Winemaking problems solved

Related titles:

Managing wine quality Volume 1: Viticulture and wine quality
(ISBN 978-1-84569-484-5)
Many aspects of both grape production and winemaking influence wine flavour, aroma, colour and mouthfeel. Factors affecting wine sensory attributes have not always been properly understood, limiting the ability of winemakers to manage these variables. In recent years, advances in research have provided new understanding of the scientific basis of quality variation in wine, promoting developments in viticulture and oenology. With chapters by leading experts, Volume 1 reviews major wine quality attributes, such as colour and mouthfeel, the measurement of grape and wine properties and the effects of viticultural practices on the final product. The focus is on recent developments, advanced methods and likely future technologies.

Managing wine quality Volume 2: Oenology and wine quality
(ISBN 978-1-84569-798-3)
Experts from major winemaking centres worldwide review oenology and wine quality in the second volume of this important collection. Chapters in Part I cover advances in winemaking technologies. Chapters in Part II focus on quality management, with chapters on flavour modulation, flavour deterioration and common taints among other topics.

Brewing: Science and practice
(ISBN 978-1-85573-490-6)
Brewing: Science and practice updates and revises the previous work of this distinguished team of authors, producing what is the standard work in its field. The book covers all stages of brewing from raw materials, including the chemistry of hops and the biology of yeasts, through individual processes such as mashing and wort separation to packaging, storage and distribution. Key quality issues are discussed such as flavour and the chemical and physical properties of finished beers.

Brewing: New technologies
(ISBN 978-1-84569-003-8)
This book describes the brewing of beer, from raw materials supply, through processing, to the final product and its analysis. It does not describe the fundamental science and technology of brewing (the necessary underpinning information to set the scene for the non-specialist or student is provided in Chapter 1), rather, each chapter contains a critical description of current technology and its applications, an outline of new and emerging technologies and views on how they may develop in the future. As such the volume can be used by professional brewers (or those intending to become same) as a source of the latest thinking in brewing strategies.

Details of these books and a complete list of titles from Woodhead Publishing can be obtained by:

- visiting our web site at www.woodheadpublishing.com
- contacting Customer Services (e-mail: sales@woodheadpublishing.com; fax: +44 (0) 1223 832819; tel.: +44 (0) 1223 499140 ext. 130; address: Woodhead Publishing Limited, 80 High Street, Sawston, Cambridge CB22 3HJ, UK)
- in North America, contacting our US office (e-mail: usmarketing@woodheadpublishing.com; tel.: (215) 928 9112; address: Woodhead Publishing, 1518 Walnut Street, Suite 1100, Philadelphia, PA 19102-3406, USA)

Woodhead Publishing Series in Food Science, Technology and Nutrition:
Number 193

Winemaking problems solved

Edited by
Christian E. Butzke

Oxford   Cambridge   Philadelphia   New Delhi

© Woodhead Publishing Limited, 2010

Published by Woodhead Publishing Limited, 80 High Street, Sawston,
Cambridge CB22 3HJ, UK
www.woodheadpublishing.com
www.woodheadpublishingonline.com

Woodhead Publishing, 1518 Walnut Street, Suite 1100, Philadelphia, PA 19102-3406, USA

Woodhead Publishing India Private Limited, 303, Vardaan House, 7/28 Ansari Road,
Daryaganj, New Delhi – 110002, India
www.woodheadpublishingindia.com

First published 2010, Woodhead Publishing Limited; reprinted 2013
© Woodhead Publishing Limited, 2010. Note: the publisher has made every effort to ensure that permission for copyright material has been obtained by authors wishing to use such material. The authors and the publisher will be glad to hear from any copyright holder it has not been possible to contact.
The authors have asserted their moral rights.

This book contains information obtained from authentic and highly regarded sources. Reprinted material is quoted with permission, and sources are indicated. Reasonable efforts have been made to publish reliable data and information, but the authors and the publisher cannot assume responsibility for the validity of all materials. Neither the authors nor the publisher, nor anyone else associated with this publication, shall be liable for any loss, damage or liability directly or indirectly caused or alleged to be caused by this book.

Neither this book nor any part may be reproduced or transmitted in any form or by any means, electronic or mechanical, including photocopying, microfilming and recording, or by any information storage or retrieval system, without permission in writing from Woodhead Publishing Limited.

The consent of Woodhead Publishing Limited does not extend to copying for general distribution, for promotion, for creating new works, or for resale. Specific permission must be obtained in writing from Woodhead Publishing Limited for such copying.

Trademark notice: Product or corporate names may be trademarks or registered trademarks, and are used only for identification and explanation, without intent to infringe.

British Library Cataloguing in Publication Data
A catalogue record for this book is available from the British Library.

ISBN 978-1-84569-475-3 (print)
ISBN 978-1-84569-018-8 (online)

The publisher's policy is to use permanent paper from mills that operate a sustainable forestry policy, and which has been manufactured from pulp which is processed using acid-free and elemental chlorine-free practices. Furthermore, the publisher ensures that the text paper and cover board used have met acceptable environmental accreditation standards.

Typeset by Godiva Publishing Services Limited, Coventry, West Midlands, UK
Printed by Lightning Source

Contents

Contributor contact details xiii

Woodhead Publishing Series in Food Science, Technology and
Nutrition xvii

1 Grape analysis in winemaking
1.1 Why is effective grape sampling and analysis important? 1
1.2 How can I sample grapes? 4
1.3 How should I store, transport and process grape samples? ... 6
1.4 How do I measure grape berry ripeness and what equipment is needed for analysis? 9
1.5 How do I undertake sensory ripeness assessment of grape berries? 12

2 Juice and must preparation in winemaking
2.1 What size wine press should I buy? 15
2.2 How do I manage unwanted botrytis or rot in harvested grapes? 17
2.3 When should pH adjustments be made to the must rather than to the fermented wine? 19
2.4 How do I adjust a juice with high pH and high titratable acidity before fermentation? 21
2.5 I sometimes see the sugar content of grape juice and must reported as °Brix, sometimes just as Brix. What is the correct nomenclature and why does it matter? 22

2.6	How should I calculate and make additions of grape concentrate to juice, musts and wines?		24
2.7	How does skin contact affect a white wine style?		26
2.8	What are the pros and cons of using pectinase when preparing a white must?		28
2.9	Is must clarification necessary?		29
2.10	What is the best way to cold settle my white juices?		31
2.11	How should I treat a must from white grapes containing laccase?		33
2.12	How can I avoid oxygen exposure with a white must?		35
2.13	How should I calculate and make water additions to facilitate the fermentation of red musts?		37
2.14	What is saignée and how will it affect my red wine?		39
2.15	What is thermovinification and why should I use this technique?		41
2.16	Should I add enological tannin prior to the fermentation of a red must to affect its color?		45
2.17	How can I estimate when to fortify a fermenting juice to achieve desired sugar and alcohol concentrations?		47
2.18	What is the ideal temperature to press ice wine ('frozen') grapes?		50

3 Yeast fermentation in wine

3.1	What different types of fermentor are there and which should I use for red and white wines?	52
3.2	What materials are used in constructing fermentors and how does this affect fermentation and storage of wine?	55
3.3	What is a yeast 'strain'?	57
3.4	Does the yeast strain have an influence on the fermentation kinetics and on the wine aromas and flavours?	58
3.5	What is a native flora fermentation?	60
3.6	What factors are important in deciding to use a yeast inoculum versus allowing the native flora to conduct the fermentation?	61
3.7	What is yeast assimilable nitrogen (YAN) and how much is needed?	62
3.8	How do I measure yeast assimilable nitrogen (YAN)?	65
3.9	How do I interpret yeast assimilable nitrogen (YAN) data?	66
3.10	When and how do I adjust yeast assimilable nitrogen (YAN)?	67
3.11	Is it important to rehydrate active dry yeast with a precise procedure?	71
3.12	Is the yeast population homogeneous throughout the tank during fermentation?	73
3.13	What is the influence on the yeast of oxygen addition during alcoholic fermentation?	74

3.14	Is it important to inoculate the grape with yeast before a cold soak with Pinot Noir?	75
3.15	What is the difference between 'carbonic maceration' and 'whole berry fermentation'?	76

4 Malolactic fermentation in wine

4.1	How many different types of malolactic (ML) starter culture preparations are in use today?	77
4.2	What are the ideal conditions for storage and rehydration of malolactic (ML) starter cultures?	79
4.3	When should I inoculate for malolactic fermentation (MLF)?	81
4.4	What are the advantages and the risks of inoculating with malolactic (ML) starter cultures at the beginning of alcoholic fermentation?	83
4.5	What is the impact of temperature on malolactic fermentation (MLF)?	85
4.6	Is it possible to manage the diacetyl levels in a wine?	86
4.7	Are there other factors beside SO_2, pH, alcohol content and temperature that can have an impact on malolactic fermentation (MLF)?	88
4.8	Do I have to add nutrients to my malolactic fermentation (MLF)?	91
4.9	Does the yeast used for alcohol fermentation have an impact on malolactic fermentation (MLF)?	93
4.10	How can I monitor malolactic fermentation (MLF)?	95
4.11	How much residual malic acid will cause visible malolactic bacteria (MLB) growth/carbonation in the bottle?	97

5 Wine clarification, stabilization and preservation

5.1	What role does vineyard nitrogen management play in juice processing and wine instabilities?	98
5.2	How do additions of potassium sorbate, potassium metabisulfite or potassium bicarbonate impact the cold stability of my wine?	100
5.3	Why do I have to heat-stabilize my white wines with bentonite?	102
5.4	What should I use: sodium or calcium bentonite?	105
5.5	I was told to treat my wine with casein. What is the best procedure?	109
5.6	What is sorbate?	111
5.7	How much sorbate should I use in wine?	113
5.8	I mixed up potassium sorbate and citric acid together to make a sorbate addition and an acid adjustment prior to bottling. An amorphous white precipitate has formed and is floating on top of the mixture. What is it and what should I do about it?	115

5.9	What does 'contains sulfites' mean versus 'no added sulfites'?	116
5.10	What is molecular sulfur dioxide (SO_2) and how does it relate to free and total SO_2?	117
5.11	How can I protect my grapes and juice from spoilage during transport to the winery?	119
5.12	The wine's pH is 3.95. How much free SO_2 do I need to prevent malolactic bacteria (MLB) or Brett growth?	121
5.13	What is lysozyme and why is it used in winemaking?	124

6 Wine filtration

6.1	How should I interpret references to 'size' in the context of filtration?	127
6.2	What are my options in terms of filtration?	130
6.3	What is osmotic distillation?	135
6.4	What are typical types of filters?	136
6.5	How do I decide what membrane pore size to use?	138
6.6	Do sterile filter pads really exist?	139
6.7	How can I minimize filtration?	140
6.8	My wines are difficult to filter: where can I look to solve this issue?	141
6.9	Can the physical nature of wine particulates affect filtration rate and volume?	142
6.10	Does filtration affect wine quality?	143
6.11	I've heard that filtration strips my wine. Is it really necessary?	144
6.12	What operational parameters should I monitor during filtration?	145
6.13	What is integrity testing and when should it be performed?	146
6.14	After sterile filtering wine, what membrane flushing schemes are recommended?	147
6.15	When can I bottle a wine without filtering it?	149

7 Wine packaging and storage

7.1	What is the best sterilization option for the bottling line?	150
7.2	How long do I need to disinfect my bottling line if my hot water is less than 82°C (180°F)?	154
7.3	How do I steam the bottling line?	156
7.4	What does sterile bottling involve?	160
7.5	What chemical additives can I utilize as an additional source of security in helping prevent the threat of re-fermentation or microbiological instability in the bottle?	164
7.6	How significant is oxygen pick up during bottling?	170
7.7	How can I control oxygen uptake at bottling?	172
7.8	How much oxygen can I have in my bottles at filling?	176

7.9	How many corks out of a 5000-cork bale would I need to sample to assure a taint rate of less than one bottle out of five cases?	178
7.10	Will screw caps make my wine better?	180
7.11	Can synthetic closures take the place of corks?	182
7.12	What precautions do I need to take when using 'bag in box' packaging?	184
7.13	What's the best way to store a wine bottle: sideways, upside down or closure up?	186
7.14	What effects do post-bottling storage have on package performance?	189
7.15	What are the optimum environmental parameters for bottle storage and what effect do these parameters have on wine quality?	191
7.16	What temperatures can my wine be exposed to during national and global shipments and storage once it leaves the sheltered winery?	195

8 Winemaking equipment maintenance and troubleshooting

8.1	What are common problems in the stemming-crushing operation and how can these be remedied?	199
8.2	Why won't my electric motor start?	202
8.3	What other problems might I encounter with my electric motor?	204
8.4	What different types of pumps are there for wine and/or must transfer?	206
8.5	What are the strengths and weaknesses of different types of pump?	208
8.6	Why isn't my pump priming and why isn't it pumping fast enough?	211
8.7	How do I care for the winery pump?	213
8.8	What is bubble point membrane filter integrity testing?	218
8.9	How do I perform a bubble point membrane filter integrity test?	220
8.10	What's the best procedure to clean or store a used barrel?	223
8.11	How do I manage my barrels?	225
8.12	Why are my bottles underfilled?	232
8.13	What should I consider for winery preventative maintenance?	234
8.14	What is corrosion?	237
8.15	What is stainless steel?	241
8.16	What is corrosion in passivated materials?	246
8.17	How do I clean and protect stainless steel?	249
8.18	How often do I need to lubricate my destemmer, press, corker jaws, etc.?	253

9 Winery microbiology and sanitation

- 9.1 What are the essential elements for an operational sanitation program? ... 257
- 9.2 What are biofilms and are they important in winery sanitation? ... 262
- 9.3 What is cross-contamination? ... 263
- 9.4 What are viable but non-culturable organisms and should I be concerned with them during winemaking? ... 268
- 9.5 I am buying a microscope for the winery. What features should I look for and how can I utilize it best in the winery? ... 270
- 9.6 Should I use or continue to use bleach or chlorinated cleaners in my sanitation program? ... 272
- 9.7 We use chlorine bleach to clean the floors and walls of the crush area and cellar once a season, and rinse well afterwards. Is this advisable? ... 273
- 9.8 What should be used instead of chlorine bleach to clean and sanitize the winery? ... 274
- 9.9 What is environmental TCA? ... 276
- 9.10 What is ozone and how is it used in the winery? ... 277
- 9.11 How do I clean my wine tanks? ... 278
- 9.12 How do I clean my winery transfer hoses? ... 283
- 9.13 How can I determine if my sanitation program is successful? ... 288
- 9.14 What are the safety issues associated with sanitation operations? ... 289

10 *Brettanomyces* infection in wine

- 10.1 What is 'Brett'? ... 290
- 10.2 What is the history of *Brettanomyces* and where does it come from? ... 292
- 10.3 How does *Brettanomyces* grow? ... 300
- 10.4 What do *Brettanomyces* do to wines? ... 307
- 10.5 How do I sample for *Brettanomyces* testing? ... 314
- 10.6 What methods do I have available to detect *Brettanomyces* infection? ... 316
- 10.7 How can I culture the *Brettanomyces* strain in my wine? ... 323
- 10.8 How can I manage *Brettanomyces* in the cellar? ... 329
- 10.9 Can I bottle my wine unfiltered if it is infected with *Brettanomyces*? ... 336
- 10.10 How should I prepare my wine for bottling if it has *Brettanomyces*? ... 340

11 Particular wine quality issues

- 11.1 How do I know my samples are representative? ... 345
- 11.2 How do I use factors of ten to make blend trials in the lab easier? ... 349

11.3	Why did my wine's pH and titratable acidity (TA) both drop significantly after fermentation?	354
11.4	How do I improve my wine color?	356
11.5	What is wine oxidation and how can I limit it during wine transfer?	361
11.6	What is the difference between oxidative and non-oxidative browning?	363
11.7	How do I identify and treat sulfur off-odors?	364
11.8	What are some of the key steps to avoid sulfur off-odors?	366
11.9	A sulfides analysis run by our wine lab shows the presence of disulfides in my wine. Should I treat it with ascorbic acid, sulfur dioxide (SO_2) and copper?	367
11.10	Can I use silver (alloys) instead of copper to bind reduced sulfur compounds such as hydrogen sulphide (H_2S) and mercaptans?	369
11.11	My wine smells like old fish what should I do?	371
11.12	My wine is too astringent what do I do?	373
11.13	I have a lactic acid bacteria problem in my winery. How do I control it to avoid high volatile acidity?	377
11.14	What are the types of hazes that can form in a wine?	378
11.15	How much residual sugar will cause visible yeast growth and/or carbonation in the bottle?	379
11.16	What causes films to form on the surface of a wine?	380
11.17	A thin, dry, white, filamentous film has formed on the surface of my wine. What is it and how do I get rid of it?	381
11.18	I have a whitish film yeast growing on top of my wine in the tank (or barrel) that is resistant to sulfur dioxide (SO_2). What should I do?	383
11.19	I am a great fan of Burgundian bâtonnage. How often should I stir my lees to release the most mannoproteins?	385

Index *387*

11.3	Why do my wine's pH and titratable acidity (TA) both drop significantly after fermentation?	354
11.4	How do I improve my wine color?	356
11.5	What is wine oxidation and how can I limit it during wine transfer?	361
11.6	What is the difference between oxidative and non-oxidative browning?	363
11.7	How do I identify and treat sulfur off-odors?	364
11.8	What are some of the key steps to avoid sulfur off-odors?	366
11.9	A sulfides analysis run by our wine lab shows the presence of disulfides in my wine. Should I treat it with ascorbic acid, sulfur dioxide (SO_2) and copper?	367
11.10	Can I use silver (alloys) instead of copper to bind reduced sulfur compounds such as hydrogen sulphide (H_2S) and mercaptans?	369
11.11	My wine smells like old fish what should I do?	371
11.12	My wine is too astringent what do I do?	373
11.13	I have a lactic acid bacteria problem in my winery. How do I control it to avoid high volatile acidity?	377
11.14	What are the types of hazes that can form in a wine?	378
11.15	How much residual sugar will cause visible yeast growth and/or carbonation in the bottle?	379
11.16	What causes films to form on the surface of a wine?	380
11.17	A thin, dry, white, filamentous film has formed on the surface of my wine. What is it and how do I get rid of it?	381
11.18	I have a whitish film yeast growing on top of my wine in the tank (or barrel) that is resistant to sulfur dioxide (SO_2). What should I do?	383
11.19	I am a great fan of Burgundian bâtonnage. How often should I stir my lees to release the most mannoproteins?	385

Index *387*

Questions 4.1, 4.2, 4.4–4.6 and 4.8

Dr Sibylle Krieger-Weber
Lallemand
In den Seiten 53
70825 Korntal Münchingen
Germany
E-mail: skrieger@lallemand.com

Question 4.7

Dr Sibylle Krieger-Weber*
Lallemand
In den Seiten 53
70825 Korntal Münchingen
Germany
E-mail: skrieger@lallemand.com

Piet Loubser
Lallemand South Africa
31 Blousuikerbos Street
7530 Proteavalley
South Africa
E-mail: ploubser@lallemand.com

Question 4.10

Dr Sibylle Krieger-Weber*
Lallemand
In den Seiten 53
70825 Korntal Münchingen
Germany
E-mail: skrieger@lallemand.com

Samantha Kollar and Dr Neil Brown
Vinquiry Inc.
7795 Bell Road
Windsor, CA 95492
USA
E-mail: Info@vinquiry.com

Questions 4.3, 5.10, 9.4, 9.5 and 11.17

Dr James Osborne
Department of Food Science and Technology
Oregon State University
Corvallis, OR 97331-4501
USA
E-mail: James.Osborne@oregonstate.edu

Questions 5.6, 5.7, 6.14, 8.8 and 8.14–8.17

Dr Brent Trela
Department of Plant and Soil Science
Texas Tech University
Food Technology 204E
Mail Stop 42122
Lubbock, TX 79409-3121
USA
E-mail: brent.trela@ttu.edu

Questions 6.1–6.3, 6.6, 6.9, 6.11–6.13 9.1, 9.2, 9.6, 9.9, 9.10, 9.13 and 9.14

Professor K. C. Fugelsang
Department of Viticulture and Enology
California State University Fresno
2360 E. Barstow Ave.
Fresno, CA
USA
Email: kennethf@csufresno.edu

Questions 6.4, 6.5, 6.7, 6.8, 6.10, 7.3, 8.11, 9.3, 9.11, 9.12, 11.1 and 11.2

Thomas J. Payette
Winemaking Consultant
7111 B Riverside Drive
Rapidan, VA 22733
USA
E-mail: winemakingconsultant@gmail.com

Questions 7.1, 7.4, 7.5, 7.7 and 7.15

Todd E. Steiner
Enology Program Manager and Outreach Specialist
Department of Horticulture and Crop Science
The Ohio State University/OARDC
Gourley Hall
1680 Madison Avenue
Wooster, OH 44691
USA
E-mail: steiner.4@osu.edu

Questions 7.8, 7.10–7.12 and 7.14

Richard Gibson
Scorpex
PO Box 309
Semaphore
South Australia 5019
Australia
E-mail: rgibson@scorpex.net

Questions 8.1– 8.3, 8.5, 8.6 and 8.12

Howard Bursen
Howard Bursen Consulting
Winery Design Services
Winemaking
Vineyard Establishment
P.O. Box 285
Pomfret Center, CT 06259
USA
E-mail: info@wineryplan.com

Questions 8.7, 8.13 and 8.18

Dr Brent Trela* and Mike Sipowicz
Department of Plant and Soil Science
Texas Tech University
Food Technology 204E
Mail Stop 42122
Lubbock, TX 79409-3121
USA
E-mail: brent.trela@ttu.edu

Questions 10.2–10.10

Lisa Van de Water
The Vinotec Group
607 Cabot Way
Napa, CA 94559
USA
E-mail: vinotecnapa@aol.com; badwinelady@aol.com

Pacific Rim Oenology Services Limited
Private Bag 1007
Blenheim
New Zealand
E-mail: info@pros.co.nz

Questions 11.4, 11.11 and 11.12

Dr James F. Harbertson
Washington State University
IAREC
24106 N. Bunn Road
Prosser, WA 99350
USA
E-mail: jfharbertson@wsu.edu

Woodhead Publishing Series in Food Science, Technology and Nutrition

1 **Chilled foods: a comprehensive guide** *Edited by C. Dennis and M. Stringer*
2 **Yoghurt: science and technology** *A. Y. Tamime and R. K. Robinson*
3 **Food processing technology: principles and practice** *P. J. Fellows*
4 **Bender's dictionary of nutrition and food technology Sixth edition** *D. A. Bender*
5 **Determination of veterinary residues in food** *Edited by N. T. Crosby*
6 **Food contaminants: sources and surveillance** *Edited by C. Creaser and R. Purchase*
7 **Nitrates and nitrites in food and water** *Edited by M. J. Hill*
8 **Pesticide chemistry and bioscience: the food–environment challenge** *Edited by G. T. Brooks and T. Roberts*
9 **Pesticides: developments, impacts and controls** *Edited by G. A. Best and A. D. Ruthven*
10 **Dietary fibre: chemical and biological aspects** *Edited by D. A. T. Southgate, K. W. Waldron, I. T. Johnson and G. R. Fenwick*
11 **Vitamins and minerals in health and nutrition** *M. Tolonen*
12 **Technology of biscuits, crackers and cookies Second edition** *D. Manley*
13 **Instrumentation and sensors for the food industry** *Edited by E. Kress-Rogers*
14 **Food and cancer prevention: chemical and biological aspects** *Edited by K. W. Waldron, I. T. Johnson and G. R. Fenwick*
15 **Food colloids: proteins, lipids and polysaccharides** *Edited by E. Dickinson and B. Bergenstahl*
16 **Food emulsions and foams** *Edited by E. Dickinson*
17 **Maillard reactions in chemistry, food and health** *Edited by T. P. Labuza, V. Monnier, J. Baynes and J. O'Brien*
18 **The Maillard reaction in foods and medicine** *Edited by J. O'Brien, H. E. Nursten, M. J. Crabbe and J. M. Ames*
19 **Encapsulation and controlled release** *Edited by D. R. Karsa and R. A. Stephenson*
20 **Flavours and fragrances** *Edited by A. D. Swift*
21 **Feta and related cheeses** *Edited by A. Y. Tamime and R. K. Robinson*
22 **Biochemistry of milk products** *Edited by A. T. Andrews and J. R. Varley*
23 **Physical properties of foods and food processing systems** *M. J. Lewis*

24 **Food irradiation: a reference guide** *V. M. Wilkinson and G. Gould*
25 **Kent's technology of cereals: an introduction for students of food science and agriculture Fourth edition** *N. L. Kent and A. D. Evers*
26 **Biosensors for food analysis** *Edited by A. O. Scott*
27 **Separation processes in the food and biotechnology industries: principles and applications** *Edited by A. S. Grandison and M. J. Lewis*
28 **Handbook of indices of food quality and authenticity** *R. S. Singhal, P. K. Kulkarni and D. V. Rege*
29 **Principles and practices for the safe processing of foods** *D. A. Shapton and N. F. Shapton*
30 **Biscuit, cookie and cracker manufacturing manuals Volume 1: ingredients** *D. Manley*
31 **Biscuit, cookie and cracker manufacturing manuals Volume 2: biscuit doughs** *D. Manley*
32 **Biscuit, cookie and cracker manufacturing manuals Volume 3: biscuit dough piece forming** *D. Manley*
33 **Biscuit, cookie and cracker manufacturing manuals Volume 4: baking and cooling of biscuits** *D. Manley*
34 **Biscuit, cookie and cracker manufacturing manuals Volume 5: secondary processing in biscuit manufacturing** *D. Manley*
35 **Biscuit, cookie and cracker manufacturing manuals Volume 6: biscuit packaging and storage** *D. Manley*
36 **Practical dehydration Second edition** *M. Greensmith*
37 **Lawrie's meat science Sixth edition** *R. A. Lawrie*
38 **Yoghurt: science and technology Second edition** *A. Y. Tamime and R. K. Robinson*
39 **New ingredients in food processing: biochemistry and agriculture** *G. Linden and D. Lorient*
40 **Benders' dictionary of nutrition and food technology Seventh edition** *D. A. Bender and A. E. Bender*
41 **Technology of biscuits, crackers and cookies Third edition** *D. Manley*
42 **Food processing technology: principles and practice Second edition** *P. J. Fellows*
43 **Managing frozen foods** *Edited by C. J. Kennedy*
44 **Handbook of hydrocolloids** *Edited by G. O. Phillips and P. A. Williams*
45 **Food labelling** *Edited by J. R. Blanchfield*
46 **Cereal biotechnology** *Edited by P. C. Morris and J. H. Bryce*
47 **Food intolerance and the food industry** *Edited by T. Dean*
48 **The stability and shelf life of food** *Edited by D. Kilcast and P. Subramaniam*
49 **Functional foods: concept to product** *Edited by G. R. Gibson and C. M. Williams*
50 **Chilled foods: a comprehensive guide Second edition** *Edited by M. Stringer and C. Dennis*
51 **HACCP in the meat industry** *Edited by M. Brown*
52 **Biscuit, cracker and cookie recipes for the food industry** *D. Manley*
53 **Cereals processing technology** *Edited by G. Owens*
54 **Baking problems solved** *S. P. Cauvain and L. S. Young*
55 **Thermal technologies in food processing** *Edited by P. Richardson*
56 **Frying: improving quality** *Edited by J. B. Rossell*
57 **Food chemical safety Volume 1: contaminants** *Edited by D. Watson*
58 **Making the most of HACCP: learning from others' experience** *Edited by T. Mayes and S. Mortimore*
59 **Food process modelling** *Edited by L. M. M. Tijskens, M. L. A. T. M. Hertog and B. M. Nicolaï*

Woodhead Publishing Series in Food Science, Technology and Nutrition

60 **EU food law: a practical guide** *Edited by K. Goodburn*
61 **Extrusion cooking: technologies and applications** *Edited by R. Guy*
62 **Auditing in the food industry: from safety and quality to environmental and other audits** *Edited by M. Dillon and C. Griffith*
63 **Handbook of herbs and spices Volume 1** *Edited by K. V. Peter*
64 **Food product development: maximising success** *M. Earle, R. Earle and A. Anderson*
65 **Instrumentation and sensors for the food industry Second edition** *Edited by E. Kress-Rogers and C. J. B. Brimelow*
66 **Food chemical safety Volume 2: additives** *Edited by D. Watson*
67 **Fruit and vegetable biotechnology** *Edited by V. Valpuesta*
68 **Foodborne pathogens: hazards, risk analysis and control** *Edited by C. de W. Blackburn and P. J. McClure*
69 **Meat refrigeration** *S. J. James and C. James*
70 **Lockhart and Wiseman's crop husbandry Eighth edition** *H. J. S. Finch, A. M. Samuel and G. P. F. Lane*
71 **Safety and quality issues in fish processing** *Edited by H. A. Bremner*
72 **Minimal processing technologies in the food industries** *Edited by T. Ohlsson and N. Bengtsson*
73 **Fruit and vegetable processing: improving quality** *Edited by W. Jongen*
74 **The nutrition handbook for food processors** *Edited by C. J. K. Henry and C. Chapman*
75 **Colour in food: improving quality** *Edited by D MacDougall*
76 **Meat processing: improving quality** *Edited by J. P. Kerry, J. F. Kerry and D. A. Ledward*
77 **Microbiological risk assessment in food processing** *Edited by M. Brown and M. Stringer*
78 **Performance functional foods** *Edited by D. Watson*
79 **Functional dairy products Volume 1** *Edited by T. Mattila-Sandholm and M. Saarela*
80 **Taints and off-flavours in foods** *Edited by B. Baigrie*
81 **Yeasts in food** *Edited by T. Boekhout and V. Robert*
82 **Phytochemical functional foods** *Edited by I. T. Johnson and G. Williamson*
83 **Novel food packaging techniques** *Edited by R. Ahvenainen*
84 **Detecting pathogens in food** *Edited by T. A. McMeekin*
85 **Natural antimicrobials for the minimal processing of foods** *Edited by S. Roller*
86 **Texture in food Volume 1: semi-solid foods** *Edited by B. M. McKenna*
87 **Dairy processing: improving quality** *Edited by G Smit*
88 **Hygiene in food processing: principles and practice** *Edited by H. L. M. Lelieveld, M. A. Mostert, B. White and J. Holah*
89 **Rapid and on-line instrumentation for food quality assurance** *Edited by I. Tothill*
90 **Sausage manufacture: principles and practice** *E. Essien*
91 **Environmentally-friendly food processing** *Edited by B. Mattsson and U. Sonesson*
92 **Bread making: improving quality** *Edited by S. P. Cauvain*
93 **Food preservation techniques** *Edited by P. Zeuthen and L. Bøgh-Sørensen*
94 **Food authenticity and traceability** *Edited by M. Lees*
95 **Analytical methods for food additives** *R. Wood, L. Foster, A. Damant and P. Key*
96 **Handbook of herbs and spices Volume 2** *Edited by K. V. Peter*
97 **Texture in food Volume 2: solid foods** *Edited by D. Kilcast*
98 **Proteins in food processing** *Edited by R. Yada*
99 **Detecting foreign bodies in food** *Edited by M. Edwards*

100 Understanding and measuring the shelf-life of food *Edited by R. Steele*
101 Poultry meat processing and quality *Edited by G. Mead*
102 Functional foods, ageing and degenerative disease *Edited by C. Remacle and B. Reusens*
103 Mycotoxins in food: detection and control *Edited by N. Magan and M. Olsen*
104 Improving the thermal processing of foods *Edited by P. Richardson*
105 Pesticide, veterinary and other residues in food *Edited by D. Watson*
106 Starch in food: structure, functions and applications *Edited by A.-C. Eliasson*
107 Functional foods, cardiovascular disease and diabetes *Edited by A. Arnoldi*
108 Brewing: science and practice *D. E. Briggs, P. A. Brookes, R. Stevens and C. A. Boulton*
109 Using cereal science and technology for the benefit of consumers: proceedings of the 12th International ICC Cereal and Bread Congress, 24–26 May, 2004, Harrogate, UK *Edited by S. P. Cauvain, L. S. Young and S. Salmon*
110 Improving the safety of fresh meat *Edited by J. Sofos*
111 Understanding pathogen behaviour in food: virulence, stress response and resistance *Edited by M. Griffiths*
112 The microwave processing of foods *Edited by H. Schubert and M. Regier*
113 Food safety control in the poultry industry *Edited by G. Mead*
114 Improving the safety of fresh fruit and vegetables *Edited by W. Jongen*
115 Food, diet and obesity *Edited by D. Mela*
116 Handbook of hygiene control in the food industry *Edited by H. L. M. Lelieveld, M. A. Mostert and J. Holah*
117 Detecting allergens in food *Edited by S. Koppelman and S. Hefle*
118 Improving the fat content of foods *Edited by C. Williams and J. Buttriss*
119 Improving traceability in food processing and distribution *Edited by I. Smith and A. Furness*
120 Flavour in food *Edited by A. Voilley and P. Etievant*
121 The Chorleywood bread process *S. P. Cauvain and L. S. Young*
122 Food spoilage microorganisms *Edited by C. de W. Blackburn*
123 Emerging foodborne pathogens *Edited by Y. Motarjemi and M. Adams*
124 Benders' dictionary of nutrition and food technology Eighth edition *D. A. Bender*
125 Optimising sweet taste in foods *Edited by W. J. Spillane*
126 Brewing: new technologies *Edited by C. Bamforth*
127 Handbook of herbs and spices Volume 3 *Edited by K. V. Peter*
128 Lawrie's meat science Seventh edition *R. A. Lawrie in collaboration with D. A. Ledward*
129 Modifying lipids for use in food *Edited by F. Gunstone*
130 Meat products handbook: practical science and technology *G. Feiner*
131 Food consumption and disease risk: consumer–pathogen interactions *Edited by M. Potter*
132 Acrylamide and other hazardous compounds in heat-treated foods *Edited by K. Skog and J. Alexander*
133 Managing allergens in food *Edited by C. Mills, H. Wichers and K. Hoffman-Sommergruber*
134 Microbiological analysis of red meat, poultry and eggs *Edited by G. Mead*
135 Maximising the value of marine by-products *Edited by F. Shahidi*
136 Chemical migration and food contact materials *Edited by K. Barnes, R. Sinclair and D. Watson*
137 Understanding consumers of food products *Edited by L. Frewer and H. van Trijp*
138 Reducing salt in foods: practical strategies *Edited by D. Kilcast and F. Angus*
139 Modelling microrganisms in food *Edited by S. Brul, S. Van Gerwen and M. Zwietering*

Woodhead Publishing Series in Food Science, Technology and Nutrition xxi

140 **Tamime and Robinson's Yoghurt: science and technology Third edition** *A. Y. Tamime and R. K. Robinson*
141 **Handbook of waste management and co-product recovery in food processing: Volume 1** *Edited by K. W. Waldron*
142 **Improving the flavour of cheese** *Edited by B. Weimer*
143 **Novel food ingredients for weight control** *Edited by C. J. K. Henry*
144 **Consumer-led food product development** *Edited by H. MacFie*
145 **Functional dairy products Volume 2** *Edited by M. Saarela*
146 **Modifying flavour in food** *Edited by A. J. Taylor and J. Hort*
147 **Cheese problems solved** *Edited by P. L. H. McSweeney*
148 **Handbook of organic food safety and quality** *Edited by J. Cooper, C. Leifert and U. Niggli*
149 **Understanding and controlling the microstructure of complex foods** *Edited by D. J. McClements*
150 **Novel enzyme technology for food applications** *Edited by R. Rastall*
151 **Food preservation by pulsed electric fields: from research to application** *Edited by H. L. M. Lelieveld and S. W. H. de Haan*
152 **Technology of functional cereal products** *Edited by B. R. Hamaker*
153 **Case studies in food product development** *Edited by M. Earle and R. Earle*
154 **Delivery and controlled release of bioactives in foods and nutraceuticals** *Edited by N. Garti*
155 **Fruit and vegetable flavour: recent advances and future prospects** *Edited by B. Brückner and S. G. Wyllie*
156 **Food fortification and supplementation: technological, safety and regulatory aspects** *Edited by P. Berry Ottaway*
157 **Improving the health-promoting properties of fruit and vegetable products** *Edited by F. A. Tomás-Barberán and M. I. Gil*
158 **Improving seafood products for the consumer** *Edited by T. Børresen*
159 **In-pack processed foods: improving quality** *Edited by P. Richardson*
160 **Handbook of water and energy management in food processing** *Edited by J. Klemeš, R. Smith and J-K. Kim*
161 **Environmentally compatible food packaging** *Edited by E. Chiellini*
162 **Improving farmed fish quality and safety** *Edited by Ø. Lie*
163 **Carbohydrate-active enzymes** *Edited by K-H. Park*
164 **Chilled foods: a comprehensive guide Third edition** *Edited by M. Brown*
165 **Food for the ageing population** *Edited by M. M. Raats, C. P. G. M. de Groot and W. A. Van Staveren*
166 **Improving the sensory and nutritional quality of fresh meat** *Edited by J. P. Kerry and D. A. Ledward*
167 **Shellfish safety and quality** *Edited by S. E. Shumway and G. E. Rodrick*
168 **Functional and speciality beverage technology** *Edited by P. Paquin*
169 **Functional foods: principles and technology** *M. Guo*
170 **Endocrine-disrupting chemicals in food** *Edited by I. Shaw*
171 **Meals in science and practice: interdisciplinary research and business applications** *Edited by H. L. Meiselman*
172 **Food constituents and oral health: current status and future prospects** *Edited by M. Wilson*
173 **Handbook of hydrocolloids Second edition** *Edited by G. O. Phillips and P. A. Williams*
174 **Food processing technology: principles and practice Third edition** *P. J. Fellows*
175 **Science and technology of enrobed and filled chocolate, confectionery and bakery products** *Edited by G. Talbot*

176 **Foodborne pathogens: hazards, risk analysis and control Second edition** *Edited by C. de W. Blackburn and P. J. McClure*
177 **Designing functional foods: measuring and controlling food structure breakdown and absorption** *Edited by D. J. McClements and E. A. Decker*
178 **New technologies in aquaculture: improving production efficiency, quality and environmental management** *Edited by G. Burnell and G. Allan*
179 **More baking problems solved** *S. P. Cauvain and L. S. Young*
180 **Soft drink and fruit juice problems solved** *P. Ashurst and R. Hargitt*
181 **Biofilms in the food and beverage industries** *Edited by P. M. Fratamico, B. A. Annous and N. W. Gunther*
182 **Dairy-derived ingredients: food and neutraceutical uses** *Edited by M. Corredig*
183 **Handbook of waste management and co-product recovery in food processing Volume 2** *Edited by K. W. Waldron*
184 **Innovations in food labelling** *Edited by J. Albert*
185 **Delivering performance in food supply chains** *Edited by C. Mena and G. Stevens*
186 **Chemical deterioration and physical instability of food and beverages** *Edited by L. Skibsted, J. Risbo and M. Andersen*
187 **Managing wine quality Volume 1: viticulture and wine quality** *Edited by A. Reynolds*
188 **Improving the safety and quality of milk Volume 1: milk production and processing** *Edited by M. Griffiths*
189 **Improving the safety and quality of milk Volume 2: improving quality in milk products** *Edited by M. Griffiths*
190 **Cereal grains: assessing and managing quality** *Edited by C. Wrigley and I. Batey*
191 **Sensory analysis for food and beverage control: a practical guide** *Edited by D. Kilcast*
192 **Managing wine quality Volume 2: oenology and wine quality** *Edited by A. Reynolds*
193 **Winemaking problems solved** *Edited by C. Butzke*
194 **Environmental assessment and management in the food industry** *Edited by U. Sonesson, J. Berlin and F. Ziegler*
195 **Consumer-driven innovation in food and personal care products** *Edited by S. R. Jaeger and H. MacFie*
196 **Tracing pathogens in the food chain** *Edited by S. Brul, P. M. Fratamico and T. A. McMeekin*
197 **Case studies in novel food processing technologies: innovations in processing, packing and predictive modelling** *Edited by C. J. Doona, K. Kustin and F. E. Feeherry*
198 **Freeze-drying of pharmaceutical and food products** *T-C. Hua, B-L. Liu and H. Zhang*
199 **Oxidation in foods and beverages and antioxidant applications: Volume 1 Understanding mechanisms of oxidation and antioxidant activity** *Edited by E. A. Decker, R. J. Elias and D. J. McClements*
200 **Oxidation in foods and beverages and antioxidant applications: Volume 2 Management in different industry sectors** *Edited by E. A. Decker, R. J. Elias and D. J. McClements*
201 **Protective cultures, antimicrobial metabolites and bacteriophages of food and beverage biopreservation** *Edited by C. Lacroix*
202 **Separation, extraction and concentration processes in the food, beverage and nutraceutical industries** *Edited by S. S. H. Rizvi*
203 **Determining mycotoxins and mycotoxigenic fungi in food and feed** *Edited by S. De Saeger*
204 **Developing children's food products** *Edited by D. Kilcast and F. Angus*

205 **Functional foods: concept to profit Second edition** *Edited by M. Saarela*
206 **Postharvest biology and technology of tropical and subtropical fruits Volume 1** *Edited by E. M. Yahia*
207 **Postharvest biology and technology of tropical and subtropical fruits Volume 2** *Edited by E. M. Yahia*
208 **Postharvest biology and technology of tropical and subtropical fruits Volume 3** *Edited by E. M. Yahia*
209 **Food and beverage stability and shelf-life Volume 1** *Edited by D. Kilcast and P. Subramaniam*
210 **Food and beverage stability and shelf-life Volume 2** *Edited by D. Kilcast and P. Subramaniam*
211 **Processed meats: improving safety, nutrition and quality** *Edited by J. P. Kerry and J. F. Kerry*
212 **Food chain integrity: a holistic approach to food traceability, authenticity, safety and bioterrorism prevention** *Edited by J. Hoorfar, K. Jordan, F. Butler and R. Prugger*
213 **Improving the safety and quality of eggs and egg products Volume 1** *Edited by Y. Nys, M. Bain and F. Van Immerseel*
214 **Improving the safety and quality of eggs and egg products Volume 2** *Edited by Y. Nys, M. Bain and F. Van Immerseel*
215 **Feed and fodder contamination: effects on livestock and food safety** *Edited by J. Fink-Gremmels*
216 **Hygiene in the design, construction and renovation of food processing factories** *Edited by H. L. M. Lelieveld and J. Holah*
217 **Technology of biscuits, crackers and cookies Fourth edition** *Edited by D. Manley*
218 **Nanotechnology in the food, beverage and nutraceutical industries** *Edited by Q. Huang*
219 **Rice quality** *K. R. Bhattacharya*
220 **Meat, poultry and seafood packaging** *Edited by J. P. Kerry*
221 **Reducing saturated fats in foods** *Edited by G. Talbot*
222 **Handbook of food proteins** *Edited by G. O. Phillips and P. A. Williams*

*e*PROVENANCE
Assuring the provenance of fine wine

What really happens to your wine in transit?

Temperature conditions during transport and storage of wine vary widely and are seldom monitored, which can lead to undetected damage.

eProvenance helps our customers create a **Fine Wine Cold Chain**™ system across global distribution channels to improve transport conditions and protect the quality of fine wine. Our high-tech sensors and online database allow a winery, importer or shipper to easily monitor the temperature conditions of their precious cargo.

Bordeaux to Brazil:
Wild Temperature Fluctuations

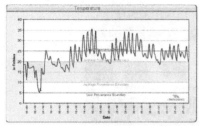

London to Hong Kong:
Optimal Conditions with
Temperature Monitoring

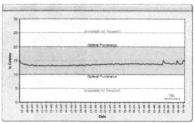

eProvenance has been endorsed by Robert Parker and Jancis Robinson

eProvenance

USA	Tel +1 617 484 2515
Bordeaux	Tel +33 (0) 5 56 11 77 30
Paris	Tel +33 (0) 9 60 45 03 00

www.eprovenance.com

Fine Wine Cold Chain is a trademark of eProvenance LLC

The largest scientifically organized and independent international wine competition in the United States

www.indyinternational.org

1
Grape analysis in winemaking

1.1 Why is effective grape sampling and analysis important?

R. Ford-Kapoor

Berry composition information derived from grape sampling and analysis is important for many reasons some of which are given below:

- Changes in berry composition during the ripening process give the grape grower an understanding of the grape berry development cycle.
- Gives current information to the grower on berry composition and quality specific to their vineyard or vineyard blocks.
- Can be used by the grower to determine harvest date based on grape quality.
- Allows the grower to make informed decisions on vineyard management strategies for improving grape quality.
- Helps determine the possible level of wine quality that can be achieved.

The point of optimal grape maturity occurs when the relative concentrations of a number of important flavor and aroma compounds are considered ideal for the production of a given wine style. However, seasonal, vineyard, vine, personnel and machinery limitations, weather, winery constraints and even berry within cluster variation mean that harvesting at optimum ripeness is impossible, unless individual berries have their own harvest dates! To combat variability, viticulturists employ practical methodology to reduce its impact such as designating harvest blocks, irrigation, vine training and canopy management.

The development cycle of the grape berry has three stages (Coombe 1992) Stage I is determined by rapid cell division, enlargement of cells and endosperm development. Stage II is signified by s slowing of growth, and the seed embryo develops at the end of this stage. The start of veraison is signified by a color change from green to red in red grape varieties and a transparent appearance in green grapes. At the beginning of Stage III veraison comes to completion and berries begin to ripen (Coombe 1980).

Important compositional changes that occur during the three stages of berry development are summarized below:

I   Tartaric acid (Ruffner 1982) and methoxypyrazine accumulation (Hashizume and Samuta 1999; Roujou de Boubee 2003).
II Malate accumulation, tartaric acid concentrations remain stable (Ruffner 1982), methoxypyrazine concentrations are at their highest (Hashizume and Samuta 1999).
III Sugar accumulation, phenolic accumulation and polymerization, terpenoid accumulation, tartaric acid levels remain stable, reduction in malate (Lakso and Kleiwer 1978; Ruffner 1982) and methoxypyrazines concentrations (Roujou de Boubee 2003), thiol precursor concentrations P-3MH remains stable but P-4MMPOH decreases in the later stages of ripening (Peyrot des Gachons *et al.* 2005).

Armed with such information about how the chemical composition of grapes changes during the ripening phase or stage III, the viticulturist and/or winemaker will notice the following changes in basic ripeness parameters; increasing soluble solids measured as °Brix, an increase in pH and reduction in titratable acidity.

Most commercial vineyards are organized into harvest blocks. When choosing a harvest block the viticulturist takes into consideration differences in the timing of maturity for different areas of the vineyard. Grape maturity of individual blocks needs to be monitored to give a holistic view of vineyard. It is upon ripeness information gathered for each vineyard block (and practicality) a harvest plan can be developed.

The viticulturist uses grape berry composition information to evaluate the effectiveness of viticultural management strategies, e.g. pruning treatments, crop load, irrigation and canopy management, for improving grape quality and vine balance.

Grape quality is a decisive factor in determining wine quality. Great vintages are those in which the range in grape variability is such that the average level of ripeness of grapes from a given region is close to optimum grape maturity. Great vintages produce wines which exhibit a synergy of outstanding palate and aromatic profiles. These are characterized by excellent balance between the sweetness of sugar and alcohol and the structural components of ripe phenolics and acids. These wines also possess a strong aromatic component of varietal, fermentation and aging bouquet.

References

COOMBE, B. G. 1980. 'Development of the grape berry. I. Effects of time on flowering and competition.' *Australian Journal of Agricultural Research* 31: 125–131.

COOMBE, B. G. 1992. 'Honorary research lecture: research on development and ripening of the grape berry.' *American Journal of Enology and Viticulture* 43(1): 101–110.
HASHIZUME, K. and T. SAMUTA. 1999. 'Grape maturity and light exposure affect berry methoxypyrazine concentration.' *American Journal of Enology and Viticulture* 50(2): 194–198.
LAKSO, A. N. and W. M. KLEIWER. 1978. 'The influence of temperature on malic acid metabolism in grape berries. II. Temperature responses of net dark CO_2 fixation and malic acid pools.' *American Journal of Enology and Viticulture* 29(3): 145–149.
PEYROT DES GACHONS, C., C. VAN LEEUWEN, T. TOMINAGA, J. SOYER, J. GAUDILLÈRE and D. DUBOURDIEU. 2005. 'Influence of water and nitrogen deficit on fruit ripening and aroma potential of *Vitis vinifera* L cv Sauvignon blanc in field conditions.' *Journal of the Science of Food and Agriculture* 85(1):73–85.
ROUJOU DE BOUBEE, D. 2003. *Research on 3-methoxy-3-isobutylpyrazine in grapes and wines*. Bordeaux: Academie Amorim.
RUFFNER, H. P. 1982. 'Metabolism of tartaric and malic acids in Vitis: A review – Part B.' *Vitis* 21:346–358.

1.2 How can I sample grapes?

R. Ford-Kapoor

Grape sampling starts about four weeks before harvest and biweekly in the two weeks prior to harvest (Krstic et al. 2003). The two most common methods for grape sampling are berry and whole cluster sampling. Berry sampling involves the collection of individual berries and cluster sampling of whole clusters for analysis.

There has been a move away from what used to be common practice for gauging field maturity by testing fruit from a single vine in a vineyard or by collecting a few bunches of grapes (Amerine and Roessler 1958). Nowadays, a more effective approach is to take a large number of randomized or systematic small samples from harvest blocks.

Each berry in a cluster develops independently (Coombe 1992). Studies have shown that there can be substantial variability between grape sample composition and that of harvested lots for processing with differences of up to 2 °Brix (Kasimatis 1984).

Berry sampling

The main advantage of berry sampling over cluster sampling is sample size. Less fruit is needed for berry sampling than for cluster sampling. Variability between clusters is similar to variability between berries (Kasimatis and Vilas 1985). However, some studies have found that berry samples do not give a true indication of ripeness in a population of grapes (Kasimatis and Vilas 1985). External 25-berry samples have been found to have higher soluble solids on average than berries from the remaining cluster samples in Zinfandel, Chenin blanc and Cabernet Sauvignon by between 0.2 and 1 °Brix (Kasimatis and Vilas 1985).

Berry sampling is a useful sampling tool when assessing differences in composition between treatments. In this situation 2 × 50-berry samples were found to have similar accuracy in detecting differences between treatments as 2 × 10 cluster samples (Kasimatis and Vilas 1985).

Berry samples are commonly taken in lots of between 30 and 100 berries. There is some disagreement about optimum sample size. However, a sample size of about 100 berries per acre (200 per hectare) in a uniform vineyard is thought to be adequate (Hamilton and Coombe 1992).

Berries should be taken from different positions on clusters, also make sure to include berries from inside the cluster. Sample collection should be done at the same time on each sampling occasion if possible.

Cluster sampling

Cluster sampling is more effective than berry sampling in determining ripeness in a harvest block (Kasimatis and Vilas 1985). When assessing pH and °Brix 20 bunch samples per harvest block is recommended (Krstic *et al.* 2003). A harvest block according to Rankine (2004) in larger vineyards is up to 2 ha (5 acre approx) in size. When collecting cluster samples take clusters from different sides of the trunk, at varying depths into the canopy, from alternating sides of the canopy (Krstic *et al.* 2003) and from different shoot numbers from the head of the vine.

Systematic sampling
For each block (<2 ha) do not include edge rows or the first two vines in each row. Walk along every second row sampling every tenth vine taking a bunch or one or two berries. Smaller vineyards should be divided up into two blocks to take into consideration causes of fruit variation including differences in aspect, slope, soil and water availability.

Random sampling
Random sampling removes human bias from the sampling process and is used for cluster sampling. Vines for sampling are determined prior to stepping in the vineyard. Numbers need to be generated for both rows and vines and some number generating methods involve drawing numbers out of a box, using the last few digits in the phone book or generating numbers using Excel (Collings 2003; Krstic *et al.* 2003).

Whatever method or combination of methods used, be consistent, methodical and keep good records.

References
AMERINE, M. A. and E. B. ROESSLER. 1958. 'Methods of determining field maturity of grapes.' *American Journal of Enology and Viticulture* 9:37–40.
COLLINGS, S. ed. 2003. *Growing quality grapes to winery specifications: quality measurement and management options for grapegrowers*. Adelaide: Winetitles.
COOMBE, B. G. 1992. 'Honorary research lecture:research on development and ripening of the grape berry.' *American Journal of Enology and Viticulture* 43(1):101–110.
HAMILTON, R. P. and B. G. COOMBE. 1992. 'Harvesting of Winegrapes.' In *Viticulture: Volume 2 Practices*, eds. B. G. Coombe and P. R. Dry. Adelaide: Winetitles.
KASIMATIS, A. N. 1984. 'Viticultural practices for varietal winemaking.' University of California-Davis Extension Short Course Series.
KASIMATIS, A. N. and E. P. VILAS. 1985. 'Sampling for degrees Brix in vineyard plots.' *American Journal of Enology and Viticulture* 36(3):207–213.
KRSTIC, M., G. MOULDS, B. PANAGIOTOPOULOS and S. WEST. 2003. *Growing Quality Grapes to Winery Specifications: quality measurement and management options for grapegrowers*. Adelaide: Winetitles.
RANKINE, B. 2004. *Making good wine: a manual of winemaking practice for Australia and New Zealand*. Sydney: Pan Macmillan Australia Ltd.

1.3 How should I store, transport and process grape samples?

R. Ford-Kapoor

Storing and transporting grape samples

When collecting berry samples the worker will need small plastic bags either sealable or using twist ties, a permanent marker for labeling, scissors or secateurs and a cooler bin with ice packs.

Prior to collecting samples label all sample bags with sample information using a permanent marker. Sample information should include date, variety, and harvest block or treatment plot. Write sample information on the bottom half of the sample bag for ease of reference especially if using twist ties to seal bags.

When taking cluster samples use scissors or secateurs as pulling clusters from vines can damage shoots. Place samples once collected in large labeled plastic shopping bags, seal and place in a cooler bin with ice packs. Be careful when grapes are close to harvest not to overfill the cooler and squash clusters on the bottom.

On returning to the laboratory place samples in the refrigerator and analyze within 24 hours.

Processing grape samples

The method for processing grape samples should always be carried out in the same way. Changing the method can result in differences in important analytical readings. The aim is not to macerate or puree berries but to liberate juice from skin and pulp. If using cluster samples remove all the berries from the cluster and place in a container immediately before processing.

Preparing berries manually

Place the sample bag flat on the work bench. Ensure most of the air is removed from the bag. Hold the bag opening end in one hand and with the palm of the other hand squash the berries in the bag (Fig. 1).

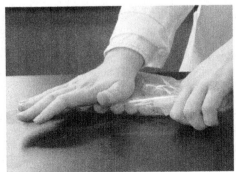

Fig. 1 Manually crushing berry samples.

Continue for approximately 1 minute until juice is liberated from the berry skin and pulp. The time it takes can be significantly influenced by berry ripeness; greener berries are harder and smaller liberating less juice. In a large well-equipped laboratory, after initial rough crushing with palm sample bags may be placed in a stomacher for a short time to achieve uniform consistency.

Using a mortar and pestle
A method outlined by Krstic *et al.* (2003) involves pressing the berries with the pestle to liberate berry juice. Take care not to crush the seeds and macerate the berry skin.

Separating juice from berry solids
A couple of methods are to use a sieve (Krstic *et al.* 2003) (Fig. 2) or muslin cloth (Fig. 3) to separate juice from solids. When using muslin or cheese cloth the used cloth is discarded. The use of pieces of muslin cloth is wasteful but saves time cleaning and rinsing the sieve. Once juice is extracted, pour contents through a piece of muslin cloth in a funnel into a small beaker. Squeeze the juice gently from the solids in the muslin. Rinse funnel between each sample with distilled water.

Fig. 2 Using a sieve to separate grape juice from solids.

Allow juice to settle for about 30 minutes before using for analysis.

8 Winemaking problems solved

Fig. 3 (a) and (b) Using muslin or cheese cloth to separate grape juice solids.

References

KRSTIC, M., G. MOULDS, B. PANAGIOTOPOULOS and S. WEST. 2003. *Growing Quality Grapes to Winery Specifications: quality measurement and management options for grapegrowers.* Adelaide: Winetitles.

1.4 How do I measure grape berry ripeness and what equipment is needed for analysis?

R. Ford-Kapoor

In viticulture and winemaking there is a lot of discussion about different types of ripeness. A grape is either ripe or it isn't, one might think. There are a number of terms thrown around about when grapes are ripe for making wine; these include physiological ripeness and phenolic ripeness. Physiological ripeness refers to all the elements of ripeness of the grape berry. Elements include sugar and acid balance, key varietal flavor and aroma compounds and phenolic composition. Phenolic ripeness refers to the phenolic composition of grapes. It is widely accepted that the elements of ripeness often develop in an asynchronous manner. This means that grapes may achieve the target sugar ripeness but lack the varietal aroma or phenolic ripeness the grower or winemaker was seeking.

Many experienced winemakers will 'pick on flavor' this means the winemaker sets harvest dates according to grape flavor. However, there would be very few that did not take into consideration objective measurements of °Brix and acid composition and if the facilities are available phenolic ripeness to support their harvest decisions.

Sugars and acids are easy and inexpensive to measure so these are the most commonly measured ripeness parameters. Phenolic ripeness and composition need more specialized equipment to assess but many wineries invest in the necessary equipment. Key varietal aroma measurements can require very expensive and specialized equipment in which most small- and medium-sized wineries would never consider investing.

Sugars

°Brix is a measurement of dissolved solids in grape juice. The dissolved or soluble solids include sugars and other non-fermentable solids. However, the non-fermentable portion of soluble solids measured using as °Brix is relatively small but is known to reach around 4–5% (Crippen and Morrison 1986). Therefore using °Brix is a reliable but not exact measurement of berry sugar concentrations. Of the dissolved solids in grape must sugars namely, glucose and fructose, are the two that are of the greatest interest. From °Brix the concentrations of sugars an approximate final wine alcohol concentration can be gleaned. Conversion rates are between 0.52 and 0.61 (Jones and Ough 1985).

Reliable equipment for field measurements of °Brix are simple to use and inexpensive. Two options are a handheld refractometer or a digital refractometer.

Sugar ripeness for grapes depends largely on the style of the wine to be made. A rough estimate of the range of ripeness for grapes:

- Sparkling wine – 18–20°Brix
- Table wine – 22–24°Brix
- Dessert wine – 26+°Brix.

Acidity

Acidity is measured in two ways in grapes to assess for harvest ripeness. One is pH and the other is titratable acidity.

pH is a logarithmic scale that measures the concentration of free hydrogen ions in grape juice. The pH of wine is very important to winemakers as it influences a wines microbial stability and resistance to oxidation during winemaking. Acid is often adjusted during winemaking to achieve a favorable pH.

Rankine (2004) gives the suitable ranges for pH for quality winemaking below:

- White wine grapes – pH 3.0–3.3
- Red wine grapes – pH 3.2–3.4.

Titratable acidity, or TA, is a measurement of how much acid is in the juice, in most countries it is measured as grams per liter of tartaric acid when grape juice is titrated to an end point of pH 8.2. The exception is France where it is measured to an endpoint of 7 and expressed as sulfuric acid instead. The range for titratable acidity in grapes in quite wide from 6 to 10 g/L.

Laboratory equipment needed to measure pH and titratable acidity includes:

- pH meter
- Beaker
- Magnetic stirrer and flea
- Dr Schilling auto-filling burette
- Bulb pipette
- Pipetter.

Phenolic composition

Anthocyanins are phenolic compounds that give a red wine its color. Red wine quality is linked to anthocyanin concentration. Measuring the anthocyanin concentration of grape berries indicates the depth and concentration of color that can be extracted during winemaking. Anthocyanin concentrations vary and fall between 0.4 and 2.6 g/L in highly colored varieties such as Australian Cabernet Sauvignon and Shiraz (Krstic *et al.* 2003). This measurement is the sum of anthocyanidins including malvidin, peonidin, petunidin, delphinidin and cyanidin that are joined to a sugar molecule to become anthocyanins giving greater color stability in the wine matrix (Rankine 2004). Phenolic compounds are viewed as important indicators of ripening (González-San José *et al.* 1991).

Laboratory equipment needed to measure anthocyanin concentration and total phenolic concentration includes:

- Spectrophotometer
- Cuvettes
- Scales
- Auto-pipette and pipette tips

- Spatula/spoon
- Homogenizer
- Centrifuge
- Centrifuge tubes.

An astute winemaker will integrate both the objective grape analysis with the subjective sensory assessment to assess berry ripeness and set harvest dates for quality wine production.

References

CRIPPEN, D. D. and J. C. MORRISON. 1986. 'The effects of sun exposure on the compositional development of Cabernet Sauvignon berries.' *American Journal of Enology and Viticulture* 37(4): 235–242.

GONZÁLEZ-SAN JOSÉ, M. L., L. J. R. BARREN, B. JUNQUERA and L. M. ROBREDO. 1991. 'Application of principal component analysis to ripening indices for wine grapes.' *Journal of Food Composition and Analysis* 4(3): 245–255.

JONES, R. S. and C. S. OUGH. 1985. 'Variation in percent ethanol (v/v) per Brix conversions of wines from different climatic regions.' *American Journal of Enology and Viticulture* 36(4): 268–270.

KRSTIC, M., G. MOULDS, B. PANAGIOTOPOULOS and S. WEST. 2003. *Growing Quality Grapes to Winery Specifications: quality measurement and management options for grapegrowers*. Adelaide: Winetitles.

RANKINE, B. 2004. *Making good wine: a manual of winemaking practice for Australia and New Zealand*. Sydney: Pan Macmillan Australia Ltd.

1.5 How do I undertake sensory ripeness assessment of grape berries?

R. Ford-Kapoor

It is common practice for winemakers to 'pick on flavor', but training the palate to perceive optimum ripeness takes practice. Studies indicate that up to 60% of winemakers and grape growers pick grapes based on how they taste (Greenspan 2006). Rosseau and Delteil (2000b) developed a method of sensory analysis that involves visual and tactile inspection of the berry connection to the pedicel, seeds and skins followed by tasting the berry pulp, skins and seeds. In addition to berry flavor and structure, assessing cluster stem ripeness can also help evaluate berry ripeness (Bisson 2001).

The Rosseau and Delteil method (Rosseau and Delteil 2000b) assesses berry ripeness on three individual berries from each sample block. A system of evaluating important aspects of berry ripeness has been developed via a numbered step-by-step method for descriptor development in regard to berry ripeness attributes via berry visual, tactile and flavor assessment.

During the sensory assessment of berry ripeness the winemaker will assess the following four areas:

- Tannin maturity
- Maturity of the grape pulp structure
- Aromatic maturity of the grape pulp
- Skin maturity.

Berry ripeness is scored from 1 to 4 in Tables 1–4 for each of the areas given above. The following method was developed based on methodology and research by Rosseau and Delteil (2000a) and Greenspan (2006):

1. Look at the berry on the cluster and note color at the point where the berry attaches to the pedicel and the ease with which the berry is removed from the pedicel.
2. Gently squeeze the berry and note deformability.
3. Place berries (3) in mouth crush against top of mouth using tongue.
4. Remove skin and seeds from mouth and keep for further assessment.
5. Note acidity and adherence of pulp to skin and seeds.
6. Note texture of pulp, juicy or gelatinous.
7. Note pulp flavors, herbaceous or fruity.
8. Chew skins same number of times for each time samples are assessed (approximately 15 times).
9. Note ease of skin break-down – skin remains intact or forms smooth paste.
10. Note tannins astringency. After swallowing or spitting note how long the dryness remains.
11. Note seed color, do not taste if there are any traces of green.
12. Crush seeds between front teeth. Note mealy, soft or biscuity crunchy texture. Note bitter, astringent or toasted character.

Table 1 Tannin maturity – color ripeness, berry deformability (Steps 1, 2, 3). Tannin flavor steps 9, 10, 11)

1	2	3	4
Green reflections (white varieties). Pink reflections (red varieties). Berries are firm and do not deform easily when squeezed. Tannins have little intensity, are acidic and astringent. Seeds are green and/or yellow.	Green reflections (white varieties). Pink reflections (red varieties). Berries are firm and deform easily when squeezed. Tannins have some acidity and astringency. Seeds have the character of green chestnuts.	Homogeneous straw yellow and amber color (white varieties). Black and dark red (red varieties). Berries are soft, deform easily when squeezed but elastic retaining shape. Grape pulp extracts color when grapes are crushed. Skin has some soft tannins but a little astringent and acidic. Seeds are yellow with no trace of green. Roasted notes with astringency.	Homogeneous amber (white varieties). Black (red varieties). Berries are soft, deform easily when squeezed and remain deformed. Tannins have a fine grain with no acidity, seeds are dark brown. Seeds have toasted, roasted flavour without a trace of astringency. Berries are easy to pick from the bunch – separate easily from the pedicel.

Table 2 Maturity of grape pulp and structure

1	2	3	4
Gelatinous. Acid dominant. Pulp adheres strongly to skin and seeds.	Gelatinous. Sugar and acid levels similar but acid dominates. Pulp has some adherence to skin and seeds.	Gelatinous/juicy. Sugar domiantes over acid, moderate acidity. Only slight pulp adherence to skin and seeds.	Juicy. Sugar dominates with weak acid. No adherence of pulp to skin and seeds.

Table 3 Aromatic maturity

1	2	3	4
Herbaceous	Neutral	Light fruit	Intense fruit, notes of jam

Table 4 Skin maturity

1	2	3	4
Hard skin with herbaceous notes. Acidic.	Quite tough texture. Neutral and/or light herbaceous notes. Acidic.	Soft texture with neutral to slightly fruity notes with a herbaceous finish. Very slight acidity.	Skin easily broken/crumbly when chewed. Forms a fine paste after chewing. Intense fruitiness with no herbaceous notes.

Samples that show synchronous berry development between the four aspects of maturity and scores of 4 in grape pulp structure, aromatic maturity of the grape pulp, skin maturity and tannin maturity will be more likely to produce high end wines. Low scores (1 and 2) at harvest for each of the parameters probably means that an assessment of viticultural methods or varieties grown needs to be evaluated (Rosseau and Delteil 2000a). The most common findings are that there is a combination of scores which equates to a wine of middle range quality. Alternatively, wines made from grapes with asynchronous development can be blended with other wines that 'fill in the gaps' or have higher scores in corresponding low score areas of development.

References

BISSON, L. 2001. 'In search of optimal grape maturity.' In *Practical Winery & Vineyard Magazine*. San Rafael, CA: PWV Incorporated.

GREENSPAN, M. 2006. 'Assessing Ripeness Through Sensory Evaluation. Many aspects of fruit ripeness can be assessed right in the vineyard through touch, taste, and sight.' In *Wine Business Monthly*. Sonoma, CA: Wine Communications Group Inc.

ROSSEAU, J. and D. DELTEIL. 2000a. *Annexe: Une grille d'interprétation synthétique permettant de caractériser le niveau de maturité du raisin et son potentiel qualitatif*. Institut Coopératif du Vin l'nologue des Vins Méditerranéens.

ROSSEAU, J. and D. DELTEIL. 2000b. *R&D Results No 2 April 2000: Evaluating maturity via grape tasting*. Institut Coopératif du Vin.

2
Juice and must preparation in winemaking

2.1 What size wine press should I buy?

C. Butzke

While there are many variations around, as much as 80% of fresh or fermented grape juice naturally separates from the skins and seeds as free-run, retrieving the remaining 20% of juice or wine can contribute significantly to the profitability of a commercial winery. The press is likely the most important and expensive piece of equipment that the winemaker has to acquire, and choosing the right size is a key investment decision as a good press will last 20+ years and depreciate relatively little.

Techniques that produce delicate white juices such as whole cluster (WC) pressing require a very gentle press. Enclosed tank presses can allow for calculated skin contact of Muscat-type varieties without additional pumping from an intermediary tank back to the press. Pressing too hard or with too much grinding action will inevitably lead to harsh wines that require extensive fining or aging.

Given the size of an estate vineyard, or the current and future tonnage of grapes that needs to be processed, the press volume must be calculated accordingly.

A US ton of whole grape cluster (2000 lbs = 908 kg), has an approximate volume of 1500 L (1.32 lbs/L), varying moderately based on the cluster structure of the variety.

Most wine presses are manufactured in Europe, and their model numbers are often referring to their capacity in either hectoliters (hL = 100 L) or in liters (L). For example, a mock membrane press model 'PressPro 50' will likely have a tank capacity of 50 hL = 5000 L = 1321 gal.

16 Winemaking problems solved

- Consequently, a 50 hL model could safely hold about 3 tons (0.06 tons/hL) of WC Chardonnay per press run. If a run – including loading the fruit and discharging the pomace – takes about four hours, this press could handle up to 10 tons of WC fruit on a busy harvest day.
- Destemmed, whole berries could be pressed at 9 tons (0.18 tons/hL), when allowing the 'free-run' juice created under the berries own weight to escape the press. This may be less when processing fruit that is high in pectin or underripe.
- Using the same press for destemmed *and* crushed white fruit, and allowing the free-run juice to leave the press immediately, one could process about four times the amount of WC fruit per load, or 12 tons (0.24 tons/hL).
- Destemmed and crushed grapes designated for skin contact inside the press could be processed at about 6 tons per load (0.12 tons/hL). Since this will put press capacity on hold for the entire maceration period, proper scheduling is essential. Depending on the temperature of the crushed fruit (>15°C/60°F), the grape's own or added commercial pectinases may break down the berry's cell walls further thus releasing slightly (ca. 5%) more free-run juice. Before opening the press, make sure the juice pan can hold the press's content, and juice pumps and hoses are ready to go!
- Frozen grapes for *ice wine* production must be pressed according to **2.18**. No more than one ton of 'frozen' fruit should be pressed in a 50 hL membrane press (0.02 t/hL), otherwise the strain on the cold plastic when rotating the press to loosen the berries may become too much and the membrane can rip.

When given the option, it is advisable to chose a bigger model as this will accommodate an unexpectedly large influx of fruit due to adverse weather conditions, as well as it will suit a future expansion of the winery. Most modern presses are quite flexible in their minimal load capacity and can handle relatively small batches of fruit just as well. When buying a press, it is also crucial that the manufacturer/supplier is capable of providing technical support and spare parts even during the busiest part of the season – and in your area – within two days.

When doing skin contact with white grapes (or reds to produce rosé), remember that the contact time is only as important as the temperature of the incoming fruit. Skin contact (cold maceration) for $t = 12$ hours at $T = 7°C$ (45°F) is roughly equivalent to 2 hours at 24°C (75°F). The aroma/color extraction kinetics $t * 1/3^{(T-7)*0.1}$ are assumed similar to the calculations in **7.2**.

2.2 How do I manage unwanted botrytis or rot in harvested grapes?

J. McAnnany

The most important elements of dealing with an unwanted *botrytis* or rot infection is making sure the juice is separated from the skins quickly and fermentations are healthy. Although time consuming, following the steps outlined below will help wines survive a botrytis infection with little impact.

Red grapes and white grapes with botrytis or rot should be handled differently. Also, the addition rate for some products will increase with the percentage of botrytis infection in the lot. Percentage of infection can be determined by a botrytis test offered by some laboratories in California.

Red grapes
1. Before the fruit is ever harvested, the winemaker should be aware of the presence of botrytis in grower vineyards. Affected fruit should be dropped before machine harvest and during hand harvest, so at least very little arrives at the winery.
2. At the winery, it is important to further sort out the infected grapes as much as possible.
3. Add SO_2 to harvest bins and to must immediately after crushing. Total destemming is recommended, even with hand harvested grapes.
4. Add a red must macerating enzyme. If grapes are lacking in concentration also add an inactive yeast product. Make sure to treat with lysozyme if any spoilage bacteria are present. Add condensed tannins.
5. Before inoculating with yeast, treat with a yeast protector to strengthen yeast for a difficult fermentation. Follow good fermentation practices by adding yeast nutrient 6–12 hours after inoculation and at 1/3 sugar depletion.
6. Separate the free run from the press run so they can be treated and evaluated separately. Protect from air during racking and treat any spoilage bacteria or volatile acidity problems now.
7. Innoculate with strong lactic acid bacteria rehydrated in bacteria nutrient.
8. Post malolactic fermentation, treat with SO_2 and top with inert gas if possible.

White grapes
1. Before the fruit is ever harvested, the winemaker should be aware of the presence of botrytis in grower vineyards. Affected fruit should be dropped before machine harvest and during hand harvest, so at least very little arrives at the winery.
2. At the winery, it is important to further sort out the infected grapes as much as possible.
3. For grapes that are mechanically harvested or if there are long delays before

crushing, add SO_2 to the harvest bins. Add a tannin suitable for white wines with botrytis problems. Add a clarification enzyme at the crusher. Make an SO_2 addition right after crushing and destemming.
4. Avoid skin contact as much as possible. If skin contact is necessary, keep contact times short and temperatures low.
5. Separate the free run from the press run so they can be treated and evaluated separately.
6. Treat wine with a beta-glucanase enzyme, if needed. Four hours after the enzyme is added treat with a casein/PVPP/bentonite fining agent. Rack after settling.
7. Add a specific inactivated yeast derivative made for white wines and a tannin suitable for white wines with botrytis problems.
8. Before inoculating with yeast, treat with a yeast protector to strengthen yeast for a difficult fermentation. Follow good fermentation practices by adding yeast nutrient 6–12 hours after inoculation and at 1/3 sugar depletion.
9. Inoculate with strong lactic acid bacteria rehydrated in bacteria nutrient.
10. Post malolactic fermentation, treat with SO_2.

Recommendations compiled from Vinquiry, Lallemand, and Martin Vialatte Oenologie.

2.3 When should pH adjustments be made to the must rather than to the fermented wine?

K. L. Wilker

The generally low pH values found in wines are an important contributor to the relatively high stability they have compared to other foods and beverages. A low pH increases the efficacy of many preservatives such as sulfur dioxide and sorbic acid. High pH wines are also generally more prone to oxidation problems.

A pH range of 2.8 to 4.0 will cover most wines. It is usually suggested that grape musts for table wine production have a pH range of 3.1 to 3.3 (Boulton *et al.*, 1996). However, must values closer to 3.5 for whites and 3.6 for reds are not uncommon. Musts for the production of sparkling wine or wine for distillation can have a pH range of 2.8 to 3.0. The pH value of a must/wine does not usually remain static throughout the course of the fermentation and maturation. Many chemical and biological reactions can occur with the net result being a rising or lowering of pH.

The most common adjustment to must pH is to lower it through the addition of acid. Determining the amount of acid one can add to a must and still reach a targeted wine titratable acidity cannot always be predicted. This is because, like pH, titratable acidity values for a must can stay the same, increase or decrease as a result of the fermentation and cold stabilization procedures depending on the makeup of the must.

Tartaric acid is the most recommended acid for must adjustments (Boulton *et al.*, 1996). It is a stronger acid than malic and citric acid and less susceptible to breakdown by microorganisms during the alcoholic and malolactic fermentations. Tartaric acid tends to precipitation as a potassium salt. This does not usually affect its ability to lower the pH of a must. The precipitation of tartaric acid can be an advantage as it will reduce the level of titratable acidity increase for a given decrease in pH. An increase in pH due to the precipitation of the potassium salt of tartaric acid is usually only a concern at pHs approaching 4.

Acid additions made to the must may result in an excessively tart wine. Decreasing acid in a wine through the use of chemical means may pose regulatory issues if the must had been previously acidified.

More precise predictions of the final effect of an acid addition on titratable acidity can usually be made following the alcoholic fermentation and malolactic fermentation. Samples can be acidified to desired pH values and then cold stabilized to determine an approximation of the final pH and acid levels.

Instances when you would consider acid additions to lower pH the must versus the wine will be influenced by a number of factors such as the following:

Wine style
Adding too much acid when correcting for a high must pH is less of a concern when working with a must that is going to be made into a sweet wine. Sugar levels can be adjusted to balance tartness. The ability to blend a batch of wine with another having a relatively low acidity level will also increase your options.

Difference between the actual and desired pH
The higher the initial pH is away from the desired range of 3.1 to 3.4 the greater will be the urgency to lower the pH.

Fruit quality
The better the fruit quality the easier it will be to decide to make pH adjustments after the fermentation. This is primarily related to the relatively low level of potential spoilage organisms associated with undamaged fruit.

Ability to process under hygienic and cool conditions in a rapid manner
This relates to concerns with microbial spoilage by organisms allowed to multiply under warm conditions, prior to the active start of the alcoholic fermentation. Oxidation is also slower at cold temperatures.

Ultimately the decision to correct a high pH at the must stage will be determined by considering several different factors and often factoring prior experiences from grapes varieties from the same or similar vineyards.

Reference
Boulton, R. B., V. L. Singleton, L. F. Bisson and R. E. Kunkee. 1996. *Principles and practices of winemaking*, New York, Chapman and Hall.

2.4 How do I adjust a juice with high pH and high titratable acidity before fermentation?

S. Spayd

Perhaps the first step is explaining how a juice, must, or wine can have both high pH and high titratable acidity (TA). What is pH? pH by definition is the inverse log of the hydrogen ion (H^+) concentration. Titratable acidity includes only those acid groups that can be neutralized using a base of known concentration. Musts and wines with high pH and high TA have high concentrations of malate and potassium (K^+). The best timing to adjust pH and TA is at the juice or must stage rather than in wine.

Although it seems contrary, the best way to adjust a high pH–high TA juice/must is by the addition of tartaric acid. To determine the amount of tartaric acid to add, prepare a 10% solution of tartaric acid in distilled water. Collect a sample of the juice to be adjusted. This is more difficult for red musts. You may wish to prepare a red must sample by blending it for 30 seconds and coarsely filtering it to remove solids. By blending, you more closely estimate the pH and TA that would be found post-fermentation on the skins. Measure exactly 100 mL of the juice to be adjusted into each of about 6 containers. Set-up an addition series by adding from 0 to 5 mL, in 1 mL increments, of the 10% tartaric acid solution to the containers. Measure the pH change. To more closely approximate the pH and TA of the wine after cold stabilization, do an overnight freeze test. Measure the pH and TA of the samples after bringing back to room temperature. Determine the amount of tartaric acid needed to achieve the desired degree of change in pH and TA. Some of the limitations to the method described above are uncertainty as to the degree of extraction of K^+ from the skins during fermentation for red wines, the freeze test does not mirror the exact conditions that occur during cold stabilization, and the solution being tested is aqueous and not alcoholic. Alcohol does influence the solubility of potassium tartarate solutions.

Nagel *et al.* (1988) recommended a combination of tartaric acid addition and treatment with Acidex® to simultaneously lower TA and pH. A 1L sample of either wine or must is treated with tartaric acid and Acidex®. The treated sample is then filtered and titrated to the desired pH with a known concentration of tartaric acid. A non-Acidex® sample is adjusted to the same pH as the deacidified sample and titrated back to the desired pH with tartaric acid. Blends of the original sample, tartaric acid adjusted sample and the Acidex® treated sample can be made to determine the proportion of wine or must that needs to be adjusted. If the wine/must is above pH 3.5, the pH will drop by 0.1 to 0.2 with cold stabilization. The advantage of this method is that it addresses the high malate concentration as well as the elevated K^+ concentration.

Reference

NAGEL, C. W., K. WELLER, and D. FILIATREAU. 1988. Adjustment of high pH-high TA musts and wines. *Proceedings Second Internation Cool Climate Viticulture and Oenology Symposium*, Auckland, New Zealand, pp. 221–224.

2.5 I sometimes see the sugar content of grape juice and must reported as °Brix, sometimes just as Brix. What is the correct nomenclature and why does it matter?

C. Butzke

Young Adi *Brix* adjusted Kalle *Balling*'s scale for the concentration of sugar solutions in water from 17.5°C (63.5°F) to a reference temperature of 15.5°C (60°F) which was later re-calculated to the currently used 20°C (68°F). One Brix equals one gram of sugar in 100 grams of sugar/water mix, i.e. the sugar measured contributes to the weight of the solution it is in, considering that it is about 1.6 times heavier than water. Note that the scale was created as a mass-to-mass ratio because it is much easier to weight a certain amount of sugar plus a certain of amount of water than it is to dissolve the sugar in water while adding more water to reach a certain volume as the reference. Practically, however, we work with volumes of juice and wine not masses. For example, the addition of sugar to chaptalize by one unit of Brix is significantly more than 1% of the juice volume in weight, or roughly 9% (1.09 g/L at around 22 Brix) due to the juice's higher-than-water specific gravity (as expressed indirectly by Ferdi *Oechsle*'s scale). On the other hand, the difference between the Balling, *Plato* and the Brix scales based on their slightly different reference temperatures is practically irrelevant.

Brix can be measured by refractometer, hydrometer or density meter, but neither of these physical measurements have a direct correlation to the concentration of sugar in water, thus a non-linear regression is used to relate the two. Even then, it is assumed that we measure fermentable sugars alone, which is incorrect as there is a variable solids fraction in juice that is non-fermentable by *Saccharomyces* yeast but contributes to the Brix assessment. That's why scales that relate a juice measurement to potential alcohol formation are highly inconsistent as grape growing conditions determine the fraction of non-fermentable solids. The ration of actual alcohol by volume in wine to Brix easily ranges from 0.55 to 0.60, another 9% difference!

An often overlooked fact is that the variation in ripeness even within one cluster at harvest can be around five Brix units. Sampling of ripe, easy-to-access, sun-exposed berries often leads to an over-estimate of block maturity, especially in cool climates. Representative, random, 200+ berry sampling of a vineyard block in question is essential to making both a good picking decision and being able to predict potential alcohol in the finished wine.

Since Brix expresses the measurement of the dissolved sugar-to-water mass ratio of a juice, it is an absolute concentration not a relative one. Therefore the use of the degree sign (°) as used, e.g., in *Fahrenheit*'s temperature scale where the measurement is relative to the warmth of a healthy Englishman's armpit or a cold winter day in London, is inappropriate. Similarly, the scale for absolute temperate, expressed in K (Kelvin) also does not use the degree symbol.

While this is not a winemaking problem *per se*, the use of concentrations during the winemaking process or quality control has often lead to false reporting and subsequent improper dosage of acids, sulfites or fining agents. For example, the titratable acidity (TA) of wine is reported by most major commercial wine laboratories historically in g/100 mL; most recent articles in enological journals, however, use g/L. While wineries are reasonably proficient in measuring TA, any confusion about the magnitude of reporting can lead to errors in measuring out tank additions for acidity adjustments.

The use of ambiguous units such as 'parts' as in ppm, ppb or ppt, should be avoided as they leave unclear if the concentration is expressed in mass/volume or mass/mass which can lead to considerable differences in dosage, as the specific gravity of water is 1.0, of sugar 1.6 and of alcohol 0.8 kg/L, respectively. Similarly, *either* metric or English units should be used throughout the winery consistently, mixing up kg and lb can obviously lead to significant over- or under-dosing.

It is recommended to buy three good scales for the winery: one to weight adjuncts up to about 500 g in 0.1 g increments for fining and blending bench trials, one to weight quantities up to about 5000 g in 1 gram increments for metabisulfite or sorbate additions, and one up to about 50 kg in 100 g increments for bentonite, DAP or sugar preparations.

Sources

http://en.wikipedia.org/wiki/Brix
BOULTON, R. B. 1996. *Principles and Practices of Winemaking.* p. 196, Springer.
BUTZKE, C. E. 2000/2001. 'Survey of Winery Laboratory Proficiency.' *Am. J. Enol. Vitic.* 53: 2: 163–169, 2002.

2.6 How should I calculate and make additions of grape concentrate to juice, musts and wines?

G. S. Ritchie

Grape juice concentrate is added to juices and musts when the growing conditions have prevented the grapes from accumulating optimal glucose and fructose concentrations for the desired wine style. It may also be used to adjust the sweetness of a table wine (usually just before bottling) or a sweet or fortified wine (after alcoholic fermentation or during maturation).

Grape juice concentrate is a very viscous material and therefore it is important to ensure thorough homogenization with the juice or wine to which it is added. Adequate homogenization can be achieved by mixing the concentrate with small volumes of the juice or wine to be adjusted, in one or more stages. A 1:1 mixture is placed in another, smaller tank and carefully mixed by circulation over the top of the tank and between the bottom and racking valves. Another equal volume of the juice or wine can then be added to the small tank and the procedure repeated until the viscosity of the mixture is closer to that of the juice or wine in the main tank. The mixture in the smaller tank can then be transferred to the main tank, followed by thorough mixing. Taking samples at different depths in the main tank and comparing their specific gravity with each other and with a mixture prepared at the laboratory scale can assess homogeneity of the complete mixture.

The actual volume to add is calculated using Pearson's Square (Rankine, 1991):

$$\text{Volume of concentrate to be added} = \frac{V(D-A)}{C-D}$$

where V = volume of juice or must, D = desired Brix, A = initial Brix, C = Brix of concentrate (1 °Brix = 1%).

For example, how much 68% grape juice concentrate do we need to add to 700 L of must to increase the Brix from 20 to 24?

$$\text{L of water required} = \frac{700(24-20)}{68-24} = 63.64 \text{ L}$$

The volume of must is calculated from the tons of grapes multiplied by the expected yield (L per ton) from the press, e.g. if the yield for the press is expected to be 650 L/ton and 3 tons are going to be crushed, the volume would be $650 \times 3 = 1950$ L.

Just as with water additions, grape heterogeneity and consistency are major sources of error in estimating the initial Brix of a must. It is best to make concentrate additions as early as possible, just before fermentation or during the early stages. Homogeneous mixing becomes very difficult once there is vigorous production of carbon dioxide.

Adding concentrate will change the equilibria in the juice or wine and so pH and titratable acidity will need to be checked and adjusted accordingly. In addition, wines will need to be re-checked for stability with respect to cold and heat and any other fining that may have occurred prior to the concentrate addition. Finally, wines will require sterile filtration to prevent unwanted fermentations occurring.

Reference

RANKINE, B. C. 1991. *Making good wine*, Australia, Pan Macmillan Publishers.

2.7 How does skin contact affect a white wine style?

K. L. Wilker

The grape skins and the underlying cell layers are where the varietal aroma compounds and their precursors are found in grapes (Ribereau-Gayon et al., 2000). Flavonoid compounds that contribute to a desirable mouthfeel are also located in this region. Grape juice extracts these compounds during skin contact. In addition to desirable compounds, this part of the grape can also contribute compounds that are responsible for bitterness and grassy aromas. Part of premium winemaking is optimizing the extraction of desirable components while reducing the undesirable.

Skin contact, prior to fermentation, is a white wine technique that evolved as a way to regain extraction benefits that were lost as wineries modernized and switched to bladder and membrane presses. These presses are much gentler to grapes than previous types and greatly reduce skin extraction. During this period there was also a move towards fermenting highly clarified grape juice. This also reduces the level of extraction from grape skins.

The increased use of machine harvested grapes, and the inherent level of extraction it provides, have reduced the need for a formal skin contact treatment (Boulton et al., 1996). Skin contact is still used somewhat to increase varietal character with some aromatic varieties. Sauvignon blanc, Chardonnay and terpene containing varieties (Muscat) seem to benefit from brief skin contact.

The actual conditions practiced vary greatly and include the following.

Duration

The extent of skin contact varies from a few hours to up to 16 hours or more. Varietal differences probably account for some of these differences. Chardonnay seems to benefit from a longer maceration. Shorter periods tend to promote fresh fruity wines while longer periods promote a wine darker in color with greater flavor (Jackson, 2000).

Temperature

The grapes should be kept cool, 10–15°C. Keeping the contact period short and temperatures cool can help to reduce the increase in bitterness and astringency.

Fruit condition

The fruit should be free of disease, of similar maturity and free of stems. The presence of laccase and spoilage organisms will potentially reduce juice quality.

Carbon dioxide

Carbon dioxide should be used to reduce oxidation. Pre-filling the tank used to hold the fruit is usually practiced. Liquid and solid carbon dioxide can be used to protect the fruit from oxygen as well as cool it.

Sulfur dioxide
Sulfur dioxide additions are avoided by some during skin contact in order to reduce the extraction of phenolic compounds (Ribereau-Gayon *et al.*, 2000). Others add sulfur dioxide in the range of 30 to 50 mg/L to control microorganisms and oxidase enzymes (Jackson, 2000). Sulfur dioxide may also be useful in increasing the rate of extraction of grape constituents (Jackson, 2000). Fruit quality and experience with different varieties will be useful factors when determining if and how much sulfur dioxide to add during skin contact operations.

Skin contact tends to decrease must acidity and increase pH. Juices with skin contact often show a greater ease in fermenting due to a higher level of amino acids. Higher levels of bentonite may be needed for stabilization because of a higher concentration of proteins.

References

BOULTON, R. B., V. L. SINGLETON, L. F. BISSON and R. E. KUNKEE. 1996. *Principles and practices of winemaking*, New York, Chapman and Hall.

JACKSON R. S. 2000. *Wine science*, San Diego, CA, Academic Press.

RIBEREAU-GAYON, P., D. DUBOURDIEU, B. DONECHE, and A. LONVAUD. 2000. *Handbook of enology Volume 1 The microbiology of wine and vinifications*, Chichester, John Wiley & Sons.

2.8 What are the pros and cons of using pectinase when preparing a white must?

S. Spayd

Grape cell walls are composed of cellulose, hemicellulose, and pectins. As grapes ripen, the integrity of the cells and their walls begins to break down and cells become leaky. The pectin that cements the cells together is also degraded. The rate of these changes is not sufficiently fast to aid juice release between crushing and press operations in white grapes. Winemakers may add commercial pectic enzymes at the crusher. Pectinases are used by winemakers to breakdown cell walls in order to increase juice yield from the pulp; and/or to increase the release of anthocyanins, the pigments responsible for red color in grapes, and/or desirable flavor components from the skins. White grapes are the primary target for pectinase since red wine fermentation on the skins is usually more than adequate for achieving desired juice yield and color and flavor extraction. In rare instances, pectic hazes may form in finished wines made from grapes that were not treated with pectinase. Pectins are insoluble in alcohol and gelling of the pectin is favored by low pH, both conditions present in wine.

Most commercial pectic enzyme solutions are made using a strain or strains of the mold *Aspergillus niger* and have multiple enzyme activities. In addition to pectinase, these preparations often contain cellulases and hemicellulases. As the names imply, cellulase catalyzes the breakdown of cellulose to more simple carbohydrates, while hemicellulases do the same to hemicellulose. In addition to these activities, other enzymes may also be present in commercial preparations. Some preparations may contain β-glucosidase that cleaves glucose molecules from other compounds such as flavor volatiles. Some preparations may also contain enzyme activities with undesirable consequences. Polyphenoloxidase, perxoidase and anthocyanase activities have been purported in some enzyme preparations. All three enzymes catalyze the degradation of phenolic compounds and cause browning and/or loss of red color. Pectinases also vary with regard to optimum temperature for maximum effect. Some solutions are selected for moderate temperature conditions, room temperature and below, while others are most effective around 120–140°F. Most solutions are sold as a highly concentrated form of the enzyme that is diluted as needed. Although activity of the solution declines with time, if refrigerated, the concentrate can be used for several years. Pectinase dose level is usually about 100 mg/L (ppm).

2.9 Is must clarification necessary?

K. L. Wilker

Must clarification is a necessary part of white wine production that has been shown to improve the aroma of white wines. Juice clarification reduces oxidative enzymes (found on grape skin and pulp), mold, wild flora, elemental sulfur and pesticide residues (Boulton et al., 1996). Wines produced from insufficiently clarified juice tend to have more vegetal and reduced odors and bitter tastes than those produced from properly clarified juice (Ribereau-Gayon et al., 2000).

Juice clarification can be achieved by natural settling when working with relatively clean juice (obtained from the free run and early pressing with a pneumatic press). The juice is cooled and after a rapid initial settling and racking it is allowed to settle again until it reaches an opalescent state. Cooling the juice to 5 to 10°C during settling will inhibit microbial activity and slow oxidation. The press fraction may require the use of a commercial pectinase to improve settling. Pectinases can quickly produce very clear juices. Not all juices will clarify at the same rate.

The rate of juice clarification will be affected by such factors as grape variety, disease, maturity and processing methods (Ribereau-Gayon et al., 2000). Usually the greater the maturity of grapes the easier the juice will be to settle due to a lowering of pectin concentrations. Pectins create a suspension in the juice hindering settling. Rotten fruit can cause extra clarification problems due to the presence of polysaccharides produced by *Botrytis cinerea*. Commercial enzymes intended for this problem may be of use. Processing methods that tend to tear or grind the grape tissue will increase the suspended solids content of the juice and decrease clarity.

While the use of turbid juice to produce white wine will decrease quality, using over clarified juice can also impair wine quality. Fermenting clear juice can lead to a slow, difficult fermentation, decreased fruity aromas and higher acetic acid levels. The use of pectic enzymes and cold settling can lead to over clarification.

A turbidimeter is used to accurately measure the clarity of juice in NTU units. The suggested range for the fermentation of a white wine is between 100 and 250 NTU. Sample for testing turbidity should be taken from the middle of the settling tank. Once a desired level of turbidity has been achieved the juice should be racked. Fine lees obtained in the second settling period can be used to increase the NTU level of an over clarified juice. Juices with varying turbidities can be blended to achieve a desired NTU level. Prior to blending juices, care should be taken to insure incorporation of the fine lees which may have settled after the second racking.

The use of suspended solids such as cellulose powder can improve the fermentation rate of over clarified juice (Ribereau-Gayon et al., 2000). This may be a result of providing nucleation sites for gas bubbles and reducing the inhibitory effects on carbon dioxide on yeast. The use of these additives does not, however, duplicate all of the advantages of the correct level of naturally occurring solids. The solids found naturally in the juice appear to also have the ability to provide nutrients to yeast as well as absorb toxic fatty acids.

References

BOULTON, R. B., V. L. SINGLETON, L. F. BISSON and R. E. KUNKEE. 1996. *Principles and practices of winemaking*, New York, Chapman and Hall.

RIBEREAU-GAYON, P., D. DUBOURDIEU, B. DONECHE and A. LONVAUD. 2000. *Handbook of enology Volume 1 The microbiology of wine and vinifications*, Chichester, John Wiley & Sons.

2.10 What's the best way to cold settle my white juices?
C. Butzke

It is in general advisable to settle the solids of pressed juices for white wine production as much as possible prior to fermentation. Removal of solids limits a number of problems, including the formation of reduced sulfur volatiles, such as H_2S and mercaptans, and extraction of phenolics from grape particles. Traditionally, juices are cold settled by gravity for a number of hours or days, practically overnight or over the weekend. Forced methods like centrifugation or the more recently introduced floatation induced via injected gas, require investment in equipment, but can save much time, energy and capacity in the long run.

The question is what temperature (e.g., 4°C or 15°C) should one chill the must or juice down to, to accomplish a quick settling without negative side effects. Proponents of warmer settling temperatures would suggest that it save refrigeration energy as well as tank capacity, and results in faster settling rates due to reduced viscosity and density of the juice.

Besides biological activity (yeast, SO_2-resistant botrytis laccase, grape's own pectinases, etc.), one has to consider extraction of phenolics from grape solids during warmer cold settling temperatures. When considering extraction from skin particles, the contact time is as important as the temperature of the incoming fruit. As with general skin contact in whites, contact time for $t = 24$ hours at $T = 4°C$ (39°F) is roughly equivalent to 7 hours at 15°C (59°F). The tannin extraction kinetics $t * 1/3^{(T-4)*0.1}$ are assumed similar to the calculations in **7.2**.

Even given a good dose of free SO_2 (e.g. 100 mg/L), some recently released *Saccharomyces bayanus* strains are advertised as extreme cold-hardy fermentors ('Has excellent cold temperature properties and has been known to ferment in conditions as low as 5°C.'). While *Saccharomyces* is historically the yeast of choice to ferment grape juice due to its SO_2-resistance, higher doses of free SO_2 in the juice will extend the lag phase during which the cells are multiplying and preparing for the subsequent alcoholic fermentation. But even if they would not grow much within a few hours of settling, any CO_2 formation from the yeast inoculum could off-set and disturb the juice settling.

As discussed in **11.19** about 'Bâtonnage', the settling rate for particles in wine or juice is independent of container shape and size. It is influenced, however, by the particle size and density, viscosity (i.e. temperature) and colloidal content of the juice. A 100 μm grape skin particle, takes just 15 minutes to fall 1 meter in a juice while a smaller particle or yeast cell, about 10 μm in diameter, takes roughly 24 hours to fall 1 meter in a juice. For example, in a 4000 L settling tank that is 2 m high, it will take 2 days to settle out the smallest solids by gravity alone. Again, this depends on grape variety and its pectin content, but one should consider that the more settled and smaller the lees are, the less a racking loss is incurred.

If the winemaker chooses to use pectolytic enzyme preparations to speed up the settling by cutting though the juice colloids, a few issues must be addressed.

One main concern about enzyme use is that they are usually mold-derived and therefore have a maximum activity at around 50°C (122°F) and practical zero activity at 4°C (39°F). Ask the enzyme supplier specifically about the products activity at your typical juice temperatures, especially when you harvest at cold fall temperatures!

Besides philosophical concerns about adding foreign materials to juice and wine, some winemaker might argue that the destruction of all grape pectins – even those soluble when alcohol is present – does affect the mouthfeel of the resulting wine.

The approximate density difference between a juice at 4°C and one at 15°C is approximately 25% (based on water's behavior and directly, inversely proportional to the settling speed), i.e. it would take 18 hours vs. 24 hours in comparison for the same settling effect, a good time saving. This is based just on viscosity differences, not considering synergistic changes in colloidal content/size due to increased enzyme activity, an additional benefit (or detriment if laccase is present) of warmer settling temperatures. However, as the phenolics extraction may still be 2.5 times higher at 15°C vs. 4°C (see calculation above), it would require extremely gentle fruit handling and processing to avoid a harsher wine that needs additional fining.

2.11 How should I treat a must from white grapes containing laccase?

K. L. Wilker

Grapes naturally contain a phenol oxidase enzyme that leads to rapid oxygen consumption and browning in juices. It is relatively easily controlled in clarified juice by the addition of 25 to 75 mg/L sulfur dioxide (Boulton et al., 1996). Turbid juices will require higher levels of sulfur dioxide. Grape phenol oxidase tends to be denatured or lost during the winemaking process and is usually not a factor in wines (Jackson, 2000).

Musts produced from grapes infected with the mold *Botrytis cinerea* will contain the enzyme laccase. This enzyme also causes rapid consumption and browning in juice as well as in wine. Laccase is capable of oxidizing a wider range of phenolic compounds than phenol oxidase. It is much more soluble in juice and wine than phenol oxidase which is more closely associated with pieces of grape pulp. Laccase is not inhibited by the levels of sulfur dioxide commonly used in winemaking (Boulton et al., 1996). This fact can be used to determine laccase activity in juice. If after the addition of 50 to 75 mg/L sulfur dioxide the juice still shows oxygen consumption with an oxygen meter then laccase is likely present. If an oxygen meter is not available one can watch for darkening of the juice. Laccase is difficult to remove and can remain a potential cause of browning through all stages of wine production and after bottling. It can be destroyed at approximately 60°C. This temperature needs to be reached rather rapidly because the enzyme will remain active until the liquid is close to the temperature needed for destruction. The equipment needed for this type of treatment is out of reach for many wineries.

While many consider it not possible to destroy laccase in wine with sulfur dioxide at levels commonly used in winemaking, some indicate that it is possible. The addition of 50 to 100 mg/L of sulfur dioxide to red wines at run off has been reported to be capable of eliminating or greatly reducing laccase activity after a period of several weeks (Ribereau-Gayon et al., 2000).

The musts and wines made from grapes containing laccase require special handling. The sulfited juice should be cooled to 10 to 12°C to reduce enzyme activity. This will then allow for the incorporation of 4 to 8 mg/L oxygen needed for the yeast inoculum. Low oxygen levels at inoculation can lead to problems with yeast viability near the end of fermentation due to low levels of survival factors (Boulton et al., 1996). Commercial yeast supplements containing survival factors can also be used to ensure yeast viability when inoculating low oxygen musts. Musts from grapes infected with *Botrytis* will likely benefit from a complete yeast supplement containing thiamine. Once the wines have finished fermenting they should continue to be kept at temperatures of 10°C or less. Extra caution should be practiced to avoid oxygen pickup during movement, storage and bottling. The use of inert gases would be advised.

Gallic and proanthocyanidin tannins have been reported to inhibit lacasse (Marquette and Trione, 1998). These tannins are available as commercial products for use with wine.

References

BOULTON, R. B., V. L. SINGLETON, L. F. BISSON and R. E. KUNKEE. 1996. *Principles and practices of winemaking*, New York, Chapman and Hall.

JACKSON, R. S. 2000. *Wine science*, San Diego, CA, Academic Press.

MARQUETTE, B. and D. TRIONE. 1998. The Tannins. In: *ASVO seminar, phenolics and extraction*. M. Allen, and G. Wall (Eds.) pp. 42–43. Adelaide, Australian Society of Viticulture and Oenology.

RIBEREAU-GAYON. P., D. DUBOURDIEU, B. DONECHE and A. LONVAUD. 2000. *Handbook of enology Volume 1 the microbiology of wine and vinifications*, Chichester, John Wiley & Sons.

2.12 How can I avoid oxygen exposure with a white must?

K. L. Wilker

Failing to protect white grape juice from oxygen will lead to oxidation. This is caused by the grape enzyme polyphenol oxidase utilizing dissolved oxygen and juice phenolic substrates. This forms brown pigments which are usually lost during the fermentation. Deliberately incorporating oxygen into unprotected juice (without sulfur dioxide additions) is known as hyperoxidation. This is sometimes practiced to remove potential browning substances so they won't undergo chemical oxidation later (in the wine) when they are less easily removed. This technique appears to be little used due to concerns over aroma losses with such high levels of oxidation. Sauvignon blanc musts are especially susceptible to aroma loss due to oxidation. The thiols responsible for some of aroma of Sauvignon blanc wines are extremely sensitive to oxidation.

Protecting must completely from the process of oxidation is usually not practical. During the crushing of grapes the enzymes present will utilize any dissolved oxygen present within minutes (Boulton et al., 1996). However, oxygen exposure during processing can be greatly reduced by blanketing with inert gas. Usually carbon dioxide is preferred because it is heavier than air and will blanket the surface of the must. It can be used in either a liquid or solid form. Some believe that if a must is protected from oxygen in an overly zealous manner the resulting wine will be more sensitive to oxidation and browning as a wine (Ribereau-Gayon et al., 2000).

Yeast starters may be adversely affected when oxygen is eliminated from a must by either exclusion or by allowing it to be readily consumed by oxidase enzymes. An oxygen concentration of 4 to 8 mg/L is considered necessary for optimum yeast health throughout the alcoholic fermentation (Boulton et al., 1996).

Practical protection of grape must is likely a middle ground between oxygen exclusion and hyperoxidation. An initial of exposure to air during grape processing allows the oxidation of compounds most readily susceptible to browning. The addition of 50 mg/L sulfur dioxide to the freshly pressed juice will usually be enough to control polyphenol oxidase enzymes. This will reduce the loss of aroma compounds and allow for the accumulation of oxygen for a yeast inoculum. Once the juice is expressed, the sulfur dioxide should be quickly added in a single dose. Adding sulfur dioxide in multiple doses is not as effective as adding all at once (Ribereau-Gayon et al., 2000).

Cooling the fruit prior to processing will help to slow down oxidase enzymes and reduce oxidation prior to the addition of sulfur dioxide. If laccase is present, due to the growth of *Botrytis* on the fruit, keeping the must cool is especially important as sulfur dioxide is much less effective in controlling this enzyme.

Clarification of juices is also important for reducing the oxidation potential by polyphenol oxidase enzymes as they are closely associated with grape tissue.

References

BOULTON, R. B., V. L. SINGLETON, L. F. BISSON and R. E. KUNKEE. 1996. *Principles and practices of winemaking*, New York, Chapman and Hall.

RIBEREAU-GAYON. P., D. DUBOURDIEU, B. DONECHE and A. LONVAUD. 2000. *Handbook of enology Volume 1 The microbiology of wine and vinifications*, Chichester, John Wiley & Sons.

2.13 How should I calculate and make water additions to facilitate the fermentation of red musts?

G. S. Ritchie

Red grapes can sometimes have sugar contents that may be high enough to cause stuck or sluggish fermentations. At harvest, grapes may have a Brix >26 and include dehydrated or raisined berries. During fermentation, the alcohol that is produced can be sufficient to kill the yeast or severely restrict its ability to take up glucose and fructose and hence prevent the fermentation from completing to the desired level. The addition of water to the must can prevent such problems and is legal in some regions (Anon, 2002).

The amount that needs to be added is determined by a balance between the maximum Brix at which alcoholic fermentation can proceed in a timely fashion, the wine style to be created and the government regulations governing water additions. Once the amount has been established, we can calculate the actual volume using Pearsons Square (Rankine, 1991):

$$\text{Volume of water to be added} = \frac{V(D-A)}{C-D}$$

where V = volume of must, D = desired Brix, A = initial Brix and C = Brix of water (i.e. zero).

For example, how much water do we need to add to lower 500 L of must from a Brix of 27 to 24?

$$\text{L of water required} = \frac{500(24-27)}{0-24} = 62.5 \text{ L or } 1.25\%$$

The volume of must is calculated from the tons of grapes multiplied by the expected yield (L per ton) from the press e.g. if the yield for the press is 650 L/ton and 3 tons are going to be crushed, the volume would be $650 \times 3 = 1950$ L.

There are several sources of error in deciding how much water to add. The biggest source of error is heterogeneity in grape ripeness and consistency. Both can lead to inaccurate estimates of the initial Brix of the must and result in an under- or overestimate of the amount required. If the growing conditions are not optimal, there can be a wide variation in the Brix of berries not only between clusters but also within berries on the same cluster (e.g., a dense canopy can lead to the exposed side of a cluster being riper than the side facing the canopy). Some varieties are more prone to heterogeneity than others (e.g., Zinfandel). The consistency for the berries may also vary. Most berries may be soft but still contain the normal amount of pulp. Others, however, may have dehydrated to the extent of being harder and raisined. The latter type actually needs to absorb liquid before the sugar in the berry is released. This can lead to the Brix of the must not decreasing or actually going up even though there are clear signs that fermentation has commenced. Therefore, in order to make an accurate estimate of the initial Brix, we must ensure that the must is thoroughly mixed and that raisined berries are removed or given time to rehydrate before making an estimate. If the heterogeneity or inconsistency of the berries is not high and you do not plan to cold soak, waiting 24 hours after crushing and making an estimate after or during a pumpover may be sufficient. If the heterogeneity and inconsistency are high, then it is better to do a 3-day cold soak to give raisined berries time to release their sugar.

There are other consequences of making water additions that we need to be aware of so that we do not create other problems. Firstly, by adding water, we are not only diluting the potential alcohol but also the desired components of our wine, i.e. flavors, phenols, acidity and pH. Secondly, the dilution of these components is not predictable by a simple linear model because of complex chemical equilibria being shifted to different extents depending on the compounds present in the must. For example, after a 5% water addition, buffering by carboxylic acids may prevent the titratable acidity from decreasing by a similar amount. Nevertheless, we can counterbalance the dilution of flavors and phenols by conducting a saignée immediately after crushing. The challenge is that the removal of juice occurs before we can make an accurate estimate of the initial Brix. Adding tartaric acid, preferably after cold soak or just before alcoholic fermentation, can ameliorate changes in titratable acidity and pH.

References

ANONYMOUS. 2002. California clarifies water regulations. *Wines and Vines* 23(2): 52–55.
RANKINE, B. C. 1991. *Making good wine*, Australia, Pan Macmillan Publishers.

2.14 What is saignée and how will it affect my red wine?

G. S. Ritchie

Saignée (or bleeding) is the removal of juice from a red must immediately after destemming/crushing. Usually, it is performed to concentrate flavors and phenols in the wine made from the remaining must, or to counterbalance water additions, made to lower the potential alcohol and allow alcoholic fermentation to proceed in a timely manner (legal in CA). Sometimes it is carried out for the sole purpose of making a Rosé wine usually using white winemaking methods.

When the juice is first released from a crushed red grape, it mainly contains water, glucose, fructose and acids. It contains very little flavor compounds, phenols or anthocyanins (color compounds) because the majority is found in the skins. After 30 minutes to one hour, however, the skins start to release anthocyanins, a few phenols and some flavor compounds.

If we wish to concentrate flavors, phenols and anthocyanins, then we must carry out the saignée as soon as possible after crushing. Some destemmer-crushers are now designed so that saignée juice may be collected as soon as it comes out of the machine. Alternatively, a sump cart with screen can be used as soon as juice will flow from the bottom valve of the tank to which the must is pumped. With small lots, the juice can be collected using buckets and sieves. If we are only interested in making a rosé wine then we might leave the juice in contact with the skins for longer, depending how much color we want in our wine.

The amount of juice to remove depends on the quality of the grapes and the purpose for using the technique (Table 1). If it has not been possible for the grapes to reach optimum maturity, then the technique should be used with caution as one could also be concentrating undesirable vegetal flavors such as bell pepper and green bean as well as accentuating bitterness and astringency.

Table 1 Possible rates of saignée to be used with grapes of varying status

Reason for saignée	Status of grapes	Percentage of juice removed
Concentration of flavors and phenols	Average size berries that have not been excessively watered and were allowed to mature to optimum extent	5–10
	Large berries with low anthocyanin content even though matured to optimum extent	10–20
	Berries with excessive water uptake	10–30
To balance water additions made to ensure alcoholic fermentation completes in a timely manner	Berries with no raisining and were allowed to mature to optimum extent	5–10
	Berries with excessive raisining	<5

In some red grapes, the glucose/fructose concentration of the must may be so high that alcoholic fermentation may be sluggish or stop completely. When there is a danger of this happening, some regions allow water additions to allow alcoholic fermentation to complete in a timely manner. Such an approach has the disadvantage of diluting the concentration of acids, flavors, phenols and anthocyanins. However, we can carry out saignée during or just after crushing to counterbalance the addition of the water. The challenge is that the saignée step occurs before the water addition. The estimate of the true glucose/fructose concentration in the must needs to be accurate so that we can estimate the percentage of juice to bleed to counterbalance the desired water addition. The biggest error in making the estimation is non-uniformity of the must just after destemming/crushing.

Irrespective of the reason for removing juice, it is important to realize that it also contains yeast assimilable nitrogen (YAN). Hence, it is advisable to determine the YAN remaining in the must and make nitrogen additions accordingly. In addition, the titratable acidity (TA) of the must should be checked, as an acid addition may be required as well.

2.15 What is thermovinification and why should I use this technique?

S. Spayd

Thermovinification is heat extraction of red grapes to 60 to 80°C for 20 to 30 minutes prior to press operations. Grapes are destemmed and crushed with an SO_2 addition as per a standard red wine fermentation. Either batch (heated-jacketed tanks) or continuous processes (see Lowe et al. 1976 and Wagener 1981 for detailed system information) can be used for thermovinification. After heating, the grapes can be pressed or fermented on the skin. In the case of no skin fermentation, free run juice is drained off, and pressed juice is added back to increase tannin content (Wagener 1981). The juice/must is cooled to fermentation temperature (12 to 16°C) and inoculated with the desired yeast strain. Fermentation of thermovinified juice is more similar to white wine fermentation than traditional red wine fermentation. Juice/must must be inoculated since the majority of microbes, particularly the yeasts, are killed during the heating process. Juice/wine yield will be high due to the high degree of cellular disruption from heating, but extraction of cell wall materials will not be the same as either traditional cold maceration or skin fermentation where cell wall polysaccharides are leached into the juice/wine as the berry skin and pulp slowly breakdown.

Thermovinification differs from production of juices for non-alcoholic products. For 'Concord' grape juice production, grapes are heated to 60°C, a heat-labile pectinase is added, and the product is then held for 20 to 30 minutes before pressing. The pomace is then hard pressed for maximum juice extraction. The juice is then pasteurized, passed through a heat exchanger to drop the temperature for storage at -2 to 0°C.

The goals of thermovinification are different from the thermal-break and hot-break methods for producing products such as tomato juice. For both of these methods the fruit is also heated to 60 to 80°C. In the case of thermal-break, fruit is chopped or crushed prior to heating; while in the case of hot-break, fruit is heated prior to disruption. The goal of both of these extraction methods is to produce a turbid product by inactivating pectolytic enzymes that would break down cell wall structures.

Thermovinification is not suitable for all wines, since it reduces the aging capacity of the wine, since some of the changes in the juice/wine brought about by thermovinification are similar to changes that occur with aging. Thermovinification is suitable for situations of low fruit color. Thermovinification increases the extraction of monomeric anthocyanins, but concentrations of polymeric pigments are lower than fermentation on the skins at 30°C (Gao et al. 2008). Monomeric anthocyanin concentrations drop rapidly (about 50%) within five days of extraction, but remain higher than in wines fermented at 30°C for at least 200 days. Polymeric pigments remained lower in thermovinified wines than the 30°C fermented wines (Gao et al. 2008). Flavonol and flavan-3-ol extraction are also increased by thermovinification, especially if the heat-treated

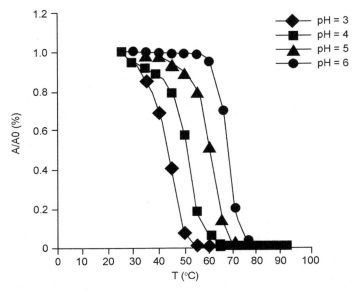

Fig. 1 Thermal stability of Victoria grape polyphenol oxidase (PPO) in a pH range of 3 to 6. Residual activity was measured after 10 minute treatment at different temperatures assay conditions: catechol 10 mM, pH 5.0, 25°C. Source: Rapeanu et al. (2005).

grapes are then fermented on the skins rather than directly pressing after heating (Netzel et al. 2003). Polymeric pigments formed from polymerization of anthocyanins with flavonols are important to color stability for aging of red wines. Long-term, more than a year, effects of thermovinification on color and color stability have not been reported.

Inactivation of enzymes, primarily oxidases, produced by molds in infected grapes is another major use for thermovinification. Inactivation of oxidases is temperature dependent with increasing temperature required for inactivation as pH increases (Fig. 1; Rapeanu et al. 2005). Grape polyphenoloxidase (PPO) inactivation is apparent as low as 55°C (Weemaes et al. 1998), but others (Lee et al. 1983; Valero et al. 1988; Wissemann and Lee 1981) report temperatures of 65°C or higher. Temperature for inactivation of PPO in grapes is apparently variety dependent. Wissemann and Lee (1981) found that for a 50% reduction in PPO activity at 75°C, 4.5 min was required for the Ravat 51 (Vignoles) grapes, while 31.6 min was required for Niagara grapes. Valero et al. (1988) reported a temperature of 75°C for 15 minutes was required to inactivate PPO in Airen grapes.

Flavor profiles are also altered by thermovinification with thermovinified wines having more cooked fruit characters than traditionally fermented red wines and less vegetal aromatic compounds due to the inactivation of lipoxygenase (Fischer et al. 2000). Lipoxygenase facilitates the oxidation of fatty acids in grapes producing C_6-alcohols (Kalua and Boss 2009). Thermovinification decreases the concentration of C_6-alcohols (Fischer et al. 2000) and 2-methoxy-

3-butylpyrazine (IBMP; Roujou de Boubeé 2004). C_6-alcohols have been reported to contribute from unpleasant-green to fruit aromas in wines (Ferreira *et al.* 2002; Guarrera *et al.* 2005), while IBMP aroma is of green bell peppers (Roujou de Boubeé 2004). Thermovinification can increase ester content.

Pros:
1. Increased color extraction from poorly colored, particularly less than optimally ripe, fruit that may not have been suitable for traditional red wine production.
2. Inactivation of oxidative enzymes, particularly in mold-infected grapes that may not have been suitable for traditional red wine production.
3. Reduction in tank space requirements for fermentation since fruit is directly pressed after heating.
4. Wine is likely to be market-ready earlier than a traditionally fermented wine.
5. Increased juice yields.
6. Reduce vegetative character.

Cons:
1. Thermovinified wines may have less aging potential than traditionally fermented wines since the heating process has already 'aged' the wines to some degree.
2. Altered flavor profile with regard to fruit character (cooked fruit).
3. Cost of equipment and energy for heating.

References

FERREIRA V., N. ORTÍN, A. ESCUDERO, R. LÓPEZ and J. CACHO. 2002. 'Chemical characterization of the aroma of Grenache rose wines: Aroma extract dilution analysis, quantitative determination, and sensory reconstitution studies.' *J. Agr. Food Chem.* 50: 4048–4054.

FISCHER, U., M. STRASSER and K. GUTZLER. 2000. 'Impact of fermentation technology on the phenolic and volatile composition of German red wines.' *Intl. J. Food Sci. Technol.* 35: 81–94.

GAO, L. B. GIRARD, G. MAZZA and A. G. REYNOLDS. 2008. 'Changes in anthocyanins and color characteristics of Pinot noir wines during different vinification processes.' *J. Agr. Food Chem.* 45: 2003–2008.

GUARRERA, N., S. CAMPISI and C. N. ASMUNDO. 2005. 'Identification of the odorants of two passito wines by gas chromatography-olfactometry and sensory analysis.' *Am. J. Enol. Viticul.* 56: 394–399.

KALUA, C. M. and P. K. BOSS. 2009. 'Evolution of volatile compounds during the development of Cabernet Sauvignon grapes. (*Vitis vinifera* L.).' *J. Agr. Food Chem.* 57: 3818–3830.

LEE, C. Y., A. P. PENNESI and N. L. SMITH. 1983. 'Purification and some properties of peroxidase from de Chaunac grapes.' *Am. J. Enol. Viticul.* 34: 128–129.

LOWE, E. J., A. OEY and T. M. TURNER. 1976. 'Gasquet thermovinification system perspective after two years' operation.' *Am. J. Enol. Viticul.* 27: 130–133.

NETZEL, A., G. STRAUSS, I. BITSCH, R. KÖNITZ, M. CHRISTMANN and R. BITSCH. 2003. 'Effect of grape processing on selected antioxidant phenolics in red wine.' *Food Eng.* 56(2–3): 223–228.

RAPEANU, G., A. VAN LOEY, C. SMOUT and M. HENRICKX. 2005. 'Effect of pH on thermal and/or pressure inactivation of Victoria grape (*Vitis vinifera sativa*) polyphenol oxidase: A kinetic study.' *J. Food Sci.* 70: E301–307.

ROUJOU DE BOUBEÉ, D. 2004. 'Recherches sur le caractère vegetal-poivron vert dan les raisins et dans les vins.' *Revue de Oenologues* 31: 6–10.

VALERO, E., R. VARÓN and F. GARCÍA-CARMONA. 1988. 'Characterization of polyphenol oxidase from Airen grapes.' *J. Food Sci.* 53: 1482–1485.

WAGENER, G. W. W. 1981. 'The effect of different thermovinification systems on red wine quality.' *Am. J. Enol. Viticul.* 32: 179–184.

WEEMAES, C. A., L. R. LUDIKHUTZE, I. VAN DEN BROECK, M. E. HENDRICKX and P. P. TOBBACK. 1998. 'Activity, electrophoretic characteristics and heat inactivation of polyphenoloxidases from apples, avocados, grapes, pears, and plums.' *Lebensm.-Wiss. u.-Technol.* 31: 44–49.

WISSEMANN, K. W. and C. Y. LEE. 1981. 'Characterization of polyphenoloxidase from Ravat 51 and Niagara grapes.' *J. Food Sci.* 46: 506–514.

2.16 Should I add enological tannin prior to the fermentation of a red must to affect its color?

K. L. Wilker

Enological tannins are made up of two types of tannin, hydrolysable and condensed. They are available commercially as separate preparations or as a combination of the two types. Hydrolysable tannins are extracted from various woods and fruits. Condensed tannins are extracted from grapes. Hydrolysable tannins are useful for the removal of protein, inhibition of laccase, antioxidant properties and improving wine structure (Marquette and Trione, 1998). Condensed tannins have these properties as well as the ability to stabilize red color.

Red wine color is stabilized through the polymerization of anthocyanins with tannins. Anthocyanins attached to tannins are less likely to undergo reactions that result in their loss. Anthocyanins attach to tannins directly or through a linkage with acetaldehyde. Acetaldehyde is naturally present during the alcoholic fermentation and during wine maturation as a result of oxygen exposure.

Anthocyanins are located in the skins of red wine grapes and are extracted during skin contact with juice. Tannins are found in grape skins and seeds. Extraction of skin tannin starts at the beginning of maceration and increases with the presence of ethanol (Ribereau-Gayon et al., 2000). Seed tannin requires the presence of ethanol and the breakdown of the seed cuticle before it can extracted. An extended period of maceration is helpful for the extraction seed tannin.

During the maceration period of a red wine fermentation anthocyanins and tannins increase in concentration due to extraction as well as decrease due to precipitation with proteins and absorption onto grape solids and yeast. It is thought that early additions of enological tannin (highly reactive) will precipitate a large portion of these proteins and reduce the precipitation of grape tannins as they are extracted from the skins (Jones et al., 2007). The addition of condensed tannin to musts deficient in extractable tannin is also thought to improve the retention of anthocyanins during maceration by polymerizing with them (Jones et al., 2007). This may be important when working with grapes that have problems producing wines with sufficient color. Enological tannin preparations containing both hydrolysable and condensed tannins are available for the purpose of stabilizing color and increasing the retention of grape tannins which are less aggressive.

A study adding commercial seed tannin to a Norton must (high in anthocyanins and low in tannin) at crush increased tannin but did not increase polymeric pigment (Wilker and Zhang, 2007). A similar addition made after fermentation and pressing, increased both polymeric pigment and tannin. Tannin concentrations were highest when additions were made after pressing. It did not appear that tannin added prior to fermentation improved color stability for these wines. It is unclear if a variety known for its problems with color stability such as Pinot noir would behave the same way.

In varieties with ample anthocyanins, it would appear that tannin additions made after pressing can result in increased polymeric pigment formation and greater tannin retention than additions made at crush. Further work needs to be carried out to determine the sensory effect of early (at crush) versus later (after pressing) tannin additions.

References and further reading

JACKSON, R. S. 2000. *Wine science*, San Diego, Academic Press.

JONES, M., J. JUST, M. KARRER, A. KYNE, S. RATHBU and R. SWANSON. 2007. *2007 Fermentation Handbook*, Petuluma, Scott Laboratories Inc.

MARQUETTE, B. and D. TRIONE. 1998. The Tannins. In: *ASVO seminar, phenolics and extraction*. M. Allen, and G. Wall (Eds.) pp. 42–43. Adelaide, Australian Society of Viticulture and Oenology.

RIBEREAU-GAYON, P., D. DUBOURDIEU, B. DONECHE and A. LONVAUD. 2000. *Handbook of enology Volume 1 The microbiology of wine and vinifications*, Chichester, John Wiley & Sons.

WILKER, K.L. and J. ZHANG. 2007. Formation of polymeric pigments in wine with a low ratio of tannin to anthocyanin. Abstr. ASEV 58th annual Meeting *Am. J. Enol. Vitic.* 58: 419A.

2.17 How can I estimate when to fortify a fermenting juice to achieve desired sugar and alcohol concentrations?

G. S. Ritchie

Fortified wines, such as Port-style wines, can be made by choosing to ferment the juice or must to a desired sugar concentration and then adding a sufficient amount of a distilled spirit to arrest the fermentation and raise the alcohol to a predetermined level.

In order to carry this out, we have to know the amount of sugar left in the juice at any one moment in time. In small wineries, a hydrometer is often used for this purpose, because it gives an instant answer and is economical to buy. However, using a hydrometer to estimate sugar concentration is complicated by the presence of alcohol in the juice and causes the hydrometer to sink more, thus indicating a lower Brix than is actually present. One percent v/v alcohol obscures 0.47 Brix (Rankine, 1991). The true Brix is given by:

True °Brix = Obscured °Brix + Measurable °Brix

Measurable Brix is that which is detected by a hydrometer. True Brix is the targeted sweetness of the resulting fortified wine.

Calculations for fortification have two stages. Initially we have to calculate the *measurable* Brix that is associated with the *true* Brix at which we wish to fortify (the fortification point). Then we have to calculate the gallons of distilled spirit to add using Pearson's Square (Rankine, 1991). The calculation will be illustrated by considering the fortification of one barrel of wine to 19% alcohol at a true Brix of 8, using 97% distilled spirit and where the Brix of the grapes at harvest was 34.

Calculating the measurable Brix at the fortification point

Step 1. *Calculate the amount of alcohol produced by the time the required sugar concentration has been reached (the fortification point):*

% v/v Alcohol produced at the fortification point =
(Initial °Brix at harvest − °True Brix at fortification) × sugar : alcohol conversion factor[1]
= (34 − 8) × 0.60 = 15.6

---

1. To calculate the fortification point, we have to make an assumption about the efficiency of conversion of sugar to alcohol. In theory, 1 g sugar is converted into 0.511 g ethanol (Boulton *et al.*, 1996). In practice, production of compounds other than alcohol and carbon dioxide, and evaporation change the observed conversion factor to between 0.55 and 0.63 (Boulton *et al.*, 1996), making it difficult to predict. Some wineries overcome this problem by collecting data over several years and estimating an average value for their white and red varieties. Here, I have assumed it to be 0.60.

Step 2. *Calculate obscuration at the fortification point:*

Obscuration of °Brix = Alcohol at fortification point × 0.47°Brix
= 15.6 × 0.47 = 7.33

Step 3. *Calculate the measurable Brix at the fortification point:*

Measurable Brix at fortification point = True Brix − Obscured Brix
= 8 − 7.33 = 0.67

Therefore, the must should be fortified at a measurable Brix of 0.67 when it will contain 15.6% v/v alcohol.

Calculating the volume of spirit to add
The actual volume to add at the fortification point is calculated using Pearson's Square (Rankine, 1991):

$$\text{Volume of distilled spirit to be added} = \frac{V(D-A)}{C-D}$$

where V = volume of juice or wine, D = desired % v/v alcohol, A = % v/v alcohol at fortification point and C = % v/v alcohol of distilled spirit.

For our example of fortifying one barrel of wine (220 L) to 19% v/v alcohol with 95% v/v distilled spirit:

$$\text{L of distilled spirit required} = \frac{220(19-15.6)}{95-19} = 9.84 \text{ L}$$

In conclusion, the calculation indicates that the barrel needs to be fortified with 9.84 L of 95% v/v distilled spirit at a measurable Brix of 0.67. Figure 1 illustrates how a spreadsheet may be set up to help with the calculations.

N.B. Adding alcohol will change the equilibria in the juice or wine and so pH and titratable acidity will need to be checked and adjusted accordingly.

References

BOULTON, R. B., V. L. SINGLETON, L. F. BISSON and R. E. KUNKEE. 1996. *Principles and Practices of Winemaking*, New York, Chapman & Hall.
RANKINE, B. C. 1991. *Making good wine*, Australia, Pan Macmillan Publishers.

Fortification calculations using Pearson's Square

Pearson's Square

$$\text{Litres of distilled spirit required} = \text{Liters of wine} \times \frac{\text{Desired Alc} - \text{Initial Alc}}{\text{Alc in distilled spirit} - \text{Desired Alc}}$$

Estimating Brix at which to fortify

% alc produced before fortification = (Brix at Harvest × 0.60)
　　　　　　　　　　　　　　　　− (True Brix at fortification × 0.60)

1% alcohol obscures 0.47 Brix of sugar
Obscured Brix = alcohol × 0.47
Targeted Brix = obscured Brix + measurable Brix
Assume sugar conversion to alcohol is 0.60

User-provided data

Liters of must/wine =	220
Brix at harvest =	34
Targeted Brix at fortification =	8
%v/v alc in distilled spirit =	95
Desired %v/v alcohol =	19

Calculations

Step 1	% alc produced before fortification =	15.60	=(D14*0.6)-(D15*0.6)
Step 2	Obscured Brix at fortification =	7.33	=D19*0.47
Step 3	Measurable Brix at fortification =	0.67	=D15-D20
	Liters of distilled spirit to add =	9.84	=$D$13*($D$17-$D$19)/($D$16-D$17)
	Final sugar concentration (g/L) =	76.6	=10*$D$16*$D$14/($D$14+$D$25)

Fig. 1 Spreadsheet equations for calculating fortification using Pearson's Square.

2.18 What is the ideal temperature to press ice wine ('frozen') grapes?

C. Butzke

Traditionally, there have been three major options for winemakers to naturally concentrate the aromas, sugars, and acids of grapes prior to fermentation: freezing them partially to leave the frozen water behind after pressing; letting the noble rot *Botrytis cinerea* penetrate the grape skins and thereby dehydrate the berries; or letting the grapes partially dry on racks post-harvest, the *passito* way.

There has been a long and often politically motivated discussion about what constitutes frozen grapes and where and when the freezing takes place. Advocates for 'true' ice wines argue that the grapes have to be frozen in the vineyard and brought straight to the winery for immediate pressing. The proponents of such definition tend to reside in areas that regularly experience early frosts in relative proximity to a regular harvest. In warmer regions, or traditional ones that are getting affected by global warming, a picking of disease-free fruit at full maturity and subsequent freezing and storage in a temperature-controlled environment, is the method of choice. The latter may also allow the winemaker to conduct the tedious pressing of frozen grapes at his/her convenience at a less busy time of year.

While either production method can be legally employed, several countries have established regulatory definitions in regard to the use of the term 'ice wine' on labels. From a winemaking perspective, however, there is little evidence that the method of freezing matters to the quality of the product. It can be argued that grapes that see long hang-times and associated mold growth provide less of a fresh ice wine character and more musty botrytized 'late harvest' aromas, making the two styles less distinguishable for the consumer.

In the US, (TTB) ATF Ruling 78-4, allows the label use of the term 'ice wine' for wine made from grapes that were 'partially frozen on the vine', a rather vague definition. In Canada, ice wine grapes have to be harvested arbitrarily at or below $-8°C$ ($18°F$), conveniently the ideal temperature to press frozen grapes.

Given their high water content, grapes start to freeze at $0°C$ ($32°F$). If pressed at this point, the pressed juice would be identical to free run juice of regularly harvested grapes. Depending on their sugar and acid content, grapes will be frozen solid at about $-11°C$ ($12°F$) which makes pressing them impossible and poses a risk for serious equipment damage.

This leaves the icewinemaker with a fairly narrow operating window of between -10 to $-7°C$ (14 to $20°F$). Pressing on the colder side will take but result in higher Brix juices. Pressing at the warmer end of the window will speed up the process but lead to a more dilute juice. In general, at those temperatures and depending on the Brix at harvest, the press goal should be to have a consolidated mix of juice at the end of the press run of 36 to 42 Brix (% sugar). This will create a nicely balanced wine with modest amounts of alcohol (around

10 to 12% by volume) and sufficient residual sugar (15 to 25% by weight) that can balance the equally concentrated acidity of the wine without being cloyingly sweet.

Given the strong fermentation performance of modern wine yeasts, especially *Saccharomyces bayanus* strains, the fermentation will have to be controlled closely and stopped (usually by rapid cooling) before the alcohol content gets too high and throws the wine off balance.

It is also important to make sure that the harvested grapes will not thaw during the transport from the vineyard or at the winery while they wait to be pressed. Harvesting/storing 2 K lower to account for a rise in fruit temperature during processing is appropriate.

Reference

http://www.ttb.gov/rulings/78-4.htm

3

Yeast fermentation in wine

3.1 What different types of fermentor are there and which should I use for red and white wines?

M. R. Dharmadhikari

A fermentor is essentially a container for fermenting must. Many types of fermentors are used in winemaking and winemakers have several options available to them. For economic reasons many fermentors are also used for bulk wine storage, thus serving a dual purpose. We will assume here that a dual purpose use is intended. Here are some factors to consider when choosing fermentors.

Material of construction
When choosing a fermentor/storage container, the first thing to consider is the material it is constructed from such as: stainless steel, wood, plastic and cement concrete. Plastic fermentors are not yet widely accepted and the use of cement tanks is becoming rare. Ideally the material used for constructing fermentor/storage containers should be impervious, inert, durable, strong, and easy to clean. It should impart no undesirable flavors to the wine (Boulton *et al.* 1996). Based on these criteria, stainless steel and wooden fermentors would appear to be the most suitable fermentation vessels.

Shape and size
Fermentors are made into various shapes such as: cylindrical, cubical, conical, oval and others. Upright cylindrical shape fermentors with a diameter to height

ratio ranging from 1:2, or larger, seem to be more common. However, a red must fermentation tank with a diameter that is larger than its height gives a thinner cap, making cap management more convenient. Fermentors also come in a wide range of sizes/capacity from 50 to several thousand gallons, with open or closed tops and in some cases with specialized features for cap management during red wine production.

Type of wine to be made
The selection of fermentor will also depend on the type of wine being made, for example, white or red?

White wine fermentors
Fruity and aromatic white wines require relatively precise temperature control during fermentation. For this reason tall fermentors with a diameter to height ratio of 1:3, or higher, are preferred. This is due to the fact that this design (shape) allows greater contact between the wine and the cooling surface. For making barrel-fermented styles of white wine, one will obviously need barrels. Usually 50-gallon oak barrels are used as fermentors. The goal is to extract and integrate the desirable oak phenolics and aroma components into wine. In barrels the fermentation occurs at a slightly warmer temperature and due to larger surface to volume ratio, artificial temperature control is seldom needed. After the completion of fermentation the wine is often left in contact with the lees for an extended period. During the lees contact the yeast constituents are released into the wine, which will add to the wine's mouthfeel and complexity.

Red wine fermentors
In red winemaking the juice and solid parts of the grape (pulp, skin and seeds) are fermented together. During active fermentation the solid parts are pushed to the top, forming a cap. In order to facilitate the extraction of color, tannins and desirable flavor compounds from the solid parts, the cap needs to be broken and mixed with the fermenting juice. In small, open top fermentors the cap can be broken by punching it down with a pedal and mixing the must. In large fermentors, the cap can be too thick to be punched down. In such a case, the juice is drawn off from the bottom of the fermentor and is sprayed over the cap by a rotating irrigator. The juice permeates through the cap and promotes extraction from the solid parts. This process is called pump over. Usually the cap is pumped over about twice a day with one juice volume. There are specially designed fermentors such as auto fermenter and rotary fermentors which mechanically manage the cap to enhance the extraction of desirable compounds from the pomace during fermentation.

Other factors to consider
In choosing a fermentor, a winemaker needs to consider some additional factors other than the ones mentioned above. These include the grape cultivar used,

style of wine, the average lot size to be processed, and the price point at which the wine will be sold. The fermentor should be easy to clean and sanitize and be placed in a well-ventilated fermentation room equipped for ease in cleaning and proper temperature control. There are fermentation and storage tanks that can be used for both white and red wine fermentation. The key feature includes a relatively steep-sloped bottom and a bottom door to allow lees and pomace removal. It should also be fitted with a racking valve, racking door, sampling port, a cooling jacket and a top manway for pumpovers.

Reference

BOULTON, R. B., SINGLETON, V. L., BISSON, L.F. and KUNKEE, R. E. 1996. *Principles and practices of winemaking*, Chapman & Hall, New York.

3.2 What materials are used in constructing fermentors and how does this affect fermentation and storage of wine?

M. R. Dharmadhikari

A wide range of materials are used in making fermentors for wine production. The main materials used are stainless steel, wood, plastic and cement concrete, but include others.

Stainless steel tanks

Stainless steel closed-top fermentors are widely used in the wine industry and their use is becoming more and more common. This is due to the fact that stainless steel, as a material of construction, offers several advantages. It is inert, so nothing from the tank is extracted into the wine. It is impervious, so it is leak-proof and for aromatic white styles of wines the access of air can be limited. It rapidly transfers heat. This means better temperature control (cooling or heating) during fermentation. The fermentors are easy to clean and sanitize and they are also durable. The tanks are designed in various shapes and sizes and some are also equipped with special features to handle automatic cap management. Small size, or porta tanks, are movable and can be stacked for efficient space utilization.

Wooden containers

Wooden containers are commonly used in the wine industry for conducting fermentation and maturation of red and white table wines. The barrels are made of oak wood and hold about 50 gallons of wine. Larger capacity barrels are also used and they are called casks, ovals or butts. The vats (open top) and the tanks (closed top) of varying capacity ranging from 1000 gallons and more are also used as fermentors. They may be made of oak or redwood and they are constructed with staves that are not bent and coopered. In some cases small 50-gallon barrels are also used for making barrel fermented wines with sur lie aging. Wooden fermentors are less expensive than cement or stainless steel tanks and the loss of heat is usually greater than the cement tanks. They are, however, prone to leak and can be difficult to clean and sanitize. Wooden containers, particularly when new, can also contribute to the flavor of wine.

Plastic tanks

The use of plastic and fiberglass fermentors is becoming popular. These tanks are less expensive and light weight, as compared to cement, wood and stainless steel. However, they may transmit light and/or oxygen, which can be detrimental to some wines (Boulton *et al.* 1996). However, with the advancement in technology and innovations, we are likely to see new plastic tanks that would overcome the drawbacks mentioned. Some of the new plastic polyethylene tanks are equipped with stainless steel manways and sanitary fittings and can be used for temperature-controlled fermentation. In recent years, new plastic fermentors with proprietary names such as Flex tank and Fermenta bags have been introduced. For many small scale producers they offer some advantage. However, how well they will be accepted remains to be seen.

Cement/concrete tanks

In older wine-producing regions cement tanks were common. They are also being used in some parts of the new world wine industry, but in general the trend is moving towards the use of stainless steel as a material of choice for making wine fermentors. Generally, the cement tanks are built with reinforced concrete and covered with layers of cement coating. The inside is finished smooth. For red wine production the walls are neutralized by scouring with tartaric acid (Penaud 1984). Usually the inside is lined with stainless steel, ceramic tiles, paraffin, epoxy resin or some inert material in order to prevent direct contact between the wines and cement surface. The cement tanks are strong and durable, not prone to leakage like wood, and need little upkeep. Some winemakers believe that cement tanks maintain steady temperatures during fermentation and produce better red wines (Teichgraeber 2005). The disadvantages of cement tanks include possibility of increased calcium content in wine. The wine contains many acids which can react with calcium and form salts rich in calcium. The calcium salts can contribute to wine instability. Other issues include difficulty in cleaning and sanitizing the tank and low heat loss (Amerine and Joslyn 1970).

References

AMERINE, M. A. and JOSLYN, M. A. 1970. *Table wines: The technology of their production*, Berkeley, University of California Press.
BOULTON, R. B., SINGLETON, V. L., BISSON, L. F. and KUNKEE, R. E. 1996. *Principles and practices of winemaking*, Chapman & Hall, New York.
PENAUD, E. 1984. *Knowing and making wine*, John Wiley & Sons, New York.
TEICHGRAEBER, T. 2005. Concrete fermenters: from old school to new world, *Wines and Vines*, November, 2005.

3.3 What is a yeast 'strain'?

L. Bisson

Yeasts are generally classified into genus and species names, such as *Saccharomyces cerevisiae*. Yeasts in the same genus share physiological traits and genetic relatedness and are placed into different species if they are not capable of successfully inter-breeding because the chromosome structure is too different. Species can be further subdivided into 'races' and 'strains'. A race classification means that the organism is capable of mating with other members of the same species but possesses unique physiological traits that can be measured and observed. A strain designation means that the genotypes of the two organisms are not identical – this can mean that only a single gene differs and that difference may not be manifest in any measurable physiological way. The strain designation simply means that the two organisms are not genetically identical.

3.4 Does the yeast strain have an influence on the fermentation kinetics and on the wine aromas and flavours?

G. Specht and D. Delteil

Among the selected *Saccharomyces* available as enological active dry yeast, there are very significant differences in their kinetics, and aptitude to achieve a complete fermentation. Those differences come mainly from the yeasts nitrogen and oxygen needs, which have been characterized for some of them, and its ability to tolerate juice conditions. Another important influence on the yeast fermentation kinetics is the individual yeast's ability to compete against the microbial soup of indigenous yeast present in the juice or must. Under winery conditions, competition between yeast for the uptake of nutrients and the influence of the competitive factor is a practical key point in the choice of an enological strain.

The secondary metabolism of the yeast produces compounds having great importance in the wine analytical and sensory profile. It had been demonstrated as early as 1990 that there are significant differences among *Saccharomyces* for their production of esters, sulphur compounds, varietal compounds such as beta-damascenone, polysaccharides, for the impact of their mannoprotein on the volatility of certain compounds and on the stability of pigments and polyphenols. With such significant possible analytical differences between enological yeast, it is logical that significant sensory differences have been demonstrated as early as the mid 1980s. More recently it has been demonstrated that yeast mannoproteins released in the fermentation must have a sensory impact on the aromatic expression of varietal volatile compound such as beta-ionone or fermentation volatiles such as ethyl hexanoate. In multi parameter sensory studies with PCA (principal component analysis) of sensory analysis results, different yeast strains in red wines, can give sensory differences as important as grapes at different stages of ripeness or as maceration length; techniques generally recognized as having a big impact on the wine style. Such sensory differences due to the yeast strain were still significant years after vinification.

Further reading

DELTEIL, D. 2001. Aspects pratiques du levurage en conditions méditerranéennes. Technique d'inoculation et rapport entre population sélectionnée et population indigène. *Revue Française d'Oenologie*, 189, 14–19.

DELTEIL, D. and JARRY, J. M. 1992. Characteristic effects of two commercial yeast strains on Chardonnay wine volatiles and polysaccharides composition. *Australian & New Zealand Wine Industry Journal*, 7, 1, 29–33.

DELTEIL, D., CHALIER, P., ANGOT, B., DOCO, T. and GUNATA, Z. 2004. Interactions entre les macromolécules du vin et les composés d'arômes. Comportement de mannoprotéines isolées de 2 souches de Saccharomyces vis à vis de la perception sensorielle de certains composés volatils du vin. *Poster OIV congress*, Wien (Austria).

DUMONT, A. 1994. Sensory and cemical evaluation of Riesling, Chardonnay and Pinot noir fermented by different strains of *Saccharomyces cerevisiae*. Thesis. Oregon State University.

SABLAYROLLES, J. M., JULIEN, A., ROUSTAN, J. L. and DULAU, L. 2000. Comparison of nitrogen and oxygen demands of enological yeasts: technological consequences. *Am. J. Enol. Vitic.*, 51, 3, 215–222.

3.5 What is a native flora fermentation?

L. Bisson

Native flora fermentations are conducted by the yeast indigenous to the vineyard or the winery. These are also called 'autochthonous' yeasts, meaning they arose from where they were found and were not externally introduced. *Saccharomyces cerevisiae* is the organism that dominates wine fermentations. In inoculated fermentations a starter culture or preparation of yeast is added to the juice or must to initiate fermentation. The starter culture generally grows in the juice or must and conducts the fermentation. In a native flora fermentation no inoculation occurs and the autochthonous yeasts grow and dominate the fermentation.

3.6 What factors are important in deciding to use a yeast inoculum versus allowing the native flora to conduct the fermentation?

L. Bisson

Although *Saccharomyces* can be found on grapes and in juices following processing of the grapes, it is generally present in levels of less than 1 to 1000 cells/mL. Wild yeasts that are members of different species, such as *Kloeckera, Hanseniaspora, Candida, Metschnikowia,* and *Pichia,* are present at levels of one million cells/mL, and bacteria can be found at even higher levels. The autochthonous *Saccharomyces* cells will eventually dominate and be found in numbers as high as 100 million cells/mL, but this takes time to reach this number starting from less than 1000 cells/mL. In that interim the other yeasts and bacteria present can grow and make by-products that will influence the flavor and aroma profile of the wine. This can be valued as adding to the complexity of the wine. On the other hand, negative characters such as acetic acid may be made that detract from wine quality. Depending upon the microbes present they may produce compounds that will inhibit *Saccharomyces* leading to an arrested or protracted fermentation or may result in the production of off-characters by *Saccharomyces*. A slower, cooler *Saccharomyces* fermentation may yield better retention of grape volatile compounds in the wine which also adds to complexity. Thus a native fermentation may enhance wine complexity or it may add spoilage characters and lead to an arrest of the alcoholic fermentation. Should native flora fermentations be considered? The answer is it depends. It depends upon the nature and numbers of the other organisms present that will be encouraged to grow. If there has been significant berry and cluster damage prior to and at harvest, the bioload of organisms transferred to the juice may be too high and guarantee that the fermentation will have difficulty finishing. If the fruit is sound at harvest, this is less of an issue. One important factor to consider though when making this decision is the technical capability of the winery to determine what organisms are present during all stages of production and whether or not inoculation is warranted. Sulfite can be used to minimize the impact of potential spoilage organisms and assist the autochthonous yeast in dominating the fermentation, but this may limit the development of the complex microbial characters.

3.7 What is yeast assimilable nitrogen (YAN) and how much is needed?

B. Dukes

Yeast assimilable nitrogen (YAN) is the sum of the nitrogen content of primary amino acids and ammonium in grape juice or must, expressed as mg N/L. This measure indicates the total amount of nitrogenous compounds available to wine yeast under normal anaerobic fermentation conditions. YAN has traditionally been used as a guide to the nutritional status of a must for vinification; however, more recent research has highlighted the importance of nitrogen management as a winemaking tool which can be used to guide the aroma and flavour profiles in wine. A quality wine is the result of sound balance of numerous integrated factors, which include fruit quality; YAN and other nutrient levels; oxygen; solids levels; yeast strain and concentration; mixing; hygiene; and temperature management. All of these quality parameters need to be in balance as limitation in any factor can compromise success.

YAN is variable in must and juices – so measure it

Surveys of YAN in North America and Australia have shown high variability in the levels, indicating it requires measurement. The determination of YAN requires that both the primary amino acid nitrogen and ammonium fractions are measured because no predictive relationship exists between these fractions. The early identification of potentially problematic musts and juices is the preferred approach to the remediation of problematic fermentations or wines. Routine petiole analysis in vineyards, commonly conducted at around bloom to assist in vineyard nutrition is not suitable to predict YAN in the juice, although the data may provide an overall indication of the health of the vineyard and hence the fruit.

How much YAN is required?

Levels of 140 mg N/L in juice are generally satisfactory for vinifications to complete to dryness, while maximal rates of fermentation in musts have been observed at about 500 mg N/L. American scientists have suggested intermediate levels which relate YAN requirements to the soluble solids content of the must, i.e. higher sugar levels require higher YAN levels. The Brix adjusted YAN is a robust guide due to the large number of interacting factors required for a successful fermentation (Table 1). Experimentation with a selection of commercial yeast strains revealed an approximate one-third variation in YAN requirements. Similarly, solids levels, micro-nutrients, temperature profiles and oxygenation of the ferment have been demonstrated to have a large impact on fermentation performance.

Table 1 Brix adjusted concentrations of yeast assimilable nitrogen (drawn from Bisson and Butzke 2000)

Brix	Beaume	YAN (mg N/L)
21	11.7	200
23	12.8	250
25	13.9	300
27	15.0	350

Optimal levels of YAN – a component of good quality wine

It appears that wine fermentations have a range of YAN which is optimal for quality and fermentation performance. White wine fermentations with intermediate levels of YAN 250 to 300 mg/L result in clean fruity wines with desirable fermentation performance. We do not have as much information on levels for reds; however, similar intermediate levels are also likely to assist in quality management.

It is hypothesised that performance of red fermentations wines may be improved by the presence of the solids (skins, seeds, pulp) because these winemaking conditions may also favour the slow release of yeast assimilable amino acids, and possibly other nutrients, from the skins and seeds as the fermentation progresses. For example, in Cabernet Sauvignon berries, the total free amino acids are distributed as 8.5% in the seeds, 15% in the skin and 77% in the pulp (Stines *et al.* 2000).

Deficient YAN – poor quality

The common wine yeast *Saccharomyces cerevisiae* is eukaryotic, which means that it has a compartmentalised cellular structure. This feature allows the yeast cell to store amino acids in vacuoles, and subsequently regulate the levels of these compounds in the cytoplasm. In fermentation conditions, *S. cerevisiae* depletes most of the YAN (and other nutritional factors) in the must during the first few Brix attenuated. The yeast cells multiply in an exponential proportion to the amount of YAN assimilated, or until the YAN is not limiting.

Deficient levels of YAN restrict the yeast's ability to multiply to sufficient levels to complete the fermentation to acceptable industry standards. Sensory descriptors associated with YAN deficient fermentations may include hydrogen sulphide (rotten egg gas), vegetal, cheesy and reduced colour in red wines. No universal relationship exists between hydrogen sulphide production and YAN levels; however, the general observation is that hydrogen sulphide and mercaptans are more frequently associated with situations of nitrogen limitation. Winemakers aim to optimise fermentation performance in order to reduce the probability of the formation of sulphide like odours during fermentation, maturation and post bottling. This is particularly important when using modern screw top closures. Winemakers often choose to supplement YAN during fermentation in response to hydrogen sulphide, often with 100 to 200 mg diammonium phosphate (DAP)/L additions, which are typically made before 10 Brix, below which the yeast is not likely to be able to assimilate it.

Stuck or incomplete fermentations leave the wine with an excess of residual sugar, i.e. not meeting quality specifications. The residual sugar can be a source of down-stream microbiological instability, particularly in red wines where levels of residual sugars above 100 mg/L may provide an energy source for *Brettanomyces*, a spoilage yeast which causes medicinal, band aid and faecal like characters.

Excessive YAN – poor quality
Measurement of YAN prior to fermentation gives the winemaker the ability to adjust optimal levels, in some cases eliminating 'routine' DAP additions. Overuse of DAP incurs costs of the DAP, labour, administration and may sacrifice wine quality. Excessive YAN can increase heat generation and rates of fermentation to the detriment of wine quality. A red ferment that completes without sufficient fermentation time may not achieve desirable tannin and other sensory profiles. Similarly, a white barrel fermented style may complete before adequate time in the barrel. Consequences of increased heat generation are increased energy costs and reduced refrigeration capacity. Excessive YAN additions may also lead to elevated levels of residual nutrients in the wine, creating down-stream microbiological problems. Sensory characters associated with excessive YAN include the over production of ethyl acetate (reminiscent of nail polish remover) and acetic acid (vinegar).

References and further reading
AGENBACH, W. 1977. A study of must nitrogen content in relation to incomplete fermentations, yeast production and fermentation activity. *Proc. S. Afric. Soc. Enol. Viti.* 66–87.
BISSON, L. and BUTZKE, C. 2000. Diagnosis and rectification of stuck and sluggish fermentations. *Am. J. Enol. Vitic.* 51, 2, 168–177.
STINES, A. P., GRUBB, J., GROCKOWIAK, H., HENSCHKE, P. A., HØJ, P. B. and VAN HEESWIJCK, R. 2000. Proline and arganine accumulation in developing berries of *Vitis Vinifera* L. in Australian vineyards: Influence of vine cultivar, berry maturity and tissue type. *Aust. J. Grape and Wine Research* 6, 150–158.
UGLIANO, M., HENSCHKE, P., HERDERICH, M. and PRETORIUS, I. 2008. Nitrogen management in critical for wine flavour and style. *Aust. N.Z. Wine Ind. J.* 22, 6, 24–30.

3.8 How do I measure yeast assimilable nitrogen (YAN)?

B. Dukes

The measurement of YAN requires determination of primary amino groups and ammonium ions. Each component must be measured because no predictive relationship exists between them. The measured concentrations of each fraction are then mathematically converted to mg N/L so that winemakers have a convenient summary form of data. Other nitrogenous compounds such as amines, peptides, purines, and pyrimidines are also present in grape juices; however, these compounds are typically present at micro-molar concentrations. Amino acids and ammonium ions are present in the milli-molar range. The bottom line is that the ammonium and amino nitrogen fractions are the most abundant and the quantitative importance of the other nitrogenous fractions are insignificant in a practical winery situation. Proteins in the grape juice are not available to the yeast as a nitrogen source in normal fermentation conditions.

The ammonium fraction is generally determined by a spectrophotometric analysis using enzyme kits supplied by various companies. The basis of this *glutamate dehydrogenase* catalysed determination is that the ammonium ions are indirectly determined by measuring the amount of NADH oxidised at 340 nm. This procedure is popular because it is rapid and easy to automate. Some wineries prefer to use a selective ion electrode.

Primary amino groups are also determined by a spectrophotometric technique that is based on the derivatisation of primary amino group(s) on the amino acids, with *o*-phthaldialdehyde (OPA) and N-acetyl-L-cysteine (NAC). The primary amino groups are present in limiting quantity, in contrast to the OPA and NAC, which are present in excess. Hence, the quantity of the derivative formed is proportional to the amount of reactive amino groups. The reaction product is a stable, uv light absorbing derivative that is used to determine the number of amino groups present. The result is expressed in mg N/L based on the number of isoleucine equivalents. The strength of this NOPA technique ('nitrogen by OPA') is that proline, a secondary amino acid that is often dominant in musts but is not available to yeast under vinification conditions, is not detected.

The wet chemistry enzymatic ammonium ion and NOPA procedures are both performed using a uv- spectrophotomer found in many winery labs. These methods have been automated; however, recent advances in mid-infrared spectrometry (MIR) methods mean that YAN determination is simplified. The reference methods for the MIR are NOPA and the enzymatic determination of ammonium ions. Most winegrowing areas have commercial labs which can economically perform YAN analysis in less than a day.

Further reading

DUKES, B. and BUTZKE, C. 1998. Rapid determination of primary amino acids in grape juice using and *o*-phthaldialdehyde/N-acetyl-L-cystine spectrophotometric assay. *Am. J. Enol. Vitic.* 49, 2, 125–134.

3.9 How do I interpret yeast assimilable nitrogen (YAN) data?

B. Dukes

The amino nitrogen component of YAN is expressed as equivalents of isoleucine, the selected reference amino acid and then converted to mg N/L. As such, this technique will always underestimate the absolute amount of nitrogen present because amino acids such as arganine and glutamine have a single reactive primary amino group, yet contain four N atoms, hence the yeast will get more N per arginine assimilated.

The main point to establish is that YAN is a robust tool to indicate the macro-nutritional health of a must for a vinification. It gives no direct information about micro-nutrition; however, low levels of YAN may indicate that the general nutritional status of that must is low. In these circumstances many winemakers supplement the must with complex nutrients. These mixtures, supplied by most major yeast companies, are used in the yeast rehydration phase or during the fermentation, as per the manufacturer's instruction. Indeed, some suppliers have published data showing winemaker preferences for Sauvignon blanc wines made with these supplements. If a winemaker decides to supplement the juice/must with a complex mixture, then the winemaker may require the YAN amount and composition (free amino nitrogen and ammonium ion) contained in the propriety blend so that the YAN content can be considered to determine if DAP is required. This information can be determined by analysis or from supplier data.

It has been demonstrated that yeast supplied with a mixture of nitrogenous substrates demonstrated better growth than when supplied with a single source. This suggests that the relative amounts of ammonium supplied N and amino supplied N may affect fermentation. At present we do not have sufficient data to determine optimum target levels of these individual components. However, in the future, the composition of the YAN is likely to become a valuable tool in wine quality management.

3.10 When and how do I adjust yeast assimilable nitrogen (YAN)?

B. Dukes

YAN should be measured in juices or musts immediately prior to inoculation after all juice or must conditioning treatments. Fining agents such as bentonite and colloidal silicone dioxide are known to have a strong affinity for amino nitrogen, hence reducing its concentration in the juice. Similarly, pre-inoculation conditions such as 'cold soaks', delayed SO_2 additions, poor mixing of SO_2 and/or insufficient SO_2 may allow for indigenous microbes such as *Kloeckera apiculata* to assimilate YAN and other nutrients. Modification of process techniques can improve YAN levels in juices. For example, white press fractions are richer in YAN than the free run, hence the treatment of the different fractions to address phenolic concerns, and combination pre-fermentation is usually advantageous. Sampling during fermentations will not give a YAN value that we can interpret because of the amount of YAN removed from the medium by yeast.

Di-ammonium hydrogen phosphate (DAP) is the most common source of YAN supplement at present, possibly because it is inexpensive and easy to use. DAP is a white crystalline powder which when dissolved in aqueous solution is 21% (w/v) N in its ammonium form and 23% (w/v) as the inorganic phosphate. A useful approximation is that to increase the must or juice by 1 mg/L of N, add 5 mg/L of DAP. The DAP is weighed, dissolved in a minimum volume of water, and added to the wine tank with mixing. Fermentation responses to DAP additions may be due to a nitrogen response, a phosphate response, the interaction of the ions, or even the mixing of the tank associated with the addition. The DAP addition may also cause a small increase in the pH which is determined by the buffering capacity of the must. The use of DAP in the wine industry is subject to specific regulations in different countries (Table 1).

Winemakers who are concerned about residual phosphate levels in wine could use liquid ammonium solutions as a YAN source if permitted in their country.

Some winemakers supplement the must or juice to a 250 mg N/L to achieve adequate cell numbers (biomass) in the growth phase, followed by a second

Table 1 2008 limit for DAP additions to must varies with country

Country	Maximum addition of DAP mg/L	Maximum addition of mg N/L corresponding to the DAP
European Union	325	64
United States	960	200
Australia	Max of 400 mg inorganic phosphate/L provides an indirect limit on DAP additions	

addition of 50 mg N/L at one third to the halfway point of the fermentation. This second addition is to stimulate the rate of the fermentation. The production of adequate biomass appears to be more important than the effect of YAN on cellular activity (Coleman *et al.* 2007). Hence, it is critical to have enough N in the must at the start to achieve adequate cell numbers. DAP additions after the growth phase have little effect on the population numbers (biomass), while generally having a positive effect on the rate of fermentation. Similarly, and at about the same timing of the second DAP addition, 5 to 10 mg/L of oxygen appears to provide additional advantages in assisting the completion of fermentations.

Why do I have to add DAP before the mid point of the fermentation?
Saccharomyces cerevisiae has energy requiring active transport systems for amino acids which appear to exploit the pH differential between the juice or must (pH 3 to 4) and the cells cytoplasm (pH 6 to 7). As each amino acid is transported into the cell, the transport system must expel a proton in order to prevent internal acidification, and hence denaturation of proteins. The cost of each amino acid in, and subsequent proton out is one adenosine triphospate (ATP). As the vinification progresses, increased levels of alcohol cause an increase in the stress on the plasma membrane. This in turn makes the plasma membrane more permeable to protons from the external environment. Eventually, the ATP expense becomes too great for the gain and the cell will stop actively transporting amino acids and ammonium ions.

In the winery, this means that DAP additions must be made while the yeast can still assimilate them, which is typically up until the mid-point of the fermentation. Additions later in the ferment have diminishing marginal returns. Indeed, the addition of DAP after the yeast cells have lost their ability to assimilate it, will increase the nutritional richness of the must or juice, potentially encouraging down-stream instability.

Winemaker adjustments of YAN via DAP during the fermentation require an understanding of the fermentation profile of the yeast, particularly towards the end of the fermentation. *Saccharomyces cerevisae* strains show a generally linear attenuation of the sugar; in contrast to *S. bayanus* strains that show a sigmoidal pattern, characterised by rapid initial attenuation with a slowing of the rate towards completion.

Case studies and applied response in nitrogen management in fermentations
White wine – aim is a clean fruity style with no residual sweetness
The winemaker reviews the vines before vintage, by considering soil analysis, bloom time petiole data, and visual observation. It appears that limitation of some soil nutrients is also reflected by petiole data and field observation. The low YAN, and the vineyard information suggests that other nutrients may be limiting.

The YAN was determined from a sample of the settled juice the day before its racking for fermentation so that the winery had enough time to act on the data before selecting a yeast strain and other considerations. The juice is 24 Brix and has a YAN value of 186 mg N/L.

The winemaker believes that the style objectives are best achieved via a low solids juice, with a target YAN of 300 mg N/L. The YAN plan is an early supplementation to 250 mg N/L after the first few Brix are attenuated, followed by an additional 50 mg N/L added between one-third and half-way through the vinification.

The winemaker selected a robust yeast with moderate YAN requirement and suggested inoculation rate of 250 mg/L. The inoculation rate was increased by 20% to 300 mg/L of the freeze dried yeast. It was hydrated in a complex nutrient mixture according to the manufacturer's instructions. The low YAN in the juice and the vineyard observations suggested that other nutrients may be low.

The calculation for the first DAP was:

$$\text{Addition required} = (\text{Target YAN} - \text{Actual YAN}) \text{ mg N/L}$$

$$= (250 - 186) \text{ mg N/L}$$

$$= 64 \text{ mg N/L}$$

Mass of DAP required is 64 mg N/L × 5 mg DAP/mg N/L = 320 mg DAP/L, i.e. 0.32 g DAP/L.

This 0.32 g/L DAP addition was made after the first few Brix had been attenuated. This was to ensure that most of the endogenous YAN had been consumed by the eukaryotic yeast before the DAP was added. The addition of ammonium ions has been shown to suppress the assimilation of the amino fractions, hence the DAP is added after the must is 'stripped' of nutrients by the yeast.

The fermentation rate and performance at one-third progress were good, hence the addition of the remaining 50 mg N/L was delayed to the mid-point of the fermentation so as to discourage excessive rate and heat generation. The mid-point DAP addition of 50 mg N/L or 0.25 g DAP/L was in-combination with 10 mg/L of oxygenation, achieved by a micro-oxygenation system.

The fermentation was also mixed without oxygenation at 6 Brix, and adjusted to 20°C to allow 'gentle' conditions for the vinification to complete.

Post vintage, the winemaker worked on strategies to get the fruit to specification in the vineyard due to the belief that a naturally balanced juice was better for wine quality.

Red must
Must 205 mg N/L, 25 Brix

Vineyard inspections, soils and petiole data suggest that the vineyard is healthy. The winemaker decided to adjust to a YAN 250 mg N/L at the mid-point of the fermentation, but decided that complex nutrients were not required.

The complete solids, warmer fermentation temperatures, acceptable ripeness and the hypothesis that reds allow a gradual release of YAN from the must, allowed the winemaker to be comfortable with the target of 250 mg N/L as YAN. The addition of 45 mg N/L (0.225 g DAP/L) was made at the mid point of the fermentation, along with oxygenation at about 10 mg/L, and coinciding with a pump over. The fermentation progressed to dryness with the appropriate kinetics.

References and further reading

BELY, M., SABLAYROLLES, J-M. and BARRE, P. 1991. Automatic detection and correction of assimilable nitrogen deficiencies during alcoholic fermentation in enological conditions. In: *Proceedings of the International Symposium on nitrogen in grapes and wine*. Seattle, Washington, USA. Ed. Rantz, J. Published by the American Society for Enology and Viticulture, 211–214.

COLEMAN, M., FISH, R. and BLOCK, D. 2007. Temperature-dependent kinetic model for nitrogen-limited wine fermentations. *App. Environ. Microbiology* 73, 18, 5875–5884.

3.11 Is it important to rehydrate active dry yeast with a precise procedure?

G. Specht and D. Delteil

Proper rehydration is perhaps the most critical phase in using active dried wine yeast (ADY). When selected wine yeasts are produced in an active dried form, the goal is to get a very high and viable cell population prior to a series of water removal steps until 5–8% moisture remains in the yeast powder. This low moisture level is necessary to ensure a good shelf-life in order to conserve the yeasts potential activity for more than 36 months. These drying stages remove not only extra-cellular water, but also most of the water within the cell and bound to the cell's organelles causing the yeast cells to shrink and desiccate. With extremely low water activity, there is almost no metabolic activity. To be functional again, the dried yeast cells must reabsorb all of their water. When the dried yeast comes into contact with water (or any other liquid) the cells literally act like dried sponges and suck up the needed water in seconds.

In winemaking, the grape juice is a very hostile medium for *Saccharomyces*: high osmotic pressure, low pH, often high SO_2. Most selected enological yeasts resist these conditions when their membrane and their intracellular components are in good physiological conditions. Therefore, before inoculating a grape juice, it is absolutely necessary to bring back vital water to the ADY cells in order for their membrane and intracellular components to reorganize properly.

Not only will yeast cells not disperse very well if not properly rehydrated, they can lose a large amount of cytoplasm, reducing the efficiency of oxygen and nutrient transfer to the cells. This impedes yeast growth and activity leading to sluggish and stuck fermentations. Proper yeast rehydration gets them off to a good start and ensures they are healthy and that they have good fermentation characteristics.

Recent studies have investigated rehydration protocols on the recovery of fermentative activity of different yeasts. Some differences were observed among yeasts using different rehydration protocols; however, these differences were not significant enough, suggesting that the best advice is to follow the yeast producer's rehydration instructions. Other recent studies have shown the influence on yeast viability and vitality of rehydrating ADY in the presence of GoFerm or Natstep micronutrient and/or sterol and unsaturated fatty acid enriched inactivated yeast suspension during rehydration. Higher maximum yeast cell density and shorter overall fermentation lengths were observed when using these types of rehydration nutrients especially under high sugar concentrations.

Further reading

KONTKANEN, D., INGLIS, D. L., PICKERING, G. J. and REYNOLDS, A. 2004. Effect of yeast inoculation rate, acclimatization, and nutrient addition on icewine fermentation. *Am. J. Enol. Vitic.* 55, 4, 363–370.

SABLAYROLLES, J. M., SOUBEYRAND, V. and JULIEN, A. 2006. Rehydration protocols for active dry wine yeasts and the search for early indicators of yeast activity. *Am. J. Enol. Vitic.* 57, 4, 474–480.

SOUBEYRAND, V., SALMON, J., LUPARIA, V., WILLIAMS, P., DOCO, T., VERNHET, H. and JULIEN, A. 2005. Formation of micella containing solubilized sterols during rehydration of active dry yeasts improves their fermenting capacity. *Journal of Ag. and Food Chemistry*, 53, 20, 8025–8032.

3.12 Is the yeast population homogeneous throughout the tank during fermentation?

G. Specht and D. Delteil

During fermentation, *Saccharomyces* produces a lot of CO_2, saturating the juice or must and creating bubbles that rise to the surface. This results in liquid movements and keeps a part of the yeast cells population in suspension. However, even during a very active fermentation, some cells settle at the bottom of the fermenter. As soon as a 1 mm layer of yeast cells are piling upon each other at the bottom of the tank, the cells at the bottom of this layer have no access to sugar and die from starvation. These dead yeasts produce very reductive conditions favouring the production of sulphur off compounds. The yeast cells in suspension will become stressed due to a higher workload to assimilate more sugar per cell in order to finish the fermentation. The slower the fermentation, the more the yeast cells settle out with fewer cells in contact with sugar, slowing down the fermentation even further, amplifying the problem. In tall fermentation tanks this can be a big concern since the yeast at the bottom of the fermenter have to support a greater liquid column pressure which alters their physiology even more. In practice, it is recommended to resuspend the yeast regularly as soon as the fermentation is active. There are different ways to do this: stirring by hand in barrel, deep punching down in open tanks, delestage, propeller agitator in bigger tanks, nitrogen or air injection, etc. In practice, regular agitation gives more regular and more complete fermentation and lower negative sulphur compounds.

Further reading
DELTEIL D. 2008. Buone pratiche di vinificazione del Pinot Noir. *Lallemand Italia conference*, Bolzano (Italy), 29 May.

3.13 What is the influence on the yeast of oxygen addition during alcoholic fermentation?

G. Specht and D. Delteil

During the early stages of alcoholic fermentation, *Saccharomyces* is able to use oxygen for the synthesis of sterols. These sterols help in keeping the yeast membrane fluidity and in resisting osmotic shock and ethanol toxicity. The practical consequences are lower acetic acid production during yeast growth and a healthier yeast population resulting in a more reliable fermentation finish. Even though there is an important variation among the different selected yeasts in their oxygen requirement, yeasts scavenge oxygen very quickly. There are no risks of juice oxidation when aerating or adding oxygen at the recommended times and quantities, even in very fragile juice from varieties such as Sauvignon or Viognier.

There are two optimal times to aerate or add oxygen (4 to 6 mg/L) to the fermenting must. First, during the beginning of yeast cell growth as soon as the fermentation is active to help the yeast resist the osmotic shock and limit acetic acid production. In practice, this corresponds to the consumption of 15–30 g/L sugar. Then it is recommended to aerate or add oxygen a second time at the end of yeast cell growth to ensure a better fermentation finish. This corresponds to the consumption of about one-third of the sugar which is when the must is colonized with the maximum yeast cell density.

Use a special device such as a macro-oxygenator to bring tiny enough bubbles of oxygen to the yeast at the right time and the right dosage. Pumping over can bring some oxygen but the juice is not homogeneous in the tank which means that not all of the yeast population will have access to the oxygen. An additional recommended method of providing sterols to the yeast is during the yeast rehydration step by adding inactivated yeast based products that are rich in sterols to help protect the yeast from the initial osmotic shock. Finally, at one third of fermentation the addition of inactivated yeast-based complex nutrients provides sterols along with a balanced source of nitrogen. This helps the yeast to resist alcohol toxicity at the end of fermentation especially in high alcohol potential juices or musts.

In practice, there is a significant lowering of the potential for sulphur off flavour formation when oxygen addition and these inactivated yeast-based protectant and nutrient products are used. This is due to maintaining a better yeast membrane physiology and balanced nutrition to keep the yeast from becoming stressed.

Further reading

DELTEIL, D. 2006. Pilotaje sensorial en la elaboración de grandes vinos blancos y tintos. *Acts of A.C.E. conference, Barcelona.*

SABLAYROLLES, J. M., DUBOIS, C., MANGINOT, C., ROUSTAN, J. L. and BARRE, P. 1996. Effectiveness of combined ammoniacal nitrogen and oxygen additions for completion of sluggish and stuck fermentations. *J. Ferm. Bioeng.* 82, 377–381.

3.14 Is it important to inoculate the grape with yeast before a cold soak with Pinot Noir?

G. Specht and D. Delteil

Cold soaking, also known as prefermentative cold maceration, is a technique widely used for Pinot Noir in most winemaking countries. Classically the grapes are kept below 12°C for 4–6 days with 30–40 ppm SO_2 and covered with CO_2. Higher additions of SO_2 would lead to problems during fermentation and aging: higher sulphur off flavours produced during alcoholic fermentation, higher acetaldehyde production and more difficult management of molecular SO_2 during aging.

During cold maceration, at this temperature and with such SO_2 level, apiculated yeast such as *Kloekera apiculata* or *Hanseniaspora uvarum* can develop significantly and produce high amounts of ethyl acetate which has a very pungent vinegar aroma. At low levels of ethyl acetate it emphasizes dryness and burning sensations in the mouth during tasting. There are no practical known techniques to then eliminate ethyl acetate during winemaking and aging.

When the grapes are immediately inoculated with a selected enological *Saccharomyces*, at about 4 million live cells per mL, the selected yeast will begin its lag phase and develop slowly. Even at the low temperature, the selected *Saccharomyces* population will begin to use the nutrients in the must making them unavailable to the indigenous apiculated yeast population. By following the practice of proper selected yeast inoculation at the beginning of cold soak, the danger of a significant spoilage by the apiculated yeast is greatly reduced. As the juice is very cold for *Saccharomyces* recently rehydrated, it is recommended to rehydrate the enological yeast with inactivated yeast based micronutrients, sterols and unsaturated fatty acids in order to help it support the low temperature and limit lag phase duration. It is also recommended to adapt the yeast to be inoculated to temperature, lowering the rehydration suspension temperature with cold must (10°C gap during a 5-minute period) until the gap between rehydration solution and the must is smaller than 10°C.

Further reading

DELTEIL, D. 2008. Buone pratiche di vinificazione del Pinot Noir. *Acts of Lallemand Italia conference, Bolzano (Italy)*, 29 May.

SABLAYROLLES, J. M., SOUBEYRAND, V. and JULIEN, A. 2006. Rehydration protocols for active dry wine yeasts and the search for early indicators of yeast activity. *Am. J. Enol. Vitic.* 57, 4, 474–480.

3.15 What is the difference between 'carbonic maceration' and 'whole berry fermentation'?

L. Bisson

Carbonic maceration is a technique developed in France for the production of light fruity wines that can be consumed while young. It was discovered by accident during analysis of methods by which grapes could be stored over long periods of time. In carbonic maceration intact clusters are held in a sealed container. The weight of the clusters usually leads to the expression of some juice that encourages the growth of microorganisms that in turn consume the available oxygen. Under these conditions the berry tissues are still living and will ferment internal sugar producing ethanol and carbon dioxide within the intact berry. During this anaerobic metabolism phase unique flavors are produced and many of the unripe characters are degraded yielding fruit that is rich in red berry flavors and aromas with a note of hay or silage. The grapes are left in the carbon dioxide atmosphere for a period of days to weeks depending upon the temperature of storage. Following the carbonic maceration period the grapes are crushed and the partially fermented juice allowed to complete fermentation. The wines produced are distinctive with limited ageabiltiy.

In whole berry fermentation intact berries are included in the juice during fermentation. There may be some initial reactions that are similar to those that occur during carbonic maceration but the fruit is in a liquid environment not a gaseous one and the ethanol produced by the yeast quickly penetrates the intact berries arresting metabolism. There is a slower leakage of berry constituents than would occur had all of the fruit been directly crushed. This practice serves to retain fruit characters as they are protected from evaporative loss by being held in the berry. At the discretion of the winemaker at some point during the fermentation the must is pressed crushing the intact fruit and releasing fruit components. This is usually performed once the ethanol has accumulated to a level that slows further fermentation. The new sugar that is released is fermented by the yeast but more slowly, minimizing the sweeping effect of carbon dioxide gas at removing volatile components. Wines made this way retain native fruit characters but have much more ageability than wines made using carbonic maceration.

4
Malolactic fermentation in wine

4.1 How many different types of malolactic (ML) starter culture preparations are in use today?

S. Krieger-Weber

The history of controlled malolactic fermentation (MLF) is short. Despite the early discovery by Müller-Thurgau in 1891 of lactic acid bacteria (LAB) contributing to the acid reduction in wine, by degrading malic acid to lactic acid and CO_2, bacteria starter cultures have been only introduced for winemaking in the beginning of the 1980s. Most common *Oenococcus oeni* starter cultures are used; however, there are also some *lactobacilli* preparations in use.

- *Liquid ML cultures* were available and used for a long time. They have to be freshly prepared and used within two days, when kept at room temperature or within two weeks when stored at 4°C. A pre-culture in a wine/juice solution supplemented with 1 g/L yeast extract for 3–7 days is necessary to achieve a 10-fold expansion. The pre-culture is inoculated into the final wine volume at a ratio of 2–5% or when using finished wine only for pre-culturing at a ratio of 5–10%.
- *Frozen malolactic bacteria starter cultures* were developed in the early 1980s. Frozen starter cultures can be stored at −26°C for 120 days and −50°C for one year in a non-defrosting freezer. Once thawed these cultures have to be used immediately. Re-freezing is not possible because it will cause cell damage and loss of vitality. Some frozen cultures need a 48 pre-acclimatization in a juice/wine mix supplemented with yeast extract. Other frozen starter cultures can be inoculated directly into the wine. Usage rates vary from 1 g/hL for red wines and 3–8 g/hL in difficult white wines.

- *Traditional freeze-dried (STANDARD) starter cultures* were introduced at the end of the 1980s. They can be stored up to 18 months at 4°C or up to 30 months at −18°C. Once opened STANDARD cultures have to be used immediately and they need several expansion steps. Depending on the wine to be inoculated the time for starter culture preparation varies between 3 and 14 days including a per-culture and one or two *pied-de-cuve* acclimatization steps. The addition of proprietary ML nutrients is recommended during pre-culturing. The inoculation ratio is about 1 g/hL.
- *Direct inoculation freeze-dried ML starter cultures* were introduced in the early 1990s and their use has virtually revolutionized the control and predictability of malolactic fermentation in wine. They benefit from storage under refrigerated (30 months at −18°C) and/or frozen conditions (18 month at 4°C). Depending on the origin of production they can be added directly to the wine, but better distribution will be achieved, when rehydrating for no longer than 15 minutes in cold water. Addition of nutrients is recommended under more challenging MLF conditions.
- *Quick 1-step build-up cultures* were recently developed as an economic solution for larger wine volumes. The starter culture kit combines a highly effective malolactic starter culture with an activator. They differ from the direct inoculation starter by needing an acclimatization step within 18–24 h in a water/wine mix supplemented with the activator. After this step the culture is not only acclimatized to the wine but is also 100% metabolically active to induce fast malolactic fermentation after being inoculated into the final wine volume.

4.2 What are the ideal conditions for storage and rehydration of malolactic (ML) starter cultures?

S. Krieger-Weber

Most starter cultures available for winemaking benefit from storage under refrigerated and/or frozen conditions in their original, unopened package and the container should not be opened until just before use. In addition, the freeze-dried bacteria should avoid contact with oxygen, excess moisture and high temperature as these conditions will be detrimental to survival of the bacteria. In order to obtain the maximum effect from ML bacteria starter cultures always follow the bacteria producers' recommendations for handling and storage.

The listing below gives an overview on the most common instructions for ML starter culture preparations.

Frozen ML starter cultures
(a) Thaw in room temperature water and not in the refrigerator. Mix 3 liters water, 3 liters grape juice and 30 g yeast extract. Adjust pH to 4.0 with calcium carbonate or other permitted buffer and mix thoroughly. Add 170 g of thawed culture, seal carboy and mix thoroughly. Hold at 18–24°C for 48 hours before inoculation.
(b) Direct addition of the frozen pellets to the wine.

Liquid ML starter culture suspensions
- Clean settled juice without added SO_2; if possible heat the juice to 60°C. Adjust sugar level to 180 g/L with water (if juice is not available, substitute with a mix of 50% finished wine (<10 ppm free SO_2 and low total SO_2), 25% water and 25% apple juice). Adjust pH 3.5–3.6 with calcium carbonate. If inoculating wine at pH <3.2, adjust pH again to 3.4 as an intermediate step.
- Add culture and maintain temperature at 22–26°C.
- Monitor to 100% malic acid degradation then expand again as a 10% inoculum at each build-up stage or inoculate.
- If finished wine was used to prepare the starter then expand culture by doubling the starter volume with wine until the starter volume is 5–10% of the amount to be inoculated.

Direct inoculation starter cultures (MBR®)
A special preparation is not required but may be suspended in clean chlorine free water at 20°C for no longer than 15 minutes to help in handling.

Quick build-up starter cultures (1-STEP® Kit)
- Rehydration phase: Mix and dissolve content of the activator mix in 100 liters of potable water at 18°C and 25°C. Add content of the bacteria sachet and dissolve carefully by gentle stirring. Wait for 20 minutes.

- Acclimatization phase: Mix the bacteria/activator solution with 100 liters of wine, pH > 3.5; temperature between 20°C and 25°C. Wait between 18 and 24 hours.
- Transfer the activated culture to the 1000 hL of wine.

Traditional freeze-dried STANDARD starter cultures
- Rehydrate in a 50:50 water:wine mix. Wine should be pH > 3.3 and total SO_2 < 30 mg/L.
- Monitor malic acid drop and when ~2/3 is converted to lactic acid, expand as a 5% inoculum into wine. Make sure pH > 3.3 and alcohol < 12.5%.
- Monitor malic acid drop and when ~2/3 is converted to lactic acid, expand as a 4% inoculum into wine.

4.3 When should I inoculate for malolactic fermentation (MLF)?

J. P. Osborne

The malolactic fermentation (MLF) is an important aspect of winemaking, in particular if you are producing red wines or cool climate white wines that may have excessive acidity. The MLF occurs as a result of metabolic activity by certain lactic acid bacteria and results in the conversion of malic acid to lactic acid. The bacteria may also impact the flavor and aroma of the wine. Although spontaneous MLF may occur due to bacteria naturally present in musts and wines, specific starter cultures of bacteria are now commonly used as they allow more control over the process with more reliable results. However, the use of these starter cultures has led to the discussion of when is the best time to inoculate your wine (or must) for the MLF.

Traditionally, it has been recommended that wines should be inoculated with malolactic bacteria (MLB) after the completion of the alcoholic fermentation. This is because of the risks associated with inoculating with MLB during the alcoholic fermentation or at the beginning of the alcoholic fermentation in conjunction with yeast inoculation (known as simultaneous fermentation). In particular, there is concern that the MLB may produce acetic acid and D-lactic acid which is referred to as 'piqure lactique'. This may occur as the lactic acid bacteria commonly used for MLF, *Oenococcus oeni*, can produce acetic acid via metabolism of sugars. The concern is that strong bacterial growth in grape musts when high sugars levels are present may result in elevated acetic acid levels in the wine. However, not all research has agreed with this finding. MLB will begin to consume sugars only when the degradation of organic acids such as malic, citric, and fumaric acid is complete. Therefore, during a simultaneous fermentation very little sugar may remain in the wine by the time the MLB have consumed all the organic acids, reducing the risk of excessive acetic acid production. A number of researchers have reported on successful simultaneous fermentations with little to no increase in acetic acid concentrations.

An additional concern is that bacterial growth may inhibit the growth of the yeast and cause a stuck alcoholic fermentation. However, there is little evidence that *O. oeni* can cause the inhibition of wine yeast (although other lactic acid bacteria such as some *Lactobacillus* sp. have been shown to cause stuck alcoholic fermentations). On the other hand, some wine yeasts have been shown to cause inhibition of the MLB, particularly during mid-alcoholic fermentation. Therefore care should be taken when inoculating for MLF mid-alcoholic fermentation versus before or after.

The advantages in inoculating for MLF at the beginning of the alcoholic fermentation are firstly that overall fermentation durations will be shortened given that the MLF is occurring simultaneously with the alcoholic fermentation. Secondly, the bacteria can slowly become acclimatized to increasing alcohol levels during simultaneous fermentation versus being immediately subjected to high alcohol levels if inoculated post-alcoholic fermentation. In addition, higher nutrient levels will be present in the must compared to the wine which may allow improved bacterial survival. The sensory profile of the wine will also be impacted differently with a simultaneous fermentation. The production of the buttery compound diacetyl is reduced if the MLF occurs quickly under the reductive conditions generated by active yeast cells. Diacetyl is reduced to acetoin, a compound that does not contribute to wine flavor. Studies have shown that simultaneous fermentations may result in a more fruit-driven wine versus lactic, buttery, nutty styles that may result if MLF occurs after the completion of the alcoholic fermentation. If a simultaneous fermentation is to be considered then care must be taken when choosing the combination of yeast and malolactic bacteria strain. Research has shown that certain wine yeasts are inhibitory to the malolactic bacteria and so these yeasts should be avoided. Often wine yeast companies will make note of whether a particular yeast strain is compatible with the MLF or not. For example, strain EC1118 is known to be a high produced of SO_2 and can cause problematic malolactic fermentations.

In summary, there are a number of options available regarding when to inoculate for MLF. There is no universal 'right time' as you must consider a number of different factors when making this decision. What is the chemical composition of my grape must? What are the fermentation conditions? What style of wine am I trying to produce? The answers to these questions may help determine when you want to inoculate. For example, the use of simultaneous fermentations is not recommended in wines with a pH above 3.5 as sugars are more easily metabolized by *O. oeni* at higher pHs resulting in elevated acetic acid production. For most red wines it is recommended that MLB inoculation occur at the end of alcoholic fermentation as it has been observed that rapid MLFs in some wine types, particularly Pinot noir, can result in a reduction in pigmentation. However, for aromatic white varietals where the pH is below 3.5 the use of simultaneous fermentations shows promise.

Further reading

HENICK-KLING, T. and PARK, Y. H. 1994. 'Considerations for the use of yeast and bacterial starter cultures: SO_2 and timing of inoculation', *Am. J. Enol. Vitic.* 45, 464–469.

JUSSIER, D., MORNEAU, A. D. and MIRA DE ORDUNA, R. 2006. 'Effect of simultaneous inoculation with yeast and bacteria on fermentation kinetics and key wine parameters of cool-climate Chardonnay', *Appl. Environ. Microbiol.* 72, 221–227.

ROSI, I., FIA, G. and CANUTI, V. 2003. 'Influence of different pH values and inoculation times on the growth and malolactic activity of a strain of *Oenococcus oeni*', *Aust. J. of Grape and Wine Res.* 9, 194–199.

4.4 What are the advantages and risks of inoculating with malolactic starter cultures at the beginning of alcoholic fermentation?

S. Krieger-Weber

The advantages of induction of malolactic fermentation (MLF) by inoculation with selected strains of lactic acid bacteria are twofold: first, a better control over the time and speed of malic acid conversion, and second, a positive influence on wine flavor and quality. The view in Bordeaux, France is to recommend MLB inoculation after completion of the alcoholic fermentation so as to avoid the possible production of acetic acid and D-lactic acid, a situation which is referred to as 'picûre lactique'. MLF, which occurs during the alcoholic fermentation, may occasionally result in a stuck alcoholic fermentation. American experience has not confirmed the French findings of increased acetic acid production, yeast antagonism, or stuck alcoholic fermentations when early growth of lactic acid bacteria is seen. Inoculation of MLB with the yeast has been advocated because it was felt the bacteria have a better chance of growing and acclimatizing in the absence of ethanol (Beelman and Kunkee 1985).

Wine pH < 3.5
Previous experiments conducted by the Lallemand research group under low pH wine conditions (pH < 3.5) showed that acetic acid will not be produced out of sugars during growth of malolactic bacteria (MLB) and active malolactic fermentation (MLF). In these experiments, acetic acid production was observed only when half of the malic acid was degraded and the bacteria began to utilize citric acid. Trials conducted using simultaneous bacterial and yeast inoculation vs. bacterial inoculation upon completion of the alcohol fermentation showed no difference in the final acetic acid concentration, but a direct relationship between citric acid degradation and an increase in acetic acid concentration was demonstrated. Simultaneous inoculation of yeast and bacteria into Riesling juice showed that the simultaneous inoculation of yeast and bacteria had no influence on the yeast fermentation, but resulted in a much faster MLF compared to the inoculation after AF. The timing of inoculation had an important impact on the sensory profile of the wine. In the co-inoculation experiment, this reductive environment generated by the yeast prevented the formation of buttery or lactic aromas, and the wine retained the varietal flavors and aromas. The technique of inducing AF and MLF simultaneously in low pH white musts has proven very satisfactory.

Wine pH > 3.5
Radler (1963) reported sugar consumption of 0.2–2 g during growth in the three phases. Showing that only in phase three, when the degradation of organic acids is complete, will the bacteria begin to consume sugars in excess. Under high pH

conditions the degradation of sugars is followed by organic acid consumption, which will result in a significant increase in volatile acidity. This could be the case when yeast fermentation is sluggish and malic acid degradation is finished far before alcoholic fermentation. Contrary early inoculation with selected *Oenococcus oeni* strains may reduce the risk of spontaneous MLF during alcoholic fermentation, suppressing wild bacteria, and at the same time conduct a more controlled MLF avoiding production of potential spoilage compounds. Early inoculation techniques (inoculation of the bacteria 24 h after the yeast) had been investigated in high pH high alcohol red wines during the last five northern and southern hemisphere vintages. Under high pH wine conditions early inoculation with a selected bacteria strain allows the dominance of the selected strain and a better control over the spontaneous bacteria populations without any negative effect on the yeast population and performance of AF. Early inoculation with MLB starter cultures, especially the simultaneous inoculation of selected yeasts and bacteria demonstrated faster MLF while avoiding significant acetic acid and biogenic amine formation. Nevertheless, under these high pH conditions apply good winemaking practices such as the selection of a compatible and alcohol tolerant yeast strain and good nutrition strategies for the yeast to avoid sluggish or stuck alcoholic fermentation.

References

BEELMAN, R. B. and KUNKEE, R. E. 1985. Inducing simultaneous malolactic-alcoholic fermentation in red table wines. *Proc. Aust. Soc. Vitic. Oenol. Sem. on Malolactic Fermentation*, 97–112.

RADLER, F. 1963. Über die Milchsäurebakterien des Weines und den biologischen Säureabbau. Übersicht. II. Physiologie und Ökologie der Bakterien. *Vitis* 3, 207–236.

4.5 What is the impact of temperature on malolactic fermentation (MLF)?

S. Krieger-Weber

Environmental wine conditions such as pH, temperature, alcohol level, nutritional status and sulphur dioxide (SO_2) play a significant role in a successful induction of malolactic fermentation. The optimum growth temperature for LAB is between 25 and 35°C, and the rate of malic acid degradation by non-growing cells is highest at approximately the same temperatures. The rate of growth of malolactic bacteria and the speed of the MLF are inhibited by low temperatures. This can be problematic, particularly during the production of white wines, which tend to be fermented at lower temperatures. But low temperature will not cause lethal defects to the ML bacteria. Tolerance to cool temperatures is dependent on the *Oenococcus oeni* strain. Temperature resistant direct inoculation starter cultures will tolerate in general temperatures above 12°C, more sensitive strains will perform best between 18 and 22°C.

Although the temperature optimum for the growth of wine lactic acid bacteria is around 25°C, depending upon the wine alcohol content, higher wine temperatures can also be inhibitory to the development and activity of ML bacteria especially in the combination with high alcohol and high SO_2 levels. A general guideline to avoid the inhibitory effects is:

Wine alcohol content (% v/v)	Temperature for MLF should not exceed:
Less than 14.5% ⇒	28°C
Greater than 14.5% ⇒	23°C

Wine total SO_2 content	Temperature for MLF should not exceed:
Less than 45 ppm ⇒	28°C
Greater than 45 ppm ⇒	23°C

Further reading

HENICK-KLING, T. 1988. Yeast and bacterial control in winemaking. In: Linskens, H. F. and J. F. Jackson (Eds). *Modern Methods of Plant Analysis, New Series, Vol. 6*. Springer-Verlag, 276–316.

LAFON-LAFOURCADE, S. 1970. Étude de la dégradation de l'acide L-malique par les bactéries lactiques non-prolifératines isolées des vins. *Ann. Technol. Agric.* 19, 141–154.

4.6 Is it possible to manage the diacetyl levels in a wine?

S. Krieger-Weber

Diacetyl is one of the main aromatic compounds produced during malolactic fermentation. Diacetyl is produced by *Oenococcus oeni* as an intermediate in the metabolism of citric acid. In this pathway, the intermediate pyruvic acid is reductively decarboxylated to diacetyl via α-acetolactate. Since diacetyl is chemically unstable, it can be further reduced by active cells of *Oenococcus oeni* or by yeast to the less flavor active end products, acetoin and 2,3 butanediol. In addition to citric acid, the metabolism of diacetyl by LAB is closely associated with the metabolism of sugars and malic acid. During the course of MLF, the maximal production of diacetyl generally coincides with the completion of L-malic acid degradation, and the sequential catabolism of citrate (Bartowsky *et al.* 2002).

The production of diacetyl is influenced by a wide range of microbiological, environmental and winemaking factors. Various studies by Henick-Kling *et al.* (1993) and Bartowsky and Henschke (2004) have shown that, due notably to their production of diacetyl, different *O. oeni* starters could result in completely different aromatic profiles. Each bacteria starter's potential for producing and reducing diacetyl is a criterion to consider when choosing a malolactic culture.

In addition to the bacterial strain involved, other winemaking factors can also have a significant impact on diacetyl content including wine type, contact with yeast and bacterial lees, the time required to complete MLF, contact of wine with air during MLF, addition of sulfur dioxide, concentrations of citric acid and fermentable sugar, temperature and pH (Bartowsky and Henschke 2004). Some of the criteria suggested to obtain a high diacetyl content in wine include using a bacteria strain with a high potential for diacetyl production, lower bacterial inoculation rates, lower pH and temperature, shorter contact with yeast and bacteria lees, higher redox potential, and stabilizing wine immediately following the disappearance of malic and citric acids. Conversely, criteria indicated to minimize diacetyl production include the use of a neutral bacterial strain, higher inoculation rates, extended contact with yeast and bacterial lees and lower redox potential (Martineau *et al.* 1995).

The timing of inoculation is also a tool for diacetyl management. Experiments have demonstrated that co-inoculation – the simultaneous inoculation of yeast and bacteria – does not influence the alcoholic fermentation or increase volatile acidity, but it does reduce the overall MLF duration. The co-inoculated wines did not have buttery or milky aromas associated with MLF, but did have a high intensity of varietal fruit aromas. The diacetyl produced under such reducing conditions during the alcoholic fermentation was immediately transformed into the less aroma-active butanediol. The same wines inoculated for MLF after alcoholic fermentation had more typical malolactic character with dominant notes of butter, yogurt and/or hazelnut while the fruit was diminished.

References

BARTOWSKY, E., COSTELLO, P. and HENSCHKE, P. 2002. Management of the malolactic fermentation wine flavour manipulation. *Aust. & N.Z. Grapegrower & Winemaker* 461a, 7–8, 10–12.

BARTOWSKY, E. and HENSCHKE, P. 2004. The 'buttery' attribute of wine – diacetyl – desirability, spoilage and beyond. *Inter. J. Food Microbiology* 96, 235–252.

HENICK-KLING, T., ACREE, T. E., KRIEGER, S. A. and LAURENT, M.H. 1993. Sensory aspects of malolactic fermentation. Stockley, C.S., Johnston, R.S., Leske, P.A. and Lee, T.H. (Eds) Proceedings of the eighth Australian wine industry technical conference; 26–28 October 1992, Melbourne, Vic, Winetitles, SA, 148–152.

MARTINEAU, B., HENICK-KLING, T. and ACREE, T. E. 1995. Reassessment of the influence of malolactic fermentation on the concentration of diacetyl in wines. *Am. J. Enol. Vitic.* 46, 385–388.

4.7 Are there other factors beside SO_2, pH, alcohol content and temperature that can have an impact on malolactic fermentation (MLF)?

S. Krieger-Weber and P. Loubser

In recent years, very reliable commercial bacterial cultures to induce a MLF became available to winemakers. These commercial bacterial cultures are selected according to very strict criteria and are able to function under extremely harsh conditions. A number of factors, well known and lesser known, play a role in the successful course of malolactic fermentation. The fact that some factors are lesser known does not mean that their impact is less significant. These factors include the following.

Tannins
Recent research has shown that certain grape tannins can have a negative influence on malolactic bacteria, and consequently on the course of MLF. For this reason it is clear that certain red cultivars, such as Merlot, for example, can have more difficulty undergoing a successful MLF. Latest results within the Lallemand group indicate that phenolic acids influence the growth of certain bacterial strains in laboratory growth media.

Selected yeast strain
It has been known for some time that certain yeast selected to conduct the alcoholic fermentation interact better with certain bacteria for the successful achievement of malolactic fermentation. Under specific conditions, certain yeast strains may produce high concentrations of SO_2 which has a negative influence on the growth and survival of the malolactic bacteria. Similarly, yeast strains that exhibit an inordinate need for nutrients could exhaust the medium to such an extent that no reserve nutrients are available for the bacteria. Implementing a specific nutrition strategy for the particular yeast in the early stages of alcoholic fermentation can largely surmount this problem.

Lees compaction
As a result of hydrostatic pressure, the lees found at the bottom of a tank can be compacted to such an extent that yeast, bacteria and nutrients are 'captured' and cannot function properly. It has recently been observed that lager tank sizes may correlate with increasing delays in the initiation of the MLF. The inhibition of the start of the MLF in larger tanks can be overcome by pumping over either on the day of inoculation or on the second day after inoculation of the bacteria.

Residual lysozyme activity
If lysozyme is used during the production of wine, it is possible that residual levels of this enzyme may impact the time course of the subsequent MLF. Care

must be taken to follow the supplier's recommendations with regard to the required time delay between the addition of lysozyme and the inoculation of the commercial MLF culture. In most cases racking the wine off the gross lees is recommended.

Excessive amounts of oxygen
Malolactic bacteria have been shown to be sensitive to excessive amounts of oxygen. This means that exposure of the bacteria to undue amounts of oxygen after the completion of alcoholic fermentation should be avoided. It was noted that micro-oxygenation might have a positive impact on the course of the MLF. This phenomenon may not be related to oxygen itself, but may be attributable to the gentle stirring action that is associated with the micro-oxygenation process itself, which keeps the bacteria and nutrients in suspension.

Fungicide residues
Certain fungicide and pesticide residues, especially the former, may have a detrimental effect upon the functioning of malolactic bacteria. Most effective, in a negative sense, are residues of systemic pesticides which are often used in humid years to control the botrytis fungus. Careful precautions should be taken in years with high incidence of botrytis contamination. Wine producers must be familiar with the spraying programs and products used, and they must adhere to the prescribed withholding periods required for the various antifungal products used.

Initial malic acid concentrations
Malic acid concentrations differ between grape cultivars and may also differ from year to year in the same grape cultivar. It's especially difficult to induce a MLF in wines with levels of malic acid which are below 0.8 g/L. In this case it's recommended to use ML starter cultures with a high malate permease activity or a short activation protocol.

Yeast derived fatty acids and peptides
Medium chain fatty acids can have a negative impact on the course of MLF. Alexandre *et al.* (2004) and Edwards *et al.* (1990) indicated that the antagonism between yeast and lactic acid bacteria could be explained by the production of certain medium chain fatty acids (C6 to C12), which are derived from yeast metabolism. More recently Nehme (2008) reported the production of inhibitory peptides of a molecular weight between 5 and 10 KDa by certain yeast strain reducing the speed of malic acid degradation by 5–11%.

References

ALEXANDRE, H., COSTELLO, P. J., REMIZE, F., GUZZO, J. and GUILLOUX-BENATIER, M. 2004. *Saccharomyces cerevisiae–Oenococcus oeni* interactions in wine: current knowledge and perspectives. *Int. J. Food Microbiol.* 93, 141–154.

EDWARDS, C. G., BEELMAN, R. B., BARTEY, C. E. and MCCONNELL, A. 1990. Production of decanoic acid and other volatile compounds and the growth of yeast and malolactic bacteria during vinification. *Am. J. Enol. Vitic.* 43, 233–238.

NEHME, N. 2008. Etude des interactions entre *Saccharomyces cerevisiae* et *Oenococcus oeni*: impact sur la réalisation de la fermentation malolactique en cultures séquentielles et mixtes. PHD thesis, Institute National Polytechnique de Toulouse.

4.8 Do I have to add nutrients to my malolactic fermentation (MLF)?

S. Krieger-Weber

Nutritional requirements of malolactic bacteria

Growth conditions during the malolactic fermentation in wine are very difficult for lactic acid bacteria. Lactic acid bacteria require an extensive complement of amino acids, purines, pyrimidines, vitamins and minerals (Radler 1963), which means that the nitrogen content of wine will be significantly altered during the MLF. The requirements for specific amino acids vary between strains of LAB. *Leuconostoc oenos* and *Pediococcus pentosaceus* isolated from wine may require up to 16 different amino acids (Weiller and Radler 1972), but they cannot use inorganic nitrogen sources, such as diammonium phosphate (DAP). The availability of amino acids has not been found to be growth limiting, since proteins and peptides in wine can also be used as a source of amino acids (Feuillat *et al.* 1977). It has been shown that bacterial growth and malolactic fermentation are stimulated by the presence of yeast-derived nitrogenous compounds such as peptides and cell wall mannoproteins. In addition to amino nitrogen, lactic acid bacteria of wine must be supplied with purines and pyrimidines or their derivatives. Several B-group vitamins are essential for growth and all strains apparently require folic acid and nicotinic acid (Radler 1966). The requirement for biotin, riboflavin, pantothenic acid, and pyridoxine is strain-dependent.

Nutrient availability is dictated by winery practices

Clarification of juice and wine can not only physically remove a large portion of LAB, but can also reduce the amount of bacterial growth obtainable. In addition, clarification will remove nutrients and suspended particles, which are stimulatory to bacterial growth, further impacting the MLF. Similarly, wines made using the process of thermo-vinification have been reported as being less able to support a malolactic fermentation. The availability of nutrients will be affected by interactions among wine microorganisms. In winemaking, there is always the possibility of interactions occurring between lactic acid bacteria and yeast, fungi, acetic acid bacteria and even bacteriophages. Moreover, there also may be interactions between species and strains of LAB. The antagonistic effect of yeast upon MLB has been explained through competition for nutrients and production of substances that inhibit bacterial growth. Conversely, yeast may support the growth of LAB in wine and stimulate the progress of MLF. During the process of yeast autolysis, vitamins and amino acids are released into wine, and the associated extended lees contact enriches the wine with micro-nutrients which stimulate malolactic fermentation.

Nutrient deficiencies
In order to successfully complete the malolactic fermentation, sufficient nutrition for the bacteria is of the utmost importance. The concept of nutrient addition is not a simple one. The importance of good sanitation practices and the addition of selected malolactic cultures are fundamentally linked, and are vital to the concept of providing the inoculated bacteria with critical nutrients. It is important to ensure that the MLF is always conducted under hygienic conditions, and that sufficient SO_2 is present in the wine to control indigenous bacterial populations. It is important to ensure that the wine is inoculated with the recommended level of a defined, bacteria starter culture, so that sufficient 'good' bacteria are present for the MLF to be completed successfully when the nutrient is added. In fact, the addition of nutrients is never recommended when spontaneous, naturally occurring populations of malolactic bacteria are relied upon to conduct the MLF.

References
FEUILLAT, M., BIDAN, P. and ROSIER, Y. 1977. Croissance de bactéries lactiques à partir des princoaux constitunts azotes du vin. *Ann. Technol. Agric.* 26, 435–447.
RADLER, F. 1963. Über die Milchsäurebakterien des Weines und den biologischen Säureabbau. Übersicht. II. Physiologie und Ökologie der Bakterien. *Vitis* 3, 207–236.
RADLER, F. 1966. Die mikrobiologischen Grundlagen des Säureabbaus in Wein. *Zentralbl. Bakteriol. Parasiten Abt. II.* 120, 237–287.
WEILLER, H.G. and RADLER, F. 1972. Vitamin- und Aminosäurebedarf von Milchsäurebakterien aus Wein und von Rebenblättern. *Mitt. Höheren Bundeslehr- und Versuchsanstalt Wein- und Obstbau Klosterneuburg* 22, 4–18.

4.9 Does the yeast used for alcohol fermentation have an impact on malolactic fermentation (MLF)?

S. Krieger-Weber

Wine is the result of a complex of microbial fermentations involving the sequential development of various species of yeasts and lactic acid bacteria. The winemaker is able to ensure that the desired fermentative yeasts and malolactic bacteria strains dominate these fermentations. Very often, starter culture failures are due to improper preparation and inoculation procedures. In some cases, starter culture failure may be due to antagonistic interactions between yeast and bacteria. One possible explanation for this difficulty is that the wine may be lacking essential nutrients for the lactic acid bacteria due to their consumption by the yeast.

The utilization of nutrients by *Saccharomyces* can lead to amino acid deficiencies that inhibit malolactic fermentation (MLF). The selected wine yeasts can be categorized into six groups, e.g. according to the amount of nitrogen used to ferment 1 g of sugar in a must deficient in nitrogen (100 mg/L of yeast-assimilable nitrogen). The nitrogen needs of the tested yeasts vary from extremely low (<0.4 mg N/g of sugar used) to extremely high (>1.2 mg N/g of sugar used). Information on the relative nitrogen needs of specific yeasts should be available from the yeast suppliers.

Free amino acids and small peptides are the most frequent sources of nitrogen for lactic bacteria growth. Depending on the strain, *Oenococcus oeni* have different needs for amino acids: each amino acid can be essential, nonessential or have no effect. For example, glutamic acid is essential for *O. oeni*, which are incapable of synthesizing it. However, the absence of proline, glycine and alanine has little effect on growth.

Other potential yeast interactions with the malolactic bacteria are *inhibitory compounds produced by the yeast* that accumulate in wine:

- Ethanol, which is produced by yeast during alcoholic fermentation (AF), affects the capacity of lactic bacteria to grow and will affect the MLF progress, especially at very high alcohol amounts.
- Sulfur dioxide is considered among the most significant limiting compounds influencing the growth of malolactic bacteria in wine. Its concentration in wine is a result of adding SO_2 to the must and the production of SO_2 by *Saccharomyces cerevisiae*. Henick-Kling and Park (1994) demonstrated the inhibition of malolactic starter cultures by active growing yeasts due to the production of high levels of SO_2 during the early stage of alcoholic fermentation. Although most yeast strains produce less than 30 mg/L of SO_2, some release more than 100 mg/L (Rankine and Pocock 1969). Some yeasts have already been characterized into one of three categories of yeasts according to their level of SO_2 production: low (10–40 mg/l), medium (41–70 mg/l), and high (> 70 mg/l).
- Peptides or antibacterial protein production by *S. cerevisiae* are an inhibiting yeast peptide which may act in synergy with SO_2 at high concentrations and

inhibit MLF (Dick *et al.* 1992, Osborne and Edwards 2007 and Nehme 2008).
- Medium-chain fatty acids are toxic to both the yeast and the malolactic bacteria, limiting bacterial growth and reducing the capacity of the bacteria to metabolize malic acid.
- Certain aromatic compounds, such as β-phenylethanol or succinic acid have also been described as MLF inhibitors (Lonvaud-Funel *et al.* 1988).

Stimulation of malolactic bacteria by yeasts
The autolytic activity of yeast at the end of AF modifies the concentration of amino acids, peptides and proteins in the wine. During this autolysis, glucans and mannoproteins are released. The levels released depend on the yeast strain (Rosi *et al.* 1998) and winemaking practices. These mannoproteins stimulate the bacteria through two mechanisms: adsorption of medium-chain fatty acids – thus, detoxification of the medium and secondly as a nutrient source of amino acids for the malolactic bacteria.

Some yeast suppliers have characterized their main yeast strains with regard to their MLF compatibility and made this information available. A general recommendation for best compatible yeast strains is medium fermenter, good autolysis, low to moderate nutrient demand and a good complex yeast nutrition strategy.

References
DICK, K. J., MOLAN, P. C. and ESCHENBRUCH, R. 1992. The isolation from *Saccharomyces cerevisiae* of two antibacterial cationic proteins that inhibit malolactic bacteria. *Vitis* 31, 105–116.
HENICK-KLING, T. and PARK, Y. H. 1994. Considerations for the use of yeast and starter cultures: SO_2 and timing of inoculation. *Am. J. Enol. Vitic.* 45, 464–469.
LONVAUD FUNEL, A., JOYEUX, A. and DESENS, C. 1988. Inhibition of malolactic fermentation of wines by products of yeast metabolism. *J. Sci. Food Agric.* 44, 183–191.
NEHME, N. 2008. Etude des interactions entre *Saccharomyces cerevisiae* et *Oenococcus oeni*: impact sur la réalisation de la fermentation malolactique en cultures séquentielles et mixtes. PhD thesis, L'Institut National Polytechnique de Toulouse.
OSBORNE, J. P. and EDWARDS C. G. 2007. Inhibition of malolactic fermentation by a peptide produced by *Saccharomyces cerevisiae* during alcoholic fermentation. *Int. J. Food Microbiol.* 118, 27–34.
RANKINE, B. C. and POCOCK, K. F. 1969. Influence of yeast strain on binding sulfur dioxide in wines, and on its formation during fermentation. *J. Sci. Food Agric.* 20, 104–109.
ROSI, I., GHERI, A., DOMIZIO, P. and FIA, G. 1998. Production des macromolécules pariétales de *Saccharomyces cerevisiae* au cours de la fermentation et leur influence sur la fermentation malolactique. *Re. Oenol.* 94, 18–20.

4.10 How can I monitor malolactic fermentation (MLF)?

S. Krieger-Weber, S. Kollar and N. Brown

During the course of the winemaking process, it is vital that the progress of the MLF be monitored for its primary chemical action, the conversion of L-malic acid into L-lactic acid and also for the microbial flora which is present in the wine during the conversion. Historically, winemakers have relied upon the physical senses of sight, smell, taste and, in some instances, sound to inform them as to the progress of the MLF. The industry now recognizes that these parameters, although important, are not sufficient to adequately define the progress of the malolactic fermentation. The highly refined and skilled senses of an experienced winemaker will always be the backbone of quality wine production, but can be reinforced and assisted by analytical laboratory techniques.

Chemistry

The two most important chemical parameters to monitor during the course of MLF are the depletion of malic acid and the rise in vaccenic acid (VA). VA is a very strong indicator of bacterial activity, and when produced in large amount is generally attributed to the growth of spoilage bacteria. Small amounts of VA are a natural byproduct of the growth of the malolactic bacteria and are to be expected. The level of VA is easily measured by common laboratory analyses. The progress of the malolactic fermentation can easily be determined by analyzing the degradation of malic acid or the production of lactic acid or by determining the amount of liberated CO_2. Typical methods used for organic acid analysis include paper chromatography, thin layer chromatography, enzymatic analysis, and reflectance, capillary electrophoresis (CE) and high performance liquid chromatography (HPLC). Table 1 shows a comparison of the organic acid analytical methods on a relative scale, with 1 being lower, and 4 being higher.

Table 1 Comparison of organic acid analytical methods

Method	Cost	Accuracy	Speed	Technical capability
Paper chromatography	1	1	1	1
Thin layer chrom.	1	1	3	1
Reflectance	2	3	4	2
Enzymatic analysis	3	4	3	3
CE	4	4	3	4
HPLC	4	4	2	4

Microbiology

The simplest way that microbiology is used to monitor the MLF is by making direct observations of the wine through a light microscope. Because of their very small size and their reflective index, a phase contrast microscope or a very good brightfield microscope capable of 1000× magnification is required to differentiate between wine bacteria. The microscope is a very powerful aid to the winemaker because, with a good microscope and a little experience, it is possible to instantaneously detect any substantial population of bacteria in the wine. Fortunately few bacteria can grow in wine, and because of their relatively distinct cellular appearance under the microscope, *Oenococcus*, *Pediococcus*, *Lactobacillus* and *Acetobacter* species are identifiable and their relative abundance can be estimated. The health of a population of *Oenococcus oeni* as well as the abundance of the bacterium can be estimated by direct microscopic observation. Contrary to direct microscopic observation, where the cells can actually be observed, viable culture plating relies on the ability of cells to reproduce, resulting in a visible colony on the surface of the nutrient media after a few days. Wine bacteria are usually cultured on agar media at relatively low pH containing apple juice, tomato juice, V8 juice, or something similar to stimulate the growth of the lactic acid bacteria (LAB). Membrane filtration is useful when dealing with wines that have a very low level of viable microorganisms. Spread plating is useful if there is a high microbial population. Recently fast molecular-biological methods based on PCR technology have been developed to detect lactic acid bacteria on strain level within a few hours, but sophisticated and expensive equipment is necessary.

4.11 How much residual malic acid will cause visible malolactic bacteria (MLB) growth/carbonation in the bottle?

C. Butzke

Grapes at harvest contain between 0.6 and 6 g/L of malic acid, depending on the growing climate. Malic acid and tartaric acid are the main acids in grapes and wine; however, tartaric acid cannot be metabolized by wine bacteria. Malolactic bacteria can turn malic acid into lactic acid and carbonic acid which leaves the fermentor as CO_2, thereby reducing the acidity of the wine.

Traditionally, the conversion of malic to lactic acid is monitored in the winery lab via paper chromatography. This method is semi-quantitative and the absence of a malic acid spot on the chromatogram indicates a concentration of less than 30 mg/L, a barely visible one about 200 mg/L. It is commercially assumed that if a wine completes MLF, the residual malic acid is less than 300 mg/L or 0.3 g/L.

Malic acid is stoichiometrically converted into two thirds lactic acid and one-third CO_2, i.e., 300 mg/L residual acid could produce 100 mg/L gas. The solubility of CO_2 in wine is relatively high, and cork-pushing carbonation would occur only beyond the saturation concentration of about 1400 mg/L at room/bottling temperature (20°C). A perceivable *spritz* may be tasted at 800 mg/L CO_2 which would require the malolactic fermentation of 2.4 g/L malic acid. However, bacteria will grow based on residual nutrients in the wine if malic acid is present, and a visible haze in a white wine due to the growth of *Oenococcus oeni* can be expected above 300 mg/L.

Sources

In: *Winemaking Basics*. Ough, C., p. 85, Haworth Press, New York, 1992.
In: *Wine Analysis and Production*. Zoecklein, B. *et al.*, p. 296, Springer, New York, 1995.
http://www.vinquiry.com/pdf/05FallNwsltr.pdf

5

Wine clarification, stabilization and preservation

5.1 What role does vineyard nitrogen management play in juice processing and wine instabilities?

S. Spayd

An adequate supply of nitrogen is required for the health of grapevines to achieve adequate growth and fruit production. However, the amount of nitrogen required to achieve maximum yield potential is considerably lower than the requirement to achieve maximum vegetative growth. Gamma butyric acid is the primary transport form of nitrogen in grapevines. Arginine is the primary alpha-amino acid found in grape berries. Proline is also usually found in relatively high concentration, but it is an imido-acid rather than an alpha-amino acid, and is not utilized by normal wine yeast. More information on nitrogen requirements during fermentation are presented elsewhere in this text. Nitrogen is also ultilized in the production of other compounds in grapes, including proteins. As nitrogen fertilization increases, the soluble protein concentration found in pressed grape juice also increases. Juice and wine pH may also increase with increasing nitrogen fertilization. Bentonite, a montmorillonite clay, is used to protein stabilize wines. Proteins are adsorbed onto the negatively charged sheets of bentonite clay. Hsu and Heatherbell (1987) indicated that proteins with a net isoelectric point (Ip), the pH that there is no net ionic charge, with higher Ip and of intermediate molecular weight are the first fractions removed by bentonite. A more recent study (Achaerandio *et al.* 2001) reported that there was no selection of proteins by bentonite based on molecular characteristics, rather that adsorption depended on the physical volume of the protein. Spayd and coworkers (1994) found that soluble protein concentrations increased linearly in unfined wines with increasing rate of nitrogen fertilization as did bentonite demand.

Nitrogen fertilization of 200 lbs/acre (224 kg N/ha) resulted in several wines that required addition of a concentration of bentonite greater than the legal US limit to produce a protein stabilized wine.

References

ACHAERANDIO, I., V. PACHOVA, C. GÜELL and F. LÓPEZ. 2001. Protein adsorption by bentonite in a white wine model solution: Effect of protein molecular weight and ethanol concentration. *Am. J. Enol. Viticult.* 22, 122–126.

HSU, J. C. and D. A. HEATHERBELL. 1987. Heat-unstable proteins in wine. I. Characterization and removal by bentonite fining and heat treatment. *Am. J. Enol. Viticult.* 38, 11–16.

SPAYD, S. E., R. L. WAMPLE, R. G. EVANS, R. G. STEVENS, B. J. SEYMOUR and C. W. NAGEL. 1994. Nitrogen fertilization of White Riesling grapes in Washington. Must and wine composition. *Am. J. Enol. Viticult.* 45, 34–42.

5.2 How do additions of potassium sorbate, potassium metabisulfite or potassium bicarbonate impact the cold stability of my wine?

C. Butzke

'Cold stabilizing' means to rid a wine of unstable potassium bitartrate ('cream of tartar'), the naturally occurring salt of the grape's tartaric acid, the latter being the major contributor to a wine's perceived tartness. Since potassium bitartrate is more soluble in water than in alcohol, the amount soluble in wine is smaller than the one in grape juice. However, due to the presence of colloidal materials in wine, such as mannoproteins, pectins and other polysaccharides, the unstable tartrate may not precipitate unless the wine is chilled or aged significantly in tank or barrel which might not be appropriate for a particular wine style. In litigious countries, the deposit of the harmless and tasteless potassium tartrate as visible crystals on the bottom of the bottle or wine glass may cause the consumer to mistake those 'wine diamonds' for glass splinters. While only an aesthetic issue that requires wine consumer education more than winemaker intervention, it has raised concerns about potential frivolous law suits. Thus, the winemaker stabilizes the wine to simulate what would happen if a wine consumer is putting a bottle of (white) wine into a very cold refrigerator to chill it down to, for example, 0°C.

Since the solubility of potassium bitartrate is also much reduced at lower temperatures, the winemaker usually chooses to 'cold' stabilize a wine at a temperature just above the freezing point of the individual wine. This is a giant waste of energy and a sad example of unnecessary over-processing of a natural product. The wine's freezing point is related most importantly to its alcohol content. The approximate freezing point of wine at 10% ethanol by volume is $-4°C$, at 12% it is $-5°C$, and $-6°C$ at 14%. This means that the winemaker has a maximum stabilization window of 4 to 6 K which translates into a solubility difference for potassium tartrate of between 200 and 300 mg/L, equal to 41 to 62 mg/L of potassium.

The time and efficacy of the cold stabilization depend on many factors, such as actual wine temperature reached, tank surface, use and size of seed crystals, etc. Assuming that the wine has reached maximum cold stabilizing at the temperature that was desired, how do further additions of sorbate, metabisulfite or bicarbonate impact the post-bottling stability of this wine?

1. The proper addition of potassium sorbate to suppress re-fermentation by wine yeast in the bottle is 268 mg/L (equivalent to 200 mg/L of sorbic acid). The potassium fraction of such a sorbate addition equals 70 mg/L.
2. Adjusting the sulfite concentration of the wine prior to bottling has to be carried out according to the individual pH of the wine plus to account for losses of SO_2 during the filtration/bottling process itself as well as the reaction with possible oxygen residues in the bottle headspace. Without sparging the bottle and the headspace with an inert gas, the wine would pick

up 8 mg oxygen, resulting in an SO_2 loss of 32 mg/L. An addition of an extra 50 mg/L free SO_2 (38 mg/750 mL bottle) would add another 30 mg/L potassium to the wine.

In the case of combining both treatments immediately prior to bottling, the extra 100 mg/L of potassium would clearly comprise the cold stability of the wine which at best could only keep 41 to 62 mg/L more soluble. That means the wine might be cold stable only down to 3°C instead of 0°C. An issue can arise when the bottled wine is shipped during the winter time to a cold region in a non-insulated truck.

3. If the wine's acidity has to be adjusted chemically, the addition of potassium bicarbonate is recommended as it does not cause the stability problems associated with calcium salts which do not respond well to cold stabilization. The recommended treatment limit for potassium bicarbonate is 1000 mg/L which neutralizes the equivalent of 1.5 g/L titratable acidity (TA). De-acidifying a wine with 1000 mg/L $KHCO_3$ will add 386 mg/L potassium. Even higher doses would make the wine taste salty. This substantial increase in potassium will obviously require cold-stabilization only after the treatment. While the reaction between the bicarbonate and the tartaric acid in the wine is instantaneous given proper mixing, the ensuing release of carbon dioxide gas and thus possible foam formation require a tank headspace of not less than 25% of the wine volume. The wine will keep a full saturation with CO_2, therefore a sparging with nitrogen is recommended to strip the wine of the majority of it. Otherwise, there will be a significant increase in perceived crispness associated with higher concentrations of this acidic and very soluble gas. The saturation concentration of carbon dioxide at 10°C is about 2000 mg/L (100 times that of nitrogen!) or the equivalent of 3.6 g/L TA as tartaric acid! Recommendation for residual dissolved CO_2 for wines range from 200 to 800 mg/L for reds to 700 to 1800 mg/L for whites to avoid 'flabby' tasting wines. In the United States, the legal limit for CO_2 in 'still' wine is 3920 mg/L, a concentration difficult to reach without carbonation or re-fermentation; in the OIV countries it is 1000 mg/L, about half the saturation at cellar temperature.

Further reading

BOULTON, R. B., V. L. SINGLETON, L. F. BISSON and R. E. KUNKEE. 1996. *Principles and practices of winemaking*. Springer, p. 434.

5.3 Why do I have to heat-stabilize my white wines with bentonite?

C. Butzke

The grape berry contains a large variety of nitrogen compounds mainly amino acids, peptides (short amino acid chains) and proteins (long amino acid chains). They serve various biological functions within the grape such as enzymes, cell wall components, etc. The nitrogen content of grapes varies greatly by variety, rootstock, vintage climate, pruning and crop levels, fertilization practices, etc.

Amino acids are soluble and can be used by wine yeast to grow and ferment the grape's sugars into alcohol. Peptides and proteins cannot be metabolized as yeast available nitrogen and their solubility decreases with the wine's alcohol content. This may lead to precipitation of agglomerated proteins in the form of a visible amorphous haze. This effect is accelerated or triggered by exposure to elevated temperatures, e.g. when a customer buys a bottle of wine in the tasting room and forgets it in the trunk of the car over the weekend. Protein hazes are a purely aesthetic, visual problem in wine as they cannot be tasted. However, while it is a natural effect, most consumers prefer a wine free from unappetizing-looking protein instabilities.

The winemaker's options to prevent such instabilities are limited:

- Bentonite clay in different forms (see **5.4**) can irreversibly adsorb various sizes of proteins and has been the protein fining agent of choice. It takes about six times the quantity of clay to take out the relevant amounts of protein: Protein content of wine ranges from around 10 to 300 mg/L, bentonite additions from 60 to 1800 mg/L.
- Heat exposure such as a high temperature–short time (HTST) treatment can denature proteins in unfermented juices and limit the need for additional fining of the wine. An HTST treatment, similar to a milk pasteurization at 80°C (176°F) for 5 seconds, does generally not affect the quality of the fermented wine. Such a treatment is also advisable for juices from grapes with heavy *Botrytis* infections, because its polyphenoloxidase (PPO) *laccase* – a browing enzyme – does not respond to traditional SO_2 treatments at the crusher the way grape PPOs do.
- Tannins derived from chestnut galls that contribute no color and low astringency have the potential to precipitate grape proteins as well as *Botrytis* laccase.
- Mannoproteins from yeast cell walls may act as protective colloids, bind to grape protein and prevent their flocculation.
- Regeneratable cation exchange resins that reversibly bind proteins from wine and do not create solid waste have been used only experimentally.

There are several potential problems with bentonite fining:

- Excess amounts of bentonite added to wine cannot just bind proteins but desirable aroma compounds or colloidal materials as well. Proper bench testing to determine the minimal effective amount to add is important as those will not have a detrimental effect on the wine. Each individual wine has separate dosage requirements. Routine additions of bentonite will lead to certain over- or under-dosing as the requirements may vary by more than an order of magnitude (60 to 1800 mg/L = 0.5 to 15 lb/1000 gal). In addition, oxidative damage to the wine may occur if the mixing of the bentonite slurry allows for air exposure during transfer operations, from the tank headspace or via subsequent filtration steps.

- The quickest (laziest) test for protein stability are heat tests that expose the (treated and filtered!) wine sample to a high temperature for a short period of time, e.g. 49°C (120°F) for 48 hours, 80°C (176°F) for six hours, or 90°C (194°F) for one hour followed by a period of cooling. Such tests are trying to simulate the precipitation of proteins at a proper, cool storage temperature over the lifespan of the wine. It is a rather uninformed assumption that these two scenarios are identical in their outcome. Moreover, the resulting over-stabilization of commercial wines against excessive heat exposure has significant consequences for the sensory quality of the wine. If the wine experiences extreme temperatures during shipping and storage, its aroma will be damaged but there may be no visible indicators that the wine was treated poorly. Temperature data logging and tracking of shipments can help identify sources of heat exposure within the distribution chain.
- A more representative way to assess the effect of fining treatments on the stability of a wine under normal storage and aging conditions is the Boulton ethanol assay that measures the stability of all colloidal materials. It can assess the effect of a bentonite fining treatment on a wine through a titration with successive quantities of ethanol while – just as with the heat tests – using a nephelometric turbidity meter to quantify the resulting cloudiness.
- Bentonite (juice) fining before fermentation may lead to a sluggish fermentation due to its clarification effect on the treated juice and the possibly stripping of certain growth factors, such as fatty acids, phospho-lipids, and sterols. An extended fermentation can lead to increased amounts of residual fructose in the wine. Since fructose is twice as sweet as glucose this may affect the perception of the wine's dryness.
- When treating blush/rosé wines make sure that the bench trials are also evaluated for any loss or change of color associated with the bentonite treatment.
- Proper storage of bentonite – much like filter pads – in a clean, dry environment and in a re-sealable container is crucial. Bentonite will absorb odors from the air, e.g. the cork taint component TCA, and release them into the treated wine.
- Bentonite creates a solid waste problem if separated with the lees. If flushed out of the tank it can clog drains and sewer lines. In the winery/irrigation pond, it will settle and enhance the seal of the pond bottom. Over the years, however, sludge will gradually accumulate and make the pond more shallow, thereby enhancing light penetration and algae growth.

A small amount of bentonite can be directly added to any wine and used to seal small leaks in a barrel that was previously dry-stored or has a more leakage-prone, e.g. fortified, wine in it.

Further reading

BOULTON, R. B., V. L. SINGLETON, L. F. BISSON and R. E. KUNKEE. 1996. *Principles and practices of winemaking.* Springer, p. 342.

ETS Laboratories. Technical bulletin on the interpretation of heat stability results and turbidity readings. http://www.etslabs.com/display.aspx?pageid=183

WEISS, K. C. and L. F. BISSON. 2002. Effect of bentonite treatment of grape juice on yeast fermentation. *Am. J. Enol. Vitic.* 53, 1, 28–36.

WEISS, K. C., T. T. YIP, T. W. HUTCHENS and L. F. BISSON. 1998. Rapid and sensitive fingerprinting of wine proteins by matrix-assisted laser desorption/ionization time-of-flight (MALDI-TOF) mass spectrometry. *Am. J. Enol. Vitic.* 49, 3, 231–239.

5.4 What should I use: sodium or calcium bentonite?

C. Butzke

It doesn't really matter! As long as the enologist performs an ethanol titration or one of the arbitrary heat stability bench tests on each wine and determines the smallest effective dose to satisfy the test.

Bentonite clay is the most widely used fining agent against heat-instable grape proteins in white wines. In red wines from *Vitis vinifera*, the inherent tannins usually denature these proteins enough to cause precipitation during aging. However, anthocyanin-rich but tannin-deficient red European-American varieties as well as blush/rosé wines and red *viniferas* from very cool climates should be tested for protein-stability.

There are two different forms of bentonites commercially available: sodium-rich ones and calcium-rich ones. Suppliers of sodium bentonites argue that this form has a protein fining capacity twice as high as its calcium cousin. Suppliers of calcium bentonites argue that their form swells less in water, and it creates fewer lees and a smaller loss of wine when racking.

Excess addition of sodium to wine is undesirable as its consumption contributes to high blood pressure and heart disease. For the same reason, the use of sodium metabisulfite for SO_2 additions or sodium bicarbonate (baking soda) for deacidification purposes is not permitted. The quantities added to wine by a heavy bentonite treatment can double the amount of sodium naturally present in grape juice (10 to 20 mg/L), but even then wine is still considered a 'very low sodium' beverage.

On the other hand, excess release of calcium into a wine from bentonite via exchange with grape proteins may increase the risk of calcium tartrate instability. For example, an addition of 1920 mg/L (16 lbs/1000 gal) calcium bentonite – equivalent to 960 mg/L (8 lbs/1000 gal) sodium bentonite – to a batch of protein-rich Gewürztraminer would result in an additional potential for 114 mg/L calcium tartrate. Since calcium tartrate does not respond as readily to cold stabilization as potassium bitartrate, this may mean the difference being a stable wine and a wine throwing a glass-like precipitate that may worry consumers.

Should I rehydrate my bentonite in water or in wine?

Water. Bentonite, independent of type, should be rehydrated with clean, chlorine-free hot (60°C, 140°F) water, by adding it under vigorous mixing to the water (not the other way around) and allowing it to swell for at least four hours. The lump-free slurry shouldn't sit longer than overnight either as this may encourage microbial growth. A maximum of 16.7 L may be used to dissolve each kilogram of bentonite (2 gallons of water per pound). Note that the total amount of water introduced from all processing sources during the winemaking should not exceed 1% of the wine. For bench trials in the winery lab, a mixing ratio of water to bentonite of 16 to 1 (60 g per 1 L) results in an easily pipettable 6% w/v slurry (see Table 1).

Table 1 Bentonite (6%) slurry additions for bench trials

g/hL	lbs/1000 gal	mL per liter wine	mL per 750 mL wine	mL per gal wine
12	1	2	1.5	8
24	2	4	3.0	15
36	3	6	4.5	23
48	4	8	6.0	30
60	5	10	7.5	38
72	6	12	9.0	45
84	7	14	10.5	53
96	8	16	12.0	60
108	9	18	13.5	68
120	10	20	15.0	76

Rehydrating with wine doesn't allow the bentonite to fully swell thereby reducing its fining capacity. In addition, it is a waste of wine that cannot be recovered.

However, if one uses 6% slurry at additions above 60 g/hl (5 lbs/1000 gal), the amount of water added to the wine exceeds 1%. Thus in practice, bentonite is typically dissolved at ratios of about 8 to 1 (1 kg bentonite per 8 L of water; 1 lb/gal) which allows for bentonite additions of up to 120 g/hl (10 lb/1000 gal). Above this addition level, a wine/water mix can be used for rehydration to keep the processing water addition below 1% of total wine volume. Alternatively, rehydrating with too little water will limit the amount of swelling and produce a difficult to stir, lumpy slurry.

Does a change in pH due to acidification/deacidification or cold stabilization change the protein stability of my wine?
Yes. Even small changes in pH can significantly alter the protein solubility and thus the wine's betonite requirements. Protein stability must be re-assessed after any treatment that changes the acidity of the wine in question. Especially higher pH values likely lead to increased bentonite demand. Calcium bentonites are not recommended for wines with a pH above 3.4.

Can blending two protein-stable wines compromise the stability of the blend?
Yes. Any shift in alcohol content, pH, protective colloid concentration, etc. can potentially render the entire blend unstable. A new test of the mixture must be conducted and additional fining might be needed.

Can non-grape proteins from fining agents, enzymes or lysozyme treatments, or sur lie aging influence the protein stability of my wine?
- Gelatin and other proteins that are used as fining agents against over-extracted seed or skin tannins may contribute to protein instabilities and often

be used in conjunction with other fining agents – such a silica gel (*Kieselsol*) – that can precipitate any excess of fining protein amounts that didn't bind to the wine's tannins.
- Additions of processing enzymes such as pectinases or glucosidases are usually not relevant sources for protein instabilities. However, they respond well to bentonite fining which allows the winemaker, e.g., to stop the activity of β-glucosidases. This limits the premature release of sugar-bound varietal aroma precursors such as monoterpenes and a loss of aging potential for the wine.
- Lysozyme additions to inhibit malolactic fermentation can add substantial amounts of non-grape derived protein to the wine when added at a recommended 250 to 500 mg/L, which is equal between 2 and 50 times the average concentration of natural grape protein in wine. Note that lysozyme – while potentially unstable – does not respond well to the heat tests.
- Mannoproteins, on the other hand, released from yeast cells during aging on the lees or added as commercially available adjuncts, may act as protective colloids and keep unstable grape protein from precipitating.

How fast/long should I mix?
The reaction between protein and bentonite is quick but not instantaneous. Proper mixing is crucial, and it has been shown that mixing speed, time and temperature affect the efficacy of the treatment. At least 10 to 15 minutes of vigorous mixing is recommended and the wine temperature should be above 10°C (50°F). To increase the effectiveness of the bentonite fining, the winemaker may choose to do it at a warmer temperature, and then proceed with a cold stabilization against tartrate precipitation thereafter. In this scenario, the dropout of potassium bitartrate may affect the pH of the wine and the bench test for protein stability should address this by simulating the chilling beforehand.

It is important that any bench trial conditions, especially the mixing speed and the temperature are representative of the conditions that can be achieved on a large scale in the cellar. Otherwise, an underestimation of the bentonite requirements will result.

Mixing with an inert gas fed in via the racking valve avoids potential oxidative damage due to mechanical mixing if the tank headspace contains traces of air/oxygen.

How long does it take to settle?
Allow one week (depending on tank height) to have all bentonite lees settle to the bottom by gravity alone. Limiting the contact time between wine and bentonite helps to minimize the amount of lead residues that could be extracted into the wine.

Further reading

Bentonite. Technical Information Sheet. Sadhana Brent. (https://www.gusmerenterprises.com/images/document/Bentonite.pdf) Gusmer Enterprises & The Wine Lab, Fresno, CA, 2009.

BOULTON, R. B., V. L. SINGLETON, L. F. BISSON and R. E. KUNKEE. 1996. *Principles and practices of winemaking.* Springer, pp. 284–286.

SCHNEIDER, V. 2005. *Die Winzer-Zeitschrift*, 02/05, pp. 32–34.

WEISS, K. C., L. W. LANGE and L. F. BISSON. 2001. Small-scale fining trials: effect of method of addition on efficiency of bentonite fining. *Am. J. Enol. Vitic.* 52, 3, 275–279.

5.5 I was told to treat my wine with casein. What is the best procedure?

J. McAnnany

The results are in from the lab trials to determine that casein is the appropriate product to add and the proper dosage to use is selected for improving the structure and sensory properties of the wine requiring treatment. How do I prepare and incorporate the addition of casein into my tank? While casein is on the list of known allergens since it is a protein derived by acid precipitation from milk, it is still the product application of choice for many wines. It is normally used in the form of potassium caseinate, which is a powder that is soluble in water. It also works with synergistic action when used in conjunction with polyvinylpolypyrrolidone (PVPP) and/or bentonite for treating must to eliminate phenolic precursors which may contribute to browning or pinking.

Milk may also be used as a source of casein. In the *Handbook of Enology, Volume 2: The Chemisty of Wine* (Ribereau-Gayon *et al.* 2006) it states that one liter of cow's milk contains about 30 grams of casein, along with other proteins. Owing to the presence of these other proteins and the content of milk fat, there is a greater risk of overfining if too high a dose of milk is used.

Careful preparation and handling are important with casein, as it can be challenging to mix homogeneously into the wine for a complete reaction without sinking or floating on the surface. Potassium caseinate carries a negative charge and it is characterized as a less soluble proteinaceous material compared to other protein fining agents, which allow it to readily precipitate and settle out without leaving any residual protein in the wine (Boulton *et al.* 1999). Owing to this quick precipitation, it is a 'safe' fining agent to use as overfining is not a problem.

To ensure proper mixing, prepare the selected dosage in solution by dissolving potassium caseinate in water (at a rate not more than 100 grams per Liter) (Martin Vialatte 2008). Allow the powder to dissolve completely by mixing gently with a whisk or other available stirring tool and use the solution immediately by introducing it in a continuous, steady stream into the wine with an injection pump or a venturi so that the complete dose is introduced in a short period of time. It is essential to mix the solution constantly and thoroughly into the wine to avoid partial flocculation which can lead to a less than effective treatment. The floating material is typically found in white wines where the tannin or phenolic content is lower and mixing is more difficult, allowing the less soluble casein portion to separate. It is a natural phenomenon and may require filtration to remove the floating particles since they don't settle completely to allow normal racking processes.

References

BOULTON, R. B., V. L. SINGLETON, L. F. BISSON and R. E. KUNKEE. 1999. *Principles and Practices of Winemaking*, p. 283.

MARTIN VIALATTE. 2008. Casein Technical Information, available at http://www.martinvialatte.com/catalogue_us/images/big/6.060_gb_casesol.pdf

RIBEREAU-GAYON, P., Y. GLORIES, A. MAUJEAN and D. DUBOURDIEU. 2006. *Handbook of Enology, Volume 2: The Chemisty of Wine*, p. 289.

5.6 What is sorbate?

B. Trela

Sorbate, as sorbic acid (2,4-hexadienoic acid) or its various salts (Ca, Na, K), has long been used in the food and wine industries as an antimicrobial and fungistatic agent. In wine, it is sometimes added to inhibit fermentation in wines to be bottled with residual sugar and that for whatever reason cannot be bottled under sterile conditions. Some wineries add it for 'insurance' against re-fermentation in the bottle. Sorbate inhibits many yeasts (*Saccharomyces*), but not all microbes and some, such as malolactic bacteria can metabolize it and produce sensorially objectionable products (i.e., 'geranium tone'). The use of sorbate should be viewed as a temporary stopgap assurance method until proper sanitary sterile filtration and bottling methods can be employed.

Sorbic acid is a polyunsaturated fat naturally occurring in fruits. Sensorially, it is often described as being similar to bees wax, especially when esterified with ethanol. Thresholds of detection are high with 50 percent of people being able to detect it at 50–100 mg/L in white wine, and 135 mg/L in red wines; however, some individuals can easily perceive it and find it objectionable (Ough and Ingraham 1960). It has been found to be non-toxic even when taken in large quantities, and breaks down in the body into water and carbon dioxide in the Krebs Cycle (FAO/WHO 1967).

As a fungistat it is important to note that sorbate inhibits *Saccharomyces* strains of yeast but does not kill them. This fungistatic activity does not diminish with time so it can be added to wines at any time before bottling. Efficacy is near optimum at wine pH values, and normal table wine ethanol and sulfur dioxide concentrations are synergistic to activity (Ough and Ingraham 1960). Sorbate has low efficacy against bacteria and some wine spoilage yeast, such as *Zygosaccharomyces*, are resistant to it.

Some lactic acid bacteria (common MLF bacteria [*Oenococcus*] included) can metabolize sorbate, reducing it to sorbyl alcohol that at wine pH values and in the presence of ethanol yields 2-ethoxyhexa-3,5-diene (Fig. 1), an ether having a strong objectionable geranium-like odor with a low sensory threshold of 100 ng/L (ppt) (Crowell and Guymon 1975).

Since the presence of ethanol is required for this reaction, the precursor to geranium tone may not be detected in juices or sweet reserves preserved with sorbate, but may become apparent if used as a sweetener to wine.

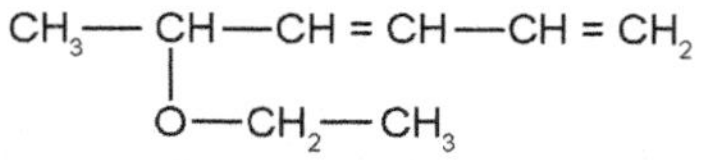

$$CH_3-CH-CH=CH-CH=CH_2$$
$$|$$
$$O-CH_2-CH_3$$

Fig. 1 Geranium tone: 2-ethoxyhexa-3,5-diene.

Advantages
- Fungistat – can help microbial stability, especially in wines with residual fermentable sugars
- Persistent – activity does not diminish with time
- Few people can detect its presence
- Its use is safe and convenient.

Disadvantages
- Some people can sensorially detect its presence and find it objectionable
- Fungistat not fungicide
- Little activity against bacteria
- Some bacteria can metabolize sorbate to produce off-odors such as geranium tone
- Some spoilage yeast such as *Zygosaccharomyces* are resistant to sorbate
- Some countries do not authorize its use at any concentration.

Health note
- Biological studies including long-term animal studies show sorbate to be quite safe, even at high doses, with health implication similar to those of various salts when the salt forms (Na, K) of sorbate were studied (FAO/WHO 1967).
- Sorbic acid had a conjugated system of double bonds, which makes it susceptible to nucleophilic attack, and therefore the potential to yield mutagenic products.

Note for use in wine
- Use the potassium salt form (potassium sorbate) of sorbic acid as it is very much easier to dissolve in wine. Sorbic acid itself is extremely insoluble in wine and not a practical choice for this purpose.

References

CROWELL, E. A. and J. F. GUYMON. 1975. Wine constituents arising from sorbic acid addition, and identification of 2-ethoxyhexa-3,5-diene as source of geranium-like off-odor. *Am. J. Enol. Vitic.* 26, 97–102.

FAO/WHO. 1967. Tenth Report of the Joint FAO/WHO Expert Committee on Food Additives, FAO Nutrition Meetings Report Series.

OUGH, C. S. and L. J. INGRAHAM. 1960. Use of sorbic acid and sulfur dioxide in sweet table wines. *Am. J. Enol. Vitic.* 11, 117–122.

5.7 How much sorbate should I use in wine?

B. Trela

Sorbic acid has very low solubility and therefore potassium sorbate (the potassium salt of sorbate) with much higher solubility is used for additions in wine. The characteristics of both forms of sorbate are provided below for reference.

Sorbic acid
- Description: White, crystalline solid with a mildly acrid odor.
- Chemical names:
    - Sorbic acid; Trans, trans-2,4-hexadienoic acid; 1,3-Pentadiene-1-carboxylic acid; Hexa-2,4-dienoic acid; (E,E)-2,4-Hexadienoic acid; 2-Propenylacrylic acid.
- Empirical formula:
    - $C_6H_8O_2$
    - $CH_3CH=CHCH=CHCOOH$
- Sorbic acid structural formula (see Fig. 1).

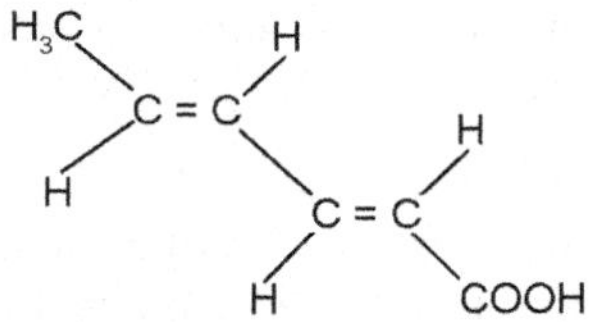

Fig. 1 Sorbic acid structural formula.

- Molecular weight: 112.13
- Solubility (aqueous): 0.15 percent, by weight (20°C)
    - Sorbic acid is very insoluble. The potassium salt of sorbic acid = potassium sorbate should be used due to its ease of dissolution.

Potassium sorbate
- Description: White or yellowish-white crystals or crystalline powder.
- Chemical names:
    - Potassium sorbate; Potassium salt of trans, trans-2,4-hexadienoic acid.
- Empirical formula:
    - $C_6H_7O_2K$
    - $CH_3CH=CHCH=CHCOOK$
- Potassium sorbate structural formula (see Fig. 2).
- Molecular weight: 150.22
- Solubility (aqueous): 58.2 percent, by weight (20°C)
    - Potassium sorbate is much more soluble than sorbic acid. Use this form for wine.

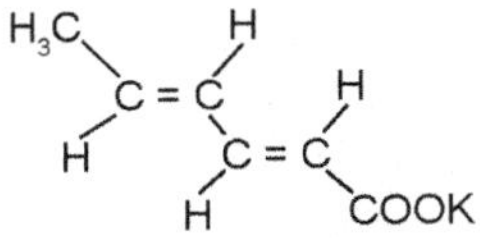

Fig. 2 Potassium sorbate structural formula.

Dose

An effective yeast inhibition dose at normal wine conditions is about 100–200 mg/L as sorbic acid regardless of the population of yeast it is expected to control (Ough and Ingraham 1960). These dosages may be within sensory perception thresholds and near legal concentration limits (200 mg/L, OIV 2005). Its use is not permitted in wines in Japan.

Potassium sorbate = 74.7% sorbic acid*

Therefore:

200 mg/L of sorbic acid = 267.7 mg/L potassium sorbate

* as a molecular weight percent of sorbic acid: 112.13/150.22 = 74.64%

Note: Sorbic acid is steam distillable and will distill over with acetic acid in volatile acidity analysis by Cash still, artificially inflating volatile acidity results.

- Sorbic acid levels can be determined by HPLC, ultraviolet analysis (255 nm), or by isolation, derivatization and colorimetric analysis.
- Correctly using the relationship: 1 g sorbic acid = 0.536 g acetic acid.
    o Subtract 0.11 from the VA value (in g/L) for every 200 mg/L sorbic acid present.

References

1. OFFICE INTERNATIONAL DE LA VIGNE ET DU VIN (OIV). 2005. International Code of Oenological Practices, pp. 11.3.3–11. OIV, Paris.
2. OUGH, C. S. and L. J. INGRAHAM. 1960. Use of sorbic acid and sulfur dioxide in sweet table wines. *Am. J. Enol. Vitic.* 11, 117–122.

5.8 I mixed up potassium sorbate and citric acid together to make a sorbate addition and an acid adjustment prior to bottling. An amorphous white precipitate has formed and is floating on top of the mixture. What is it and what should I do about it?

C. Butzke

Sorbic acid is the main antimicrobial adjunct one can use to prevent a wine's alcoholic re-fermentation in the bottle by yeast, including the common wine yeasts *Saccharomyces cerevisiae* and *bayanus*. The maximum (and required!) dose is 200 mg/L sorbic acid.

The problem is that sorbic acid, a fatty acid and a solid at room temperature, has a very low solubility in water (2500 mg/L) or wine. If one, for example, is to treat a 1000 L tank of wine with the required dose, one would have to add 200 grams of sorbic acid. If one attempts to mix up the solid acid in a 8 L bucket of water, then the maximum solubility is tenfold exceed at 25 000 mg/L. This is the reason why sorbic acid is used in the form of its highly soluble salt, potassium sorbate. A total of 268 mg of potassium sorbate equals 200 mg of sorbic acid, so the dosage needs to be adjusted accordingly, although in the OIV countries, strangely, the legal limit (200 mg/L) is the same for sorbic acid and potassium sorbate.

The potassium salt is indeed very soluble in water. However, if mixed directly into a small volume of wine, the wine's acidic pH will cause the sorbate anions to combine with hydrogen ions (protons) from the grape's tartaric and malic acids to form sorbic acid molecules. This effect can be observed as white flocks of insoluble sorbic acid floating inside and atop of the mixture.

The same holds true if potassium sorbate is added directly to a bucket of acidified water, in a situation where the winemaking goal is to make simultaneous sorbate and citric acid adjustments before bottling. In order to avoid the flocculation of sorbic acid, it is recommended to mix up the potassium sorbate separately in a bucket of clean water, then to slowly add the solution to a gently N_2-stirred tank.

Any acid adjustment should be performed separately from the sorbate addition. In case white flocks have formed, continue the stirring until they have resolved. Do not attempt to skim or filter them off as this will result in an underdosing of the required sorbic acid and will thereby increase the potential for a re-fermentation.

The effectiveness of the sorbic acid is closely related to proper pH-dependent SO_2 management of the wine and a minimal microbial load. To achieve the lowest count of residual yeast cells possible, a clarifying filtration (2 to 3 μm; Seitz K200 to K300 or equivalent) prior to sorbate additions is advised for all wines with residual sugar, followed by a sterile filtration immediately prior to bottling.

5.9 What does 'contains sulfites' mean versus 'no added sulfites'?

L. Bisson

Legislation in the United States requires disclosing that wine contains sulfite. If a wine contains more than 10 mg/L of sulfite at the point of bottling, the wine must be labeled with a declaration of sulfites, either as 'Contains sulfites' or 'Contains (a) sulfiting agent(s)'. Sulfite or SO_2 is produced during amino acid biosynthesis and the reduction of sulfate for inclusion in the sulfur-containing amino acids methionine and cysteine. These are both essential amino acids in the human diet, but yeast has the ability to synthesize both of these compounds. During sulfate reduction some sulfite may be released by the yeast. Humans contain the enzyme sulfite oxidase because we also generate sulfite in our bodies from the catabolism of sulfur containing amino acids in the diet. There are some individuals with a deficiency in this enzyme. Since sulfite is an antioxidant it can have negative effects on lung tissue especially in individuals with a deficiency in sulfite oxidase. When exposed to sulfite these individuals may display difficulty in breathing and have symptoms similar to an allergic reaction requiring immediate medical treatment. Even though sulfite sensitivity affects only a very small portion of the human population they should be warned against consuming foods that contain sulfites as consumption of such products might trigger a reaction. Sulfites in wine can be made by the yeast or can also come from the use of sulfite as an antioxidant and antimicrobial used to protect wine flavor and aroma. Sulfite also functions as a natural antioxidant within the human body so it has been adopted for use because of its ability to be metabolized by humans. Wine labels that say 'contains no added sulfites' mean that sulfite was not added as a chemical powder or solution but that sulfites naturally made by the yeast may be present. Wine below the 10 mg/L level must be certified to be below that level for labeling purposes.

5.10 What is molecular sulfur dioxide (SO₂) and how does it relate to free and total SO₂?

J. P. Osborne

Sulfur dioxide (SO$_2$) has been used for thousands of years during winemaking as an antimicrobial and antioxidant agent. It is very effective in these roles, is readily available, and is relatively cheap and easy to use. SO$_2$ can exist in many different forms in wine and these forms have different properties, especially with regards to their effectiveness as antimicrobial and antioxidant agents. For example, in aqueous solutions, SO$_2$ dissociates according to the following equations resulting in three pH dependent forms:

$$SO_2 \text{ (gas)} + H_2O \rightleftharpoons SO_2 \cdot H_2O \text{ (molecular SO}_2\text{)}$$

$$SO_2 \cdot H_2O \rightleftharpoons H^+ + HSO_3^- \text{ (bisulfite ion)}$$

$$HSO_3^- \rightleftharpoons H^+ + SO_3^= \text{ (sulfite ion)}$$

Of the three forms, (molecular, bisulfite, and sulfite) the bisulfite ion dominates in wine due to the pH (wine pH is usually between 3 and 4). In addition, the bisulfite ion can exist in a 'free' or 'bound' form due to its ability to bind with various wine components such as acetaldehyde, pyruvic acid, anthocyanins, and sugars. This binding results in bound SO$_2$ and together with the free SO$_2$ component determines the total SO$_2$ concentration. Of these various forms, the free SO$_2$ component is the most important regarding antimicrobial and antioxidant properties.

A component of free SO$_2$, molecular SO$_2$, is generally believed to be the antimicrobial form of SO$_2$. Molecular SO$_2$ is calculated based on the concentration of free SO$_2$ in the wine and the pH. The following formula can be used to calculate the molecular SO$_2$ content of a wine:

$$[\text{Molecular SO}_2] = \frac{[\text{Free SO}_2]}{[1 + 10^{pH-1.8}]}$$

In practice, it is easier to make tables based on target molecular SO$_2$ levels in a wine and base your free SO$_2$ levels on that rather than to calculate them on an individual basis using the formula. For example, if you were aiming for a target molecular SO$_2$ level of 0.6 mg/L (accepted as a level that prevents growth of most wine spoilage microorganisms) you could prepare a table such as the one shown in Table 1.

As illustrated by the table, the amount of free SO$_2$ required to maintain a given molecular SO$_2$ level is directly related to the pH and increases as pH increases. This means it is possible to have two wines with the same level of free SO$_2$ but have one that is more susceptible to microbial spoilage due to its higher pH and therefore lower level of molecular SO$_2$. Therefore just using one universal free SO$_2$ level as a target for ensuring the protection of your wine is inadequate if you don't take pH into consideration.

Table 1 Levels of free SO_2 to achieve 0.6 mg/L molecular SO_2 at different pH values

pH	Free SO_2	pH	Free SO_2
3.0	10	3.4	24
3.1	12	3.5	32
3.2	15	3.6	40
3.3	18	3.7	49

In summary, molecular SO_2 is based on the concentration of free SO_2 and the pH of the wine. It is the antimicrobial form of SO_2 and therefore an important parameter to monitor during the winemaking process. It is recommended that you use molecular SO_2 rather than free SO_2 as an indicator when you are trying to protect your wine from microbial spoilage.

Further reading

FUGELSANG, K. C. and EDWARDS, C. G. 2007. *Wine Microbiology: Practical Applications and Procedures*. Second edition. New York, Springer Science and Business Media.

ROMANO, P. and SUZZI, G. 1993. 'Sulphur dioxide and wine microorganisms', in Fleet G, *Wine Microbiology and Biotechnology*, Switzerland, Harwood Press, 373–392.

5.11 How can I protect my grapes and juice from spoilage during transport to the winery?

G. S. Ritchie

Sulfur dioxide (SO_2) can be added to harvesting bins if there is a risk of spoilage due to high temperatures, high pH, diseased or damaged fruit, long wait times until the fruit can be vinified or long transport distances to the winery. High temperatures and pH increase proliferation of microbes after infection and increase the rate of oxidation. Machine harvested or diseased and damaged berries often have ruptured skins, releasing juice which drains to the bottom of the container. Accumulation of juice in the base of the bin increases the risk of infection with microbes, as the juice is more easily accessible. The longer the journey to the winery, the greater the opportunity for spoilage.

Sulfur dioxide's main role is to prevent microbial infection of the juice and thereby prevent unwanted or spontaneous fermentations by yeasts other than that planned by the winemaker and infections by undesirable bacteria (e.g., *acetobacter, lactobacillus*). Molecular sulfur dioxide (SO_2 or H_2SO_3) is the form that inhibits microbes. The sulfite ion (SO_3^{2-}) is mainly responsible for preventing oxidation but there is not much present at the pH range of most juices and wines (Boulton *et al.* 1996). Sulfur dioxide can react with compounds other than oxygen that may be found in musts (e.g., anthocyanins, acetaldehyde, glucose) to form 'bound' SO_2, which is unable to prevent microbial spoilage or oxidation. Consequently, when we add sulfur dioxide to a juice or wine, not all will be available to protect the wine (depending on its distribution between the different forms), which complicates deciding how much to add. In practice, we have to make an estimate of how much will be in the bound form to ensure that there is sufficient molecular SO_2. At this stage of the winemaking process, we may assume that 25–30% may be in the bound form once it dissolves in juice. A further complication is that different amounts of molecular SO_2 are required to inhibit different microbes.

The simplest way to add SO_2 to harvesting bins is as the solid, potassium meta-bisulfite (KMBS). Fifty seven percent of potassium meta-bisulfite is SO_2 (some winemakers approximate this to 50%). The great advantage of KMBS is that the SO_2 is not released until it comes in contact with juice in the harvesting bin (i.e., when spoilage could begin to occur). The amount to add will vary according to the extent of the risks mentioned above. If there were more than one risk evident, then the amount would need to be increased. Table 1 illustrates how the amount added might vary according to the level of two different risks and assumes that approximately 25% may be in the bound form. If a third factor was present, the additions may increase to 60–70 mg/L. If it is hot, then some of the KMBS could volatilize as SO_2 which may kill micro-organisms on the surfaces of whole berries. However, this is a secondary mechanism of protection and the extent to which it might occur has not been investigated.

Table 1 Possible examples of the variation in SO_2 additions (mg/L) with temperature (°C) and disease level

Disease level	SO_2 additions (mg/L) at different temperatures		
	<15°C	15–25°C	>25 °C
Low	30	35	40
Medium	35	40	45
High	40	45	50

Calculating the weight of KMBS to add to achieve a specific concentration of SO_2 in the juice[1]

Grams of KMBS to add = tons of grapes × press yield × mg/L SO_2 required × 0.00175

For example, the amount of KMBS required to add 40 mg/L of SO_2 to a picking bin holding 0.5 ton of grapes (that one would expect a press yield of 650 L/ton) would be:

Grams KMBS = 0.5 × 650 × 40 × 0.00175 = 22.75 g

The KMBS can then be weighed into small, resealable plastic bag, one for each harvesting bin. In the vineyard, the solid can then be sprinkled over the base of the harvesting bin just before it is filled. Juice from machine-harvested fruit (or damaged grapes) will drain to the bottom of the bin where it dissolves the KMBS and becomes protected from spoilage. If there is more time and the bin is not being filled too quickly, the KMBS could be dissolved in a small amount of water and added in portions as the bin was filling. These small amounts of KMBS should not present a problem of pickers breathing SO_2 fumes particularly since they do not stay by the bins for long periods of time. If a sorter was being employed to sort through the bin while it was being loaded, then a gas mask should be provided for protection as a precaution.

Reference

BOULTON, R. B., SINGLETON, V. L., BISSON, L. F. and KUNKEE, R. E. 1996. *Principles and Practices of Winemaking*, New York, Chapman & Hall.

1. In order to carry out the calculation, we need to make an assumption about the amount of juice per ton (commonly referred to as the press yield, L/ton). The volume of juice is calculated from the tons of grapes multiplied the expected yield (L per ton) from the press, e.g. if the yield for the press is 650 L/ton and 3 tons are going to be crushed, the volume would be 650 × 3 = 1950 L.

5.12 The wine's pH is 3.95. How much free SO_2 do I need to prevent malolactic bacteria (MLB) or Brett growth?

C. Butzke

Depending on the alcohol content of the wine, the amount needed is between 79 and 112 mg/L (see recommendations in Table 1).

Ethanol acts synergistically and enhances the effect of molecular sulfur dioxide gas which kills bacteria, so high alcohol wines require less SO_2 protection. Only a very small fraction of the sulfites in wine is present in the molecular (gaseous) form and is *exponentially* dependent on the wine's pH value (7% at pH 3.0 vs. 0.7% at pH 4.0). The vast majority is ionized in the bisulfite form, which can destroy the vitamin thiamin, essential for the growth of *Brettanomyces* yeast and certain wine bacteria. Much of the sulfites that the winemaker adds to juice at crush or to wine post-ML, is bound up by acetaldehyde, glucose/glucosides, as well as certain metabolic acids of microbial origin. Only properly kept free SO_2 provides additional capacity to bind more products of oxidative aging, to cleave thiamin or kill unwanted, SO_2-sensitive microbes. Brettanomyces thrives at higher pH, temperatures (>13°C!), ullages and residual yeast nutrient levels, but luckily the thiamin break-up – given proper amounts of free bisulfite – also occurs faster at higher pH.

The requirements for free SO_2 are not a stylistic option for the winemaker. Neither is it a good winemaking practice to routinely and 'only' add 20 or so *parts free* to each and every wine. If the winemaker decides not to follow these charts then he or she might as well not add any SO_2 at all as it will not make very much of a difference.

In recent decades, the increased demand for wines made from very ripe fruit has led to the production of wines with very high (>3.8) pH values following Emile Peynaud's observation 'that great red wines are always low in acidity'. While the preference for wines that appear softer and younger due to a lack of acidity is understandable, the rules for SO_2 additions remain the same. Unfortunately, because of the exponential nature of the relationship between acidity – measured as proton concentration – and SO_2, high-pH wines require the addition of excessively high amounts of free sulfites. At these concentrations, the winemaker may believe that the sensory threshold for sulfur dioxide will be exceeded.

The fear that the freshly-lit-match, metallic, pungent character of SO_2 gas will overpower the wine's varietal character and mask its fruitiness, can lead the winemaker to avoid the proper dosage according to chart. Interestingly, the sensory attributes of SO_2 are related to the volatile molecular form whose concentration remains intentionally the same at any of the recommended doses. More importantly, the smell of SO_2 is very closely related to the wine's temperature, so tasting it straight from the barrel at 12°C vs. from the bottling tank at 20°C will make quite a difference, as the warmer temperature releases approximately 50% more molecular SO_2 into the tasting glass's headspace.

Table 1 Free SO$_2$ required at wine pH (mg/L)

(a) Molecular SO$_2$: 0.85 mg/L @ 12% alcohol by volume

3.0		3.1		3.2		3.3		3.4		3.5		3.6		3.7		3.8		3.9	
3.00	13	3.10	16	3.20	20	3.30	25	3.40	32	3.50	40	3.60	50	3.70	63	3.80	79	3.90	100
3.01	13	3.11	16	3.21	20	3.31	26	3.41	32	3.51	41	3.61	51	3.71	64	3.81	81	3.91	102
3.02	13	3.12	17	3.22	21	3.32	26	3.42	33	3.52	42	3.62	52	3.72	66	3.82	83	3.92	105
3.03	13	3.13	17	3.23	21	3.33	27	3.43	34	3.53	43	3.63	54	3.73	68	3.83	85	3.93	107
3.04	14	3.14	17	3.24	22	3.34	28	3.44	35	3.54	44	3.64	55	3.74	69	3.84	87	3.94	110
3.05	14	3.15	18	3.25	22	3.35	28	3.45	35	3.55	45	3.65	56	3.75	71	3.85	89	3.95	112
3.06	14	3.16	18	3.26	23	3.36	29	3.46	36	3.56	46	3.66	57	3.76	72	3.86	91	3.96	115
3.07	15	3.17	19	3.27	23	3.37	29	3.47	37	3.57	47	3.67	59	3.77	74	3.87	93	3.97	117
3.08	15	3.18	19	3.28	24	3.38	30	3.48	38	3.58	48	3.68	60	3.78	76	3.88	95	3.98	120
3.09	15	3.19	19	3.29	25	3.39	31	3.49	39	3.59	49	3.69	62	3.79	78	3.89	98	3.99	123
3.10	16	3.20	20	3.30	25	3.40	32	3.50	40	3.60	50	3.70	63	3.80	79	3.90	100	4.00	126

(b) Molecular SO$_2$: 0.60 mg/L @ 14% alcohol by volume

3.0		3.1		3.2		3.3		3.4		3.5		3.6		3.7		3.8		3.9	
3.00	9	3.10	11	3.20	14	3.30	18	3.40	22	3.50	28	3.60	35	3.70	44	3.80	56	3.90	70
3.01	9	3.11	11	3.21	14	3.31	18	3.41	23	3.51	29	3.61	36	3.71	46	3.81	57	3.91	72
3.02	9	3.12	12	3.22	15	3.32	19	3.42	23	3.52	29	3.62	37	3.72	47	3.82	59	3.92	74
3.03	10	3.13	12	3.23	15	3.33	19	3.43	24	3.53	30	3.63	38	3.73	48	3.83	60	3.93	76
3.04	10	3.14	12	3.24	15	3.34	19	3.44	24	3.54	31	3.64	39	3.74	49	3.84	61	3.94	77
3.05	10	3.15	13	3.25	16	3.35	20	3.45	25	3.55	31	3.65	40	3.75	50	3.85	63	3.95	79
3.06	10	3.16	13	3.26	16	3.36	20	3.46	26	3.56	32	3.66	41	3.76	51	3.86	64	3.96	81
3.07	10	3.17	13	3.27	17	3.37	21	3.47	26	3.57	33	3.67	42	3.77	52	3.87	66	3.97	83
3.08	11	3.18	13	3.28	17	3.38	21	3.48	27	3.58	34	3.68	42	3.78	53	3.88	67	3.98	85
3.09	11	3.19	14	3.29	17	3.39	22	3.49	27	3.59	35	3.69	43	3.79	55	3.89	69	3.99	87
3.10	11	3.20	14	3.30	18	3.40	22	3.50	28	3.60	35	3.70	44	3.80	56	3.90	70	4.00	89

Certainly though, a high pH reduces the ageability of a wine as oxygen uptake and thus the major oxidative aging reactions, particularly browning, are accelerated. This must be considered, especially if the wine is made for lengthy barrel and bottle aging prior to release as required by law in certain appellations. This should be of particular concern to regions that are experiencing elevated natural cellar temperatures due to global warming.

Ideally, the winemaker could adjust the pH of the wine without much changing its perceived acidity and mouthfeel. However, lowering of the pH without noticeably increasing the TA of the wine is only possible by using strong inorganic acids such as hydrogen chloride. This, however, is illegal. The winemaker's repertoire usually and rightfully comprises the acids naturally occurring in grapes. The dated techniques of ion exchange and electrodialysis for pH adjustments have not been widely adopted by the premium wine industry, leaving the modern winemaker with few options. These include: modest acid additions *at crush* to bring the pH below at least 3.8 based on predictive modeling; a partial, earlier picking and back-blending of more acidic lots; (field-)blending with varietals – including hybrids – that can retain more acid; suggested changes to the code that regulates local winemaking practices.

Further reading

BOULTON, R. B., V. L. SINGLETON, L. F. BISSON and R. E. KUNKEE. 1996. *Principles and practices of winemaking*. Springer, p. 452.
http://www.winenet.com.au/articles/WineNetwork_Managing-brett_MP-MK-GB03.pdf
http://www.winerysolutions.com/acid.html

5.13 What is lysozyme and why is it used in winemaking?[1]

G. S. Ritchie

Lysozyme is an enzyme that is extracted from the egg white of hens but also exists in mammalian milk, tears and saliva. For many years, it has been used as a natural antimicrobial and antiviral in the food and pharmaceutical industries (Davidson and Zivanovic 2003). It can be used in winemaking to prevent microbial spoilage by *gram positive* bacteria, delay malolactic fermentation (MLF) or delay sulfur dioxide (SO_2) additions after MLF is complete. It is usually effective within 1–3 days but does not provide a lasting protection in red wines. In contrast, lysozyme has been observed to remain active for up to six months in white wines (Bartowsky *et al.* 2004). Lysozyme cannot replace sulfur dioxide (SO_2) because it will not prevent infections by other spoilage organisms such as *Acetobacter* or *Brettanomyces*. It is very important to monitor the wine for VA regularly and for the presence of other spoilage microbes. It will not affect desirable yeasts either. Lysozyme can be used in organic wines in the USA at concentrations up to 500 mg/L.

In juices, must and wines, the main gram-positive bacteria are lactic acid bacteria (*Pediococcus, Lactobacillus* and *Oenoccocus*). We use *Oenococcus* in our wines to convert malic acid into lactic acid (MLF) to prevent this microbial action happening in an uncontrolled way at an undesirable time (e.g., after bottling), to lower the pH and or to change the flavor profile of our wine (e.g., to introduce buttery flavors). On the other hand, we do not want *Lactobacillus* or *Pediococcus* to carry out MLF because they produce undesirable compounds such as volatile acidity (VA). Problems can even arise when our wines contain *Oenococcus* if MLF finishes before alcoholic fermentation. At the end of MLF, all lactic acid bacteria may begin to consume sugars and produce high levels of VA, ruining the wine.

Spoilage of juice or musts due to lactic acid bacteria can happen when we have diseased grapes, unsanitary conditions or a stuck or sluggish alcoholic fermentation. Once alcoholic fermentation commences, we cannot control the bacterial population with SO_2 because it will also inhibit the production of alcohol by yeast. On the other hand, lysozyme additions will eliminate the lactic bacteria population without inhibiting the yeast. Hence, the addition of lysozyme can prevent unwanted increases in VA before or during alcoholic fermentation because it will not inhibit yeasts. Nevertheless, prevention is always better than cure and can be achieved by sorting out diseased fruit, keeping the winery clean and sanitizing equipment before use.

Immediately monitoring the microbial population in juices or in musts from diseased or low quality grapes (by microscopic evaluation or other tests) will indicate whether a lysozyme addition may be warranted. Tracking the rate of decrease of glucose/fructose concentration of fermenting juices or musts will

---

1. Potential users of lysozyme need to check the legal requirements for its use in their region.

alert us to a sluggish or stuck fermentation. If this occurs we need to establish if lactic acid bacteria are present so that we know that the addition of lysozyme may be beneficial.

Lysozyme additions may be useful when we wish to delay MLF (some winemakers like to delay MLF of Pinot noir for 3–6 weeks after alcoholic fermentation). Adding lysozyme at this stage may help prevent the onset of MLF and infection by undesirable lactic acid bacteria.

If lysozyme has been added to a red wine, some will react with phenols and may produce lees which can be removed by racking. There also may be some color loss in wines made from varieties that have potentially low color, such as Pinot noir (Bartowsky et al. 2004). In white and rosé wines, the lysozyme may contribute to protein hazes (Bartowsky et al. 2004, Weber et al. 2009) and so treated wines may require bentonite fining before bottling. Fining trials would need to be run to ascertain how much bentonite would be required but the presence of lysozyme has been shown to at least double the rate of bentonite required to heat stabilize Riesling and Pinot blanc (Weber et al. 2009).

Lysozyme can be purchased as a solid or as a liquid and could cost from $31–185 per 1000 L juice or wine, depending on the rate applied (150–500 mg/L), the form used and whether a discount was offered for purchasing larger quantities. The maximum rate of 500 mg/L is recommended to inhibit unwanted malolactic bacteria and/or prevent MLF, while 250–300 mg/L is considered sufficient to delay the onset of MLF. Less may be required in white and rosé wines because there are far less phenols present to potentially denature the enzyme.

Lysozyme is an allergen and the EU has decided to require declaring its presence in wine on the label. Estimates of the percentage of the population that suffers from allergic reactions to egg protein appear to be 0.09% but the role that lysozyme specifically plays is less clear (Iaconelli et al. 2008). White wines treated with the enzyme have been found to contain concentrations harmful to allergic consumers. Without bentonite fining, 34–73% of lysozyme was found to remain in Riesling, Pinot blanc and Pinot gris wines and approximately 10% in a Dornfelder red wine. On the other hand, bentonite fining decreased the amount of lysozme to negligible concentrations (Weber et al. 2009).

References and further reading

BARTOWSKY, E. J., COSTELLO, P. J., VILLA, A. and HENSCHKE, P. A. 2004. The chemical and sensorial effects of lysozyme addition to red and white wines over six months' cellar storage. *The Australian Journal of Grape and Wine Research*, 10(3), 143–150.

CIOSI, O., GERLAND, C., VILLA, A. and KOSTIC, O. 2008. Application of lysozyme in Australian winemaking. *The Australian and New Zealand Wine Industry Journal*, 23(2), 52–55.

DAVIDSON, P. S. and ZIVANOVIC, S. 2003. The use of natural antimicrobials, in *Food Preservation Techniques* (eds P. Zeuthen, and L. Bøgh-Sørensen), Woodhead Publishing Ltd, Cambridge, pp. 5–30.

GAO, Y.C., ZHANG, G., KRENTZ, S., DARIUS, S., POWER, J. and LAGARDE, G. 2002. Inhibition of spoilage lactic acid bacteria by lysozyme during wine alcoholic fermentation. *The Australian Journal of Grape and Wine Research*, 8(2), 76–83.

IACONELLI, A., FIORENTINI, L., BRUSCHI, S., ROSSI, F., MINGRONE, G. and PIVA, G. 2008. Absence of allergic reactions to egg white lysozyme additive in Grana Padano cheese. *The Journal of the American College of Nutrition*, 27(2), 326–331.

WEBER, P., KRATZIN, H., BROCKOW, K., RING, J., STEINHART, H. and PASCHKE, A. 2009. Lysozyme in wine: A risk evaluation for consumers allergic to hen's egg. *Moleluclar Nutrition and Food Research*, 53(1), 1–9.

6
Wine filtration

6.1 How should I interpret references to 'size' in the context of filtration?

K. C. Fugelsang

In classic filtration, including pad/membrane and diatomaceous earth (DE), filter porosity is defined in terms of the physical dimensions of particulates removed. Here, the reference point is the micrometer (syn: μm or micron) a unit of metric measure equivalent to 39 millionths of an inch. In terms of scale, the unaided eye can detect particulates larger than 100 μm. Recognizing this, the head of a pin is approximately 200 μm while familiar wine microorganisms, such as lactic acid bacteria and yeast, range in size from near 0.5 μm to slightly over 1 μm, respectively.

Compared with particulate dimensions, 'size' at the molecular level is relative to weight (or mass) of the component removed. Here, Daltons (Da) or kilodaltons (kDa) identify the mass (or molecular weight) of the target molecule(s) compared to that of a hydrogen atom. As an example, a molecular mass of 500 Da corresponds to solute dimensions of approximately 1nm and, extrapolating to particulate filtration, a membrane porosity of 1.0 μm is equivalent to around 500 000 Da.

Considered in other terms, macrofiltration removes microbes and larger suspended materials, depending upon the nominal/absolute porosity of the filter matrix, whereas microfiltration effects clarification in the range of 0.1–0.2 μm. Ultrafiltration (UF) removes soluble macromolecules such as pigments, tannins, polysaccharides (and combinations of the three) as well as colloidally-suspended substances in the range of 1000–10^6 Da. 'Porosity,' in the world of ultra- and nanofiltration, is defined in terms of molecular weight cut-off (MWCO). For example, a UF membrane with a MWCO of 100 000 will remove solutes/colloidal of MW > 100 000. However, a clear distinction should be made between MWCO and absolute porosity as defined in sterile membrane applications. As is the case with nominal filtration, MWCO represents an approximation and should not be taken to define a specific solute mass.

Molecular mass, by itself, does not predict ease of solute separation. Components in the wine/juice matrix react among themselves in formation of aggregates whose functional weights (size) may be many times larger than their individual formula weights. The polar properties associated with intramolecular charge distribution in water serve as one example. Here, the electonegativity associated with the oxygen creates a slight negative charge whereas the hydrogen atoms carry a slight positive charge, thereby creating molecule with two poles; positive and negative. The electrical dipole created by the charge distribution interacts with other similarly charged molecules (via hydrogen bonding) in creation of a hydration shell or 'skin'. It is estimated that each water molecule reacts with the equivalent of 500–900 Da of hydration thereby increasing the effective 'size' and, hence, physical properties of a single molecule of water by many-fold. Hydration also drives other components of the wine matrix to aggregate in formation of colloids. Thus, anthocyanins and other phenolics exist in colloidal complexes many hundreds of times larger than their formula weights would predict.

Nanofiltration serves as a loosely-defined bridge between UF and high retention ('tight') reverse osmosis. Unfortunately, a universally accepted definition, based upon solute size retention/removal, is lacking and varies depending upon whether one views the separation from a regulatory or applications mode. Section 24.248 of the Federal Register (Title 27 CFR) summarizes these distinctions in terms of size of solute removed and the maximum transmembrane pressure (ΔP) allowed to achieve the separation:

Separation mode	Molecular mass (Da)	ΔP
Ultrafiltration	500–25 000	<200
Nanofiltration	<150	<250
Reverse osmosis	<500	>200

It is clear that, from the regulatory point of view, NF is a subset of RO.

Operationally, nanoseparation refers to use of membranes which preferentially pass monovalent while excluding divalent ions. On a molecular size (mass) scale, this amounts to solutes between 500 and 1000 Da. Thus, 'nanofiltration' should be viewed as a extension of reverse osmosis and, reflecting this, is often referred to as 'loose RO.' Its primary application is to reduce levels of 4-ethyl phenol/4-ethyl guaiacol, haloanisoles and smoke taint which lie in the molecular weight range of 100–150 Da. Unfortunately, there is generally some collateral loss of wine flavor and character associated with the separation.

Because of the membrane's near impermeability to wine flavor, color and tannin, 'tight RO,' or hyperfiltration, is used for the removal of small molecules such as alcohol (MW = 46), and acetic acid (MW = 60) which pass readily into the permeate. These membranes can also be used to remove ethyl acetate (MW = 88) and even 4-ethyl phenol (MW = 122), but the passage of these molecules is slower and thus more processing cost is required.

Acknowledgements
I would like to express my sincere appreciation to Clark Smith (Vinovation, Inc.), John Giannini (Winemaker, California State University, Fresno) and Landon Donley and Marie Walker (graduate students, California State University, Fresno) for their review of my manuscript of Questions 6.1–6.3, 6.6, 6.9 and 6.11–6.13.

6.2 What are my options in terms of filtration?

K. C. Fugelsang

Depending upon one's goals, filtration can vary from little more than removal of visible debris to microbes. With the advent of cross-flow filtration, winemakers can now clarify relatively high solids and colloidally-laden wines that, historically, would not have been filterable using classic perpendicular flow technology.

Mechanistically, filtration can be divided into perpendicular-flow ('dead end') and cross- or tangential-flow filtration.

(1) Perpendicular flow filtration

As the name suggests, the particulate-laden suspension approaches the filter media head-on where those solids larger than the filter matrix's nominal (or absolute porosity) are retained. Filter media can be divided into two types: diatomaceous earth (DE or powder) and paper/pad. While the physical nature of each varies, the operational principles of clarification are the same. Based upon the physical properties of the filter matrix and particulates to be removed, three types of filters can be employed: screen or pre-filters, depth filters and membrane filters.

Screen filtration

Screen or pre-filters are occasionally employed to reduce levels of relatively large particulates *en route* to pad or DE filters. Frequently referred to as 'bug catchers', their primary application is to reduce high solids loads that would lead to premature plugging of downstream filter(s). Mechanically, screen filters trap debris on the surface of the upstream side and thus rely (primarily) on direct interception of particles larger than screen porosity. In practice, most winemakers opt to use either conventional gravity clarification or selective fining agents to achieve clarity sufficient to operate the primary filter directly rather than reliance on another tier of filtration.

Depth filtration

Depth filtration, including pad and DE, utilize the inherent properties of the filter matrix to trap and remove particulates. Those not blocked at the filter surface (surface interception) must 'navigate' through a maze of narrowing channels, turns, twists and dead ends within the filter matrix. Within the filter pad or cake, impaction may also indirectly 'catch' smaller particles through bridging resulting from multiple particles striking an opening simultaneously. In instances where particles do not entirely plug the channel, a tighter filter matrix is created bringing about an increased differential pressure (ΔP) and, eventually, plugging. Charge interaction also plays a role in retaining particulates and solutes smaller than pore/channel size. Referred to as *Zeta* potential, negatively-charged suspended debris is attracted to and retained by positively-charged filtration surfaces.

The advantages of depth filtration include retention of particles throughout its matrix, (via the interactive forces described above) rather than solely on the surface. Collectively, these forces result in significantly greater filtration or 'dirt handling capacity' relative to screen and membrane filters.

The filter matrix (pad) may be composed of cellulosic and/or polymeric fiber bound together with food-grade binder. Alternatively, diatomaceous earth or 'powder filtration' utilizes the same physical principles of retention but the cellulosic matrix is replaced by siliceous filter-aid in the form of diatomaceous earth (syn: DE, diatomite or kieselguhr/kieselgur) as a filtration surface in/upon which suspended materials are trapped.

The principal disadvantage of depth filtration is the inability, given the nature of the product, to accurately define porosity. Rather, depth filters employ an average or 'nominal' rating based upon factory laboratory testing of the particular lot. Depending upon nominal porosity, the finished wine may be broadly described as 'rough' or 'polish-filtered.' Conventional filters, either pad or DE, fundamentally act as depositories of suspended solids and have to be replaced frequently. Hence, filtration materials are expendable and, in the case of DE, increasingly a waste management concern.

Another potential disadvantage of depth filtration is 'media migration' or 'throughput' which refers to the tendency of filter media (fragments), or already-trapped debris, to slough off during filtration and end up in the filtrate. The problem is seen most often when the wine encounters the filtration surface as a surge. Thus, filter operators are advised to run pumps at low-to-moderate flow rates, without interruption and within in recommended operational pressure differentials (ΔP). Shutting down the filter for lunch or at breaks as well as ratcheting up the flow rate towards the end of filtration, although sometimes tempting, may lead to the increased likelihood of debris being forced through the pad/cake into the filtrate.

Membrane filtration

Membrane ('absolute') filtration utilizes a geometrically regular, porous filtration membrane (approx. 150 μm thick), that retains particles on its surface by direct interception/size occlusion. Pore size is controlled in the manufacturing process. Using absolute filtration, the operational integrity (porosity) of the membrane can be validated by the winemaker prior to use by methods described in **6.7**. (See also **6.13** on integrity testing.)

Depending upon porosity, membrane filtration may (or may not) produce a filtrate free of microbes. A common fallacy among winemakers is the belief that yeast removal can be achieved using a 0.6 to 0.8 μm membrane whereas tighter (0.45 μm) membrane filtration should be reserved only for lactic acid bacteria. This approach ignores the fact that recently separated yeast daughter cells (e.g., *Brettanomyces*) and physiologically stressed yeast are typically <0.6 μm and may, therefore, end up in the filtrate. It is strongly recommended in cases where biological stability of the wine is in question, that the winemaker recognize and adhere to time-honored dogma: STERILE FILTRATION IS DEFINED AS 0.45 μm (ABSOLUTE).

Because membrane filtration is based upon surface interception/retention, this family of filter media exhibits a very low dirt-handling capacity and, thus, has a finite life span dependent upon particulate challenges. Their best application is at bottling as a final in-line filtration for the removal of microbes (<10/L) remaining after pad/powder filtration.

The critical distinction between depth filters and (absolute) membranes lies in the winemaker's ability to test the physical integrity of the later. In sterile filtration applications, membrane integrity testing should precede and follow the operational cycle and, in cases of interruption, be revalidated prior to proceeding. In each case, test data should be recorded and become part of the permanent bottling record.

Winemakers are, occasionally, targets of a secondary or after-market industry that specializes in reconditioning membranes prior to resale at a fraction of the price of new. Unfortunately, unscrupulous purveyors of reconditioned membrane filters generally have no ties to the original manufacturer and, thus, there are no warrantees/guarantees associated with the purchase. Unless you are dealing directly with the manufacturer's representative, *caveat emptor*! The best approach is to decline purchase of 'rejuvenated' membranes from any source.

Further, attempting to extend the operational life of plugged membrane filters by back-flushing is not recommended. Whereas the practice is an integral part in cross-flow, it will irreparably damage the membrane in perpendicular-flow applications. Again, unless you are dealing with the manufacturer's/supplier's filtration specialist, one should not attempt regenerating plugged membranes by back-flushing or any other method. In most cases, such practices void guarantees should the filter subsequently fail.

(2) Cross- or tangential-flow filtration
Interest in and technology capable of separation at sub-micron or molecular levels was driven by the medical community's need for dialysis membranes in the early 1960s. Recognizing that conventional perpendicular flow technology resulted in almost immediate plugging of the membrane, the concept of tangential- or cross-flow filtration was born. By the 1980s, filter and membrane design had progressed to the point where cross-flow filtration could be used routinely by the winemaker to effect changes in chemical/physical properties of wine and juice ranging from removal of colloids to water.

Compared with classic filtration, the feed stream, in cross-flow, approaches the membrane tangentially, rather than head-on, and is separated, at the filter surface, into two product streams: (1) the permeate or, the component that passes through the membrane, and (2) the retentate, or concentrate, enriched by those solutes and/or suspended solids which, by their physical nature, are rejected at the barrier surface. As with any filtration, the driving force for separation is the pressure differential between the feed and permeate-side of the membrane barrier. In this case, however, directional flow of wine/juice relative to the membrane surface, creates a peripheral turbulent flow, or 'eddy-effect' which dislodges particulates thereby minimizing membrane plugging/fouling through most of the operational cycle. Those solutes/solids not passing through the membrane are swept away and returned to 'feed tank' via a circulation loop. Despite the self-cleaning effect, retentate solids levels will, eventually, become sufficiently high that the system requires regeneration. Unlike perpendicular-flow membrane filtration, back-flushing, in cross-flow applications, is standard operating procedure for dislodging trapped solids. In most units produced today, back-flushing is part of the automated filtration program.

In terms of size (dimension/mass) of components removed, contemporary cross-flow separation/filtration applications range by more than five orders of magnitude, from particulate (micropore) to sub-micron applications; ultra-, nano- and hyper- ('tight') RO.

- Micropore clarification has become increasingly popular in recent years due to its convenience as well as increasing concern, regulation and costs associated with disposal of diatomaceous earth. Additionally, health concerns stemming from inhalation of the filter aid represent an increasingly unacceptable risk. Although most winemakers now recognize micropore filtration's value as a clarification tool in whites, debate continues regarding its efficacy for highly structured reds.
- Ultrafiltration modules have developed over the last thirty years from flat sheet stock stretched across frames and stacked in a fashion similar to plate-and-frame filters to tubular, hollow fiber and spiral-wound configurations. Although each design has been successfully utilized in different processing applications, evolution has, generally, addressed the issues of available filtration area as well as strength of construction materials which can be polymeric or ceramic depending upon operational pressures and separation goals.

  In terms of MWCO, nanofiltration (NF) is theoretically positioned between RO and UF but, legally and practically, is a sub-category of RO in wine applications. Although NF and RO are similar in concept and operation, successful separation using NF is, in large part, dependent upon membrane selection. Since membrane rejection is not solely based upon size, but structure and charge of the components in solution as well, it is often difficult to predict the performance of NF membranes. Thus, a variety of application-specific membranes are available; each specifically suited to a certain application and, generally, not universally applicable.
- While RO can remove the smallest of solute molecules and ionic species, it does so at higher operational pressures. An important distinction between UF and RO is that in UF applications, the product is generally the permeate. In RO, the permeate is generally a recombined side stream and the product is the retentate. That is, the goal is to prevent color, flavor and structure from passing through the membrane, whereas in UF the reverse is true.

Despite their sub-micron filtration capabilities, cross-flow filters should be regarded as separation tools. Hence, they should be utilized as nominal and not absolute (sterile) filters. Where the permeate is biologically unstable, bottling will require the use of conventional perpendicular flow sterile membrane (0.45 μm) filtration in advance of the fillers.

6.3 What is osmotic distillation?

K. C. Fugelsang

Whereas osmosis refers to the movement of water across a membrane from an area of higher to lower concentration, osmotic distillation (OD) or, more correctly, isothermal membrane distillation, is a separation process that uses the differences in vapor pressures of the liquid phases. Here, a liquid mixture containing one or more volatile components (e.g., wine/juice) is separated from a salt solution by a non-wetting microporous membrane and the driving force for movement of volatile components across the barrier is differential vapor pressure of each component relative to the liquid phases. The temperature gradient across the membrane is typically less than 2°C making the process nearly isothermal. Membrane construction is typically either in hollow fiber or flat sheet form.

OD's primary advantage, relative to RO and other separation methods, lies in its ability to concentrate solutes to very high levels with minimal thermal or mechanical damage to (or loss of) those solutes. However, costs are still high relative to other available technology which may delay its routine use in wine/juice applications.

6.4 What are typical types of filters?

T. J. Payette

The Webster definition of 'filter' is *'A device containing a porous substance through which a liquid or gas can be passed to separate out suspended matter'*. In the majority of instances filtration is the physical removal of solid objects, in suspension, by way of a mechanical means. In most cases the wine is forced through a filter media by way of positive pressure. In only one case the author is aware a vacuum is used.

The typical types of filtration are:

- *Plate and frame* – perhaps one of the more common types of filtration one finds in the industry for large particle removal and it has the versatility, by way of changing the filter pads, to remove very small particle sizes down to 0.45 microns and potentially smaller. Pads, mostly cellulose today, are installed between frames, in the proper orientation, and the unit is tightened down on the filter pads so leaking is minimal. This type unit uses positive pressure from a pump to physically push the wine through the filter pads. A very versatile piece of equipment for the winery after the wine has been allowed to settle reasonably well in the wine vessels. For 'dirtier wines' one may consider the DE filter descried below.
- *Diatomaceous earth (DE)* – a diatomaceous earth filter uses diatomaceous earth, similar to that one may find used in the swimming pool industry, as the filter media. Stainless steel screens hold the DE on them, while positive pressure is exerted continuously, holding the DE in place to be the filter media. A dosing pump will systematically add extra DE to the wine, upstream of the filter screens, to continuously add to the filter media. If run properly one may be able to filter large amounts of wine with few to no tear downs or much clogging. The flow rates, if operated properly will not be reduced significantly. It takes time to learn how to run these filters and the media (DE) has some health concerns during its storage, use and as a filter waste product after use. Please look into these issues to make an informed decision should the choice to move forward with these types of units exist.
- *Cartridge* – these can range from large pore sizes to minute pore sizes and their surface area is not as easily adjusted as one may be able to do with a plate and frame as mentioned above. This may be the only filter a smaller winery may have but as volumes get larger in the cellar one rapidly outgrows this type of filtration, very much dependent on the style and age of the wine. One advantage these filters have added to the industry is the ability to do a final polished filtration during bottling. The author understands these filters are made for the pharmaceutical industry, they have significant cost, and it is recommended wineries obtain them for bottling and to re-sterilize them before each use. This feature provides the most effective use of these filters and makes them cost effective. They should also be tested for their integrity if one is using them for 'sterile bottling' as the industry knows it today.

- *Lees press* – this may be considered the largest pore size filter and most frequently used with juices during the crush. Diatomaceous earth may be added to juice or wine lees and forced through canvass filter sheets. Often the results are that muddy looking juices filter opulently clear, before fermentation, to provide a large economic return because the muddy juices, once discarded, are filtered, kept and fermented. It is estimated that a winery of 3000–4000 cases of wine will benefit from these types of filters and pay for themselves, with proper use, easily within the first two years. The cost of the DE is minimal and the rewards are great. The one slight disadvantage is that this becomes another process to do during harvest when many processes require the winemakers' attention. If managed properly it becomes easily juggled during harvest and worth the winemakers while not to mention the wineries financial bottom line.
- *Vacuum drum* – typically used very much like the less filter press above for 'dirty juices'. For this reason vacuum drum filters are used mostly with lees, raw juices and the same mentioned above, muddy racking bottoms. This mode of action has a coating of diatomaceous earth on a large drum that spins while partially submerged in the liquid. The liquid is sucked through the DE coating on the drum. As the drums spins a large blade or knife cuts off the clogged portion of the DE and revealing a clean layer of DE for filtration as the drum continues to spin into the juice or wine for uninterrupted continuous filtration. The clean juice or wine after filtration exits the drum by way of a central tube in the center of the drum.
- *Centrifuge* – worth mentioning briefly in this segment because it does offer clarification to a wine or juice by way of spinning the liquid, magnifying gravity, and forcing the particles out of solution. This is not a filtration but it does offer a means to reduce solids, including yeast and bacteria, from the wine or juice.
- *Cross-flow* – another up and coming filtration option for the winemaker not commonly found in wineries today. Sophisticated and proving very promising to become a mainstay in the industry. In this case the filtrate and the filter media are moving in opposite directions, not perpendicular as in most other cases, while crossing over a barrier from one wine media to another. The unfiltered wine continues to wash away the solids, known as the retentate, from the porous membrane. To continue to be current on this topic one should contact their sales representative. Especially useful for fruit wines also and it has no expendables such as DE or filter pads with little waste other than the solids that were removed.

6.5 How do I decide what membrane pore size to use?

T. J. Payette

Every winemaker struggles with this question. Once one stands at the filter and an assessment is carried out, as a before and after application, one understands certain parts of the wine dissociate and/or become lost. In some cases this is temporary and in other cases this is permanent. On the contrary, however, some wines 'clean up' after filtration.

The pore size is often determined by analytical review linked back to bacteria and/or yeast. If a wine has not undergone a malolactic fermentation to completion most winemakers will use a 0.45 micron rated absolute filtration. Absolute means all particles of that size or larger will be removed if operated properly. If a wine has a residual fermentable sugar remaining many wine makers will once again use the 0.45 micron rated 'sterile filter'. Dry wines having gone through a complete malolactic fermentation will offer a potential to review a larger pore size filtration. This is often desired because many winemakers feel it leaves more in the wine and therefore it is more pleasing to the customer. In many cases the dry wine with a complete malolactic will only need a filtration down to 0.80 microns. In this case the winemaker is targeting the next largest potential threat and that may be *Brettanomyces*.

Caution: Many winemakers try their best to filter as little as possible and this often has positive results. Some go as far as no filtration with success. The reader should be aware that a wine could go bad if no filtration is used and certain microbial risks are very large. A wine that goes bad 'on the consumer's shelf' will offer a new meaning to most winemakers and their increasing desires to filter in the future if the problem is the result of a lack of filtration.

Filtration is a wonderful tool that previous winemaking generations were not fortunate enough to have. Filtration is the number one tool to decrease cross-contamination in the wine cellar because each microbe must have an origination point. Filtered wines, if filtered tightly enough, carry less to zero wine harmful microbes and therefore a cross-contamination risk is eliminated.

6.6 Do sterile filter pads really exist?

K. C. Fugelsang

Nominal porosity ratings are based upon the filter matrix's *average* particle size removal (e.g., 0.60 μm) as determined by laboratory testing prior to packaging. However, this rating is valid only under strictly defined conditions of flow, temperature, pressure, and viscosity. Change in any parameter may well affect particle retention. Because of the nature of the filter matrix, an absolute particle retention rating (100%) is, theoretically, unattainable for any form of pad or diatomaceous earth filtration. Rather, such filters are assigned a 'nominal rating', or particle size, above which a percentage (usually 70–98%) of particulates will be retained.

So-called sterile pads are depth filters specially fabricated to a higher degree of uniform porosity. However, the most significant concern with pads, even when sold as 'sterile', is that the winemaker cannot, independently, validate porosity. Thus, a 'sterile' pad filter may be nominally rated at 0.45 μm but, in fact, porosity may vary significantly thereby calling into question the claims of 'sterile'.

Absolute filtration, by comparison, relies on surface filtration resulting almost solely from particle-size exclusion. Membrane porosity is defined at 99%+ reliability and, importantly, can be independently validated by pre-filtration integrity testing (e.g., bubble-point, etc.). Since sterile (0.45 μm) membranes have very low dirt-handling capacity, wine must be free of particulates (<10/L) prior to attempting filtration.

6.7 How can I minimize filtration?

T. J. Payette

Time and patience using traditional winemaking practices perhaps offer the best solution utilizing natural gravity to clear the wine. Rackings dislodge fermentation carbon dioxide gas from holding certain particulate matter in suspension allowing them to settle. Fining agents also afford many particulate removal benefits and these should be tried using trials in a laboratory setting before using them in larger quantities in the wine tank. Haphazard use of fining agents without laboratory proof of the positive benefits may lead toward unexpected results and/or damage to the wines' aroma and palate structure.

Excellent sanitation will be next in priority to minimize filtration. A winery that is free of spoilage microbes will most likely have many wines that are free of spoilage microbes. These wineries may more easily bottle wines with less or no filtration. Professional or internal laboratories may offer certain lab test as predictive models to a certain wine's ability to have little or no filtration. Realizing this will depend on the cleanliness of the bottling line and whether the sample of the wine offered to the lab is representative.

6.8 My wines are difficult to filter: where can I look to solve this issue?

T. J. Payette

Most often wines that are difficult to filter are so for several reasons.

Pectins

Wines that appear reasonably clear yet are difficult to filter often link back to pectins. This is a gelatinous type slimy material that can be eliminated or reduced by the use of pectic enzymes. These enzymes will break down the pectins and allow them to settle or be more easily removed during a filtration. Most winemakers use pectinase enzymes at the juice stage of the winemaking process where they are most effective. Speak with suppliers if this may be an issue with your wine.

Heavy solids load

If a wine has not settled or has not been given time to settle, the simple large amounts of solids may be the culprit. Certain obvious principles apply, so one must gauge how dirty a certain wine is and what pore size filter media to use. Dirtier wines will need a large pore size and then subsequent filtrations may allow the winemaker to step down the pore size. In many cases winemakers will use their experience and a wineglass; yet, many potential filtration lab tests do exist for those that care to pursue them as a predictive model.

Colloids and polysaccharides

Wine that has these issues are rare and perhaps difficult to identify.

6.9 Can the physical nature of wine particulates affect filtration rate and volume?

K. C. Fugelsang

The physical characteristics of particles being removed can have a significant impact on the success or failure of filtration. Structurally, wine/juice particulates can be categorized as: (1) non-deformables, whose shape *will not change* as pressure is applied; and (2) deformables which lack rigid structure. Whereas non-deformables (e.g., filteraid) are not influenced by pressure build-up during filtration, deformables (yeast and bacteria, bentonite, protein-phenol-polysaccharide complexes) respond to pressure increases by spreading out and blocking filter channels. In conventional perpendicular-flow filtration applications, these lead to rapid plugging ('blanking off' or 'fouling') of pad, filter cake or membrane. Thus, deformables should be dealt with in a proactive manner by racking and/or fining before attempting filtration. Where colloidal polysaccharides are known to be problematic, enzyme utilization may be another pre-filtration option. Unfortunately, cost and predictability of post-fermentation enzyme treatments are a consideration.

6.10 Does filtration affect wine quality?

T. J. Payette

Yes. Filtration most certainly affects wine quality. As noted earlier, some wines clean up and become better while others lose some of their palate weight or other positive characteristics. It is easy to introduce large amounts of air during most of the filtration exercises, so it is best to know how to properly operate any filtration equipment to make sure the wine is not damaged by air and oxygen alone. In the author's opinion filtration has been one of the best advancements for the winemaker because a certain amount of predictive control is placed in the winemakers' hands as a tool allowing them to use less or fewer chemicals for the wines preservation. Overall filtration has enhanced wine quality, to the consumer, because fewer wines become spoiled after bottling. Think of filtration in terms of bacteria/yeast removal and less for clarification; yet, it will assist both.

6.11 I've heard that filtration strips my wine. Is it really necessary?

K. C. Fugelsang

Despite the availability of filtration technology capable of accomplishing multiple goals, debate continues as to whether the process, in any form, compromises wine character. As evidenced by the dearth of scientific documentation supporting or refuting post-filtration changes in physical/chemical matrix, it is clear that such claims are difficult to qualify/quantify. Considering the range of soluble flavor and aroma-active compounds present in wine relative to the diameter of a 0.45 μm pore, an argument could be made that such effects may be negligible or non-existent.

Whereas soluble species are, likely, not directly removed by conventional macro- and micropore filtration, colloidally-suspended macromolecules may well be a different matter. Compared with soluble compounds, colloids are subparticulate substances dispersed throughout the wine; small enough to be homogeneously suspended but large enough not to be fully dissolved. They may be present as large aggregations of polysaccharide, mannoprotein or protein-phenolic complexes (500+ kDa) where they may play a role in the wine's textural/structural presentation. Interactions between macromolecular species and lower molecular weight volatile compounds may, in part, account for apparent aromatic changes noted after sterile filtration.

Most agree that, minimally, filtration reduces the particulate load and, in the case of sterile applications, viable microorganisms. Hence, those opting not to filter must be acutely aware of the potential for post-bottling biological instability. Despite the availability of technology to sterile package, the frequency of secondary biological activity in bottled wine is seen with increased frequency compared with a decade ago. Today, we routinely recover a range of wine microbes from *Saccharomyces* to lactic acid bacteria from wines with alcohols levels over 14% (vol/vol) and total sulfur dioxide levels exceeding 80 mg/L. One explanation is that microbes are becoming more resistant to the environmental challenges presented by wine and the winemaker. Considering the direct and indirect costs of recall and remediation, the option to filter (or not) becomes more a practical than philosophical debate. New technologies such as Velcorin™ and supercritical CO_2 injection, high pressure and high temperature short time (HTST) pasteurization are being perfected as alternatives to sterile bottling without filtration.

Those advocating utilization of management strategies rather than sterile bottling are, philosophically, divided into two camps. The first, and more 'old school' approach, relies on a proactive program of sanitation (**9.1–9.13**) directed towards minimizing build-up of microbial populations in the cellar. The other school encourages a more holistic approach to stability; striving for microbial balance leading to a 'naturally stable' wine prior to bottling. Proponents minimize utilization of sulfur dioxide during cellaring. Additionally, cellar temperatures >60°F are required to facilitate growth of competitive strains.

6.12 What operational parameters should I monitor during filtration?

K. C. Fugelsang

Without question, filtration, in any form, will have a dramatic, although when carried out correctly, typically transitory effect on the wine. However, unless the winemaker takes steps to prepare the wine and monitor the operation, there may well be consequences that do not resolve with time.

Before filtration, the winery staff should make certain that the wine is racked free of solids that may result in premature fouling of the filter surface. Free SO_2 as well as dissolved O_2 levels should be checked and adjusted as appropriate. In that oxygen is more soluble at low temperatures, this is particularly crucial in cases where the wine has been refrigerated and allowed to warm-up prior to filtration or is filtered cold and then allowed to warm to cellar temperature. Sulfur dioxide and oxygen levels should be rechecked after filtration and, again, adjusted as necessary.

During the operational cycle, cellar staff should regularly monitor differential pressure (ΔP) increases arising from particulate build-up on/within the filter matrix. During operation, it is expected that ΔP will rise to mid-range or slightly higher, depending upon solids levels. However, filter operators should never allow ΔP to exceed operational specifications. The potential for breech or throughput, while not predictable, should be of concern.

Once filtration has begun, operators should not increase the flow rate to shorten run time. The practice leads to cake/pad compression; changing the nominal porosity of the matrix and, eventually, plugging the filter. It also increases the likelihood of media migration and throughput. Likewise, once started, the filtration should, optimally, proceed without interruption. Hydraulic surges, arising from reactivation of pumps, lead to already noted problems.

Because of inherent low flow rates, tangential flow operations are often automated. Loss prevention systems are employed to detect breeches in hoses as well as out-of-spec temperatures and pressures resulting from membrane fouling and other failures which could result in wine damage or loss.

6.13 What is integrity testing and when should it be performed?

K. C. Fugelsang

In bottling operations utilizing sterile filtration, integrity testing establishes the operational readiness/condition of the filter and validates rated porosity and proper installation. Operationally, winemakers utilize relatively easily performed integrity tests before and after each bottling. The most frequently performed integrity test is the bubble point, which is based upon the observation or measurement of a volume of gas transported from one side of a wetted membrane to the other by a gas pressure gradient.

The bubble-point test is based on the fact that water is held in the pores of the filter by surface tension and capillary forces. The minimum pressure (detected as gas bubbles in the filtrate) required to force liquid out of the pores is a measure of the pore diameter. Mechanically, the filter is wetted with water and pressurized to about 80% of the expected bubble point pressure which is stated in the manufacturer's literature. Pressure is slowly increased until rapid continuous bubbling is observed at the outlet. Pressure is compared with manufacturer's literature *and recorded.*

Interpretation: A bubble point value lower than specified is an indication of one of the following: a damaged membrane or seal, an intact filter, but wrong pore size, high temperature and/or incompletely wetted membrane.

Note: the measurement is *not in-line* and care should be taken to prevent secondary contamination.

The pressure hold (aka pressure decay or pressure drop) test, is another widely-used non-destructive integrity test. In this case, gas diffusion through the membrane is detected by monitoring changes in upstream pressure. Because there is no need to measure gas flow downstream of the filter, *any risk to downstream sterility is eliminated.* Pressure hold testing is the method employed by most automated integrity test systems.

6.14 After sterile filtering wine, what membrane flushing schemes are recommended?

B. Trela

After sterile filtering wine, the following membrane flushing schemes are recommended to prolong the life of the filter.

Water regeneration

Flushing the filter with clean, softened, chlorine-free, iron- and manganese-removed, sand and rust-free, sterile-filtered water removes and dissolves particles from the membrane (regeneration). This process also reduces bio-burden contamination, which is the number of living microorganisms retained on the membrane surface. After filtering wine and while the membrane is still in the filter housing, the membrane is rinsed with cold water (to prevent denaturing and subsequent deposit of proteins), followed by a hot water flush for 20 minutes, both in the direction of the original flow of wine ('forward recirculation'). The recommended temperature for maximum cleaning efficiency is typically 82°C (180°F). Remember, no chlorinated water can be used in and around the winery. Note that it is recommended to install a separate sterile filter membrane/housing for the rinse water that enters the wine filtration system.

Chemical regeneration

When a hot water washing cycle does not properly restore flow rate or the bubble point, chemical cleaning may be recommended. Typical chemical cleaners include strong acids, and many other commercially available CIP (clean-in-place) chemicals. However, under no circumstances should the membrane be subjected to pHs above 10 or below 2.

Example of a chemical cleaning procedure:

1. 0.1% solution of tri-sodium phosphate (TSP) at 60°C (140°F) for 20 minutes at 10 L/minute flow. This removes polypeptides and proteins from the membrane. Note that excessive use of phosphate-based cleaners will cause algae growth in your irrigation pond or wherever the waste water is pumped to. Alternatively, a mild alkaline flush with a 2% sodium hydroxide (NaOH) solution at 60°C (140°F) for 20 minutes can be applied.
2. Follow with a cold water flush.
3. Re-circulate with cold 200 mg/L SO_2 in 200 g/L citric acid solution or 150 mg/L peroxyacetic acid (peracetic acid) for 20 minutes.

Back-flushing

Back-flushing of the membrane is not recommended as it may diminish the structural integrity, particularly at elevated temperatures and pressure differentials (ΔP).

Sanitization
Sanitization must be performed immediately before each production batch and after extended shutdown periods to kill microorganisms retained on the filter surface. Typical sanitation includes using one of the below methods:

- Steam: 105°C for 30 minutes.
- Hot water: 82°C for 20 minutes.
- Ozonated water (if all gaskets are ozone-safe and no air pockets exist that cannot be contacted directly).
- Peracetic acid: 100 mg/L at 40°C.

Membrane storage
Store each membrane in a 200 mg/L SO_2 in 200 g/L citric acid solution, fully submerged, inside a wall-mounted, separate and labeled housing which can be easily built out of standard 6″ PVC piping with end caps or custom-made from stainless steel. Make sure to check the membrane regularly to ensure the storage solution still has free SO_2 and the membrane is fully covered.

Further reading
ANON. 2005. Vitipore® II and Vitipore® II Plus Cartridge Filters. Lit. No. DS0137EN00 Rev. A 02/05. Millipore Corporation, Billerica, MA 01821, USA.
ANON. 2008. 3M Filtration. BevASSURE® II Membrane Technology. CUNO Incorporated, Meriden, CT 06450, USA.

6.15 When can I bottle a wine without filtering it?
C. Butzke

Proper filtration is not an evil practice that strips the wine of finesse, color and its soul. Poorly performed or unnecessary filtration can oxidize a wine, or remove colloidal materials that contribute to the wine's *mouthfeel*.

While modern filtration has made possible stable wines with residual sugar and without excessive amounts of preservative, it is not a wine processing option that should be used routinely, either. In certain but relatively rare situations, a wine may be stable enough to be bottled without any filtration at all. How can we assess such stability needs?

The following criteria must be met:

1. The wine must have finished alcoholic fermentation to *dryness*, i.e. less than 1 g/L fructose and glucose remain as fermentable sugars.
2. The wine has less than 1 g/L of other reducing sugars such as pentoses left that can act as substrate for *Brettanomyces*.
3. The juice's original nitrogen content was assessed, and no excess nitrogen or vitamin supplements were added beyond the yeast strains' needs for finishing the alcoholic fermentation.
4. The wine has completed MLF, i.e. less than 300 mg/L malic acid is left.
5. The wine has tested negative for *Brettanomyces*.
6. The wine is bottled with the pH-dependent amount of free SO_2 plus a sufficient amount to account for losses during and immediately after bottling (an additional 10 to 25 mg/L is suggested).
7. The wine has been cold-stabilized.
8. The wine has been protein-stabilized with bentonite if necessary.
9. The wine has settled completely and is free of all visibly suspended solids, including yeast and bacteria.
10. The *final blend* has been stable at cellar temperature for another six months.
11. Test bottles of the wine have been stable at room temperature for one month.
12. The winemaker is willing to take a small risk to minimize the processing of his/her wine.

Note that reverse osmosis is also a membrane filtration. Efforts to specifically remove excess ethanol (molecular mass = 46), volatile acidity (60), ethyl acetate (88), *Brettanomyces* (4-ethyl phenol, 122) or disulfide (diethyl disulfide, 122) off-odors, depend on a narrow range of the membrane's pore sizes. Just as with sterile filtration the *nominal* pore size does not reflect an absolute cut-off but is the average of a normal distribution of pores sizes, some of which can be significantly larger thus capturing larger molecules. Because many possibly desirable aroma molecules in wine such as oak lactone (156), diacetyl (86), methoxypyrazines (110), 2-phenolethanol (122), vanillin (152), CO_2 (44) or SO_2 (64) are of similar molecular mass and size, they will be partially removed. Unfortunately, there is precious little quantitative analytical or sensory information published on the effects of proprietary wine treatments.

7
Wine packaging and storage

7.1 What is the best sterilization option for the bottling line?
T. E. Steiner

Introduction
Since bottling is the last winemaking operation taking place, it is essential that all unit processes including the filtration unit and downstream are properly sanitized to prevent microbial instability in the bottle. Although there are many options of sterilization, the most efficient and proven method for the bottling line is the use of hot (>180°F) water or steam.

Benefits of steam and hot water
Although the use of chemicals such as chlorine-based compounds, iodiphors, sulfur dioxide with citric acid and ozone are effective, they have limitations involving clean-in-place (CIP) operations. This is due to the inability of water-based cleaning compounds to penetrate the submicron cracks and crevices of the bottling line and filter system that may harbor yeast or bacteria.

The sterilization protocol and identifying sources of infection
The use of hot water or steam is extremely effective because the entire bottling line system, including the cracks and crevices, reach temperatures that kill wine microbes. When using hot water or steam, it is essential that the temperature be 82°C (180°F), or above, as measured at the furthest point on the bottling line (i.e., the filler spouts). Once the water or condensate temperature has reached 82°C (or above) at the filler spouts, the sterilization period should be performed for twenty

Wine packaging and storage 151

minutes. In addition, once temperatures have been reached, the flow can then be reduced to maintain temperature and save energy costs. Temperature-sensitive crayons or other temperature indicators can be used to determine whether the stainless steel components of the bottling equipment and lines are at the appropriate temperature (Meier 1994).

After sanitation, the temperature of the filter membrane should be reduced back to room temperature with cold sterile filtered water before attempting integrity tests. When performing sterilization procedures, it is vital to recognize sources of microbial contamination during the bottling process. A study by Neradt (1982) indicated possible places of microbial contamination during bottling (Fig. 1). The sources with the highest percentage of contamination were the filler (48%), corker (28%), bottle sterilizer (10%), bottle mouth (8%) and sterilizing filter (6%). Therefore, special attention to these areas should be addressed during the cleaning and sterilization process.

Donnelly (1977) indicated other sources such as filter drip trays, wine spills and floor gutters or drains are contamination sites which release airborne microorganisms. In preventing airborne contamination, filter-pad drip trays should be emptied at regular intervals and areas with juice, and wine spills should be cleaned and sanitized. In addition, gutter or drains of the bottling room should be washed with a sterilizing agent and flushed with copious amounts of water to reduce airborne contamination. Cleaning these areas should be done on a routine basis.

The use of hand sprayers for misting a solution of 25 ppm iodophore, 90% isopropanol or 70% ethanol can help sanitize the fill spouts. Special attention should be paid in spraying/cleaning the bell rubbers and rubber spacers of the

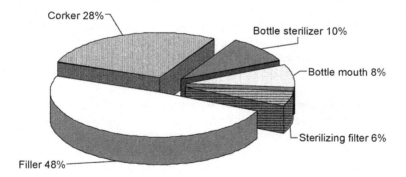

Cause of infection
Filler: infected centering bells/valves, improper sterilization, unauthorized technical modification
Corker: inadequate sterilization of centering bells/cork jaws, cork hopper not sterilized
Bottle sterilizer: mechanical failure, human error
Bottle mouth: inadequate or non-existing flaming
Sterilizing filter: damaged pressure gauge diaphragm or filter plate, unsuitable filter

Fig. 1 Sources and causes of reinfections during bottling. Adapted from Neradt (1982).

filling machine (Donnelly 1977). These solutions can also be used to spray corker jaws. It is a good practice to spray the fill spouts and corker jaws every hour while bottling and after line stoppages.

Mechanical maintenance and inspection of the bottling line such as greasing and chemical cleaning should take place the day before bottling. This includes the use of detergents for removing physical dirt and organic debris; thus making the sterilization process more effective. Sterilization of the bottling line with hot water or steam should take place both before and after bottling.

After bottling, any remaining wine should be flushed from the bottling line prior to sterilization by hot water or steam. In addition, the bottling line should be taken apart with each part being meticulously cleaned and sanitized. This sanitation process can take place in an iodophor bath at 25 mg/L followed by a thorough rinsing with sterile water and re-assembled. Although this takes time, total dismantling and thorough cleaning of the bottling line should be completed after each bottling.

Besides the use of a steam generator, a viable option for small- to medium-sized wineries may be a pressure washer/steamer or to designate a holding tank or kettle for water to be heated by propane or natural gas in a closed loop system to provide hot water sterilization at the bottling line.

Importance of water quality in sterilization of the bottling line

Sterilization efficiency may depend on the water quality. Soft water with filter sterilization is recommended for final rinsing of bottling lines. The purpose of softening the water will help prevent scale build-up from calcium and magnesium salts that can interfere with the effectiveness of detergents for cleaning the system prior to sterilization. Hard water build-up sites also act as sources for accumulation of organic build-up and microbial populations (Zoecklein 1995). These hard water deposits and scale can also accumulate in the filter system, thus causing the filter to plug up and shorten the life of the filter.

Quality control monitoring of the sterilization procedure

It is important to develop a quality control program which includes taking swab and environmental samples for microbiology analysis of the bottling line and room. This will determine the effectiveness of your sterilization procedure and identify areas of high microbial counts. Also, collecting wine samples during bottling at regular intervals (i.e., first, middle and end of bottling run) for laboratory analysis and plating procedures is recommended.

References

DONNELLY, D. M. 1977. 'Airborne microbial contamination in a winery bottling room.' *Am. J. Enol. Vitic.* 28 (3): 176–181.

MEIER, P. M. 1994. 'Influence of wine composition on filtration achieving microbiological stability through membrane filtration.' Proc. of The Ohio Grape-Wine Short Course, Wooster, OH, pp. 128–141.

NERADT, F. 1982. 'Sources of reinfections during cold-sterile bottling of wine.' *Am. J. Enol. Vitic.* 33 (3): 140–144.

ZOECKLEIN, B. W., K. C. FUGELSANG, B. H. GUMP and F. S. NURY. 1995. *Wine Analysis and Production*. New York: Chapman & Hall.

7.2 How long do I need to disinfect my bottling line if my hot water is less than 82°C (180°F)?

C. Butzke

The general assumption is that if you run 'hot' water through your bottling hose, filler bowl and fill spouts, it will kill *all* vegetative yeast and bacteria cells. This guess is wrong in two important ways. First, there is no absolute sterility when it comes to killing populations of microbes. Disinfection is a conventional attempt to significantly (by five orders of magnitude) reduce the number of spoilage microorganisms on equipment surfaces. Sterilization is an attempt to kill 100% of a population of spoilage microorganisms. However, microbiological statistics tell us that we can only reduce the percentage of germs surviving the treatment, we can theoretically never get them all.

If we have truly *hot* – 82°C (180°F) – water available, we can assume that if we reduce the number of vegetative cells present by 12 orders of magnitude (D_{12} concept), the equipment can be considered safe and 'sterile'. For example, one mL of fermenting wine with residual sugar – accidentally left in the bottling tank hose – may have 100 million (10^8) living yeast cells, so reducing their numbers by D_{12} will leave us with one cell surviving in 10 000 L of wine. If one would bottle this wine, then one in every about 13 000 bottles would be contaminated and could re-ferment, a risk acceptable to take.

This also implies that the lower the number of infectious cells is to begin with, the smaller the chance that re-growth will affect the bottling run. This is true for both thermal and chemical sterilization, i.e., adding 268 mg/L of potassium sorbate will only be safe if the wine was almost free of yeast cells. Adding it to a wine that already is still fermenting will not work, and cause a high probability that most bottles will re-ferment.

The killing power of hot water varies dramatically, i.e., exponentially depending on its temperature. For every 10°C (18°F) that the water temperature is lower than 82°C (180°F), it will take three times longer to disinfect the line. A treatment with 82°C (180°F) water for 5 minutes with kill 99.9999999999% of all yeasts present, but if your water is only 72°C (162°F) then the rinse time increases to 15 min. At 62°C (144°F), equal to a household water tank turned to the highest setting, it would take more than 45 min. Note that humans perceive water as painfully hot at temperatures just above 42°C (108°F), and scalding is always a risk working with very hot water or solutions.

The temperature of the equipment being sterilized must be taken at the *coldest* point, i.e. the part where the water leaves the system. For example, reaching 82°C at the fill spout will require that it leaves the hot water boiler significantly hotter. The boiler must be sized based on the flow requirements for the temperatures and times shown above. At a flow rate of 26 L/min (7 gal/min), the boiler capacity must be at least 130 L (35 gal) if you reach 82°C at the coldest spot, or 225 L (60 gal) if you only reach 77°C, equal to 70% more with water only 5°C colder. Multiplication factor: $3^{(82-T)*0.1}$ with T being the coldest temperature in °C.

Note that stainless steel is a poor heat conductor among the metals and it requires time for the heat from the hot water to reach all parts of the bottling line's periphery. Here is a comparison of the relative thermal conductivity of different materials in the winery: stainless steel = 1, iron 4, aluminum 17, copper 32, water 0.04, wood 0.03, air 0.002.

Biofilms that form on tank walls or in barrels are of particular concern to the winemaker as they resist traditional sanitation. They are often made of different types microorganisms that – on a suitable surface and under stress-related adaptation – form a *quorum* from multiple isolated colonies that can signal each other and channel nutrients, etc. They stick to the surface and shield themselves from heat and chemicals by producing protective muco-polysaccharides and require physical scrubbing before thermal or chemical disinfection can take place.

7.3 How do I steam the bottling line?

T. J. Payette

Steaming the bottling line
Steam is one of the most widely used methods of sanitizing a bottling line and cartridge filter prior to bottling. If done properly a 'sterile bottling' can be secured at each bottling run. Steam, for safety reasons, is a nuisance, yet most wineries still find it the bottling sanitation measure of choice. **Steam is hazardous to use and the author accepts no liability for error on behalf of the operator. Please be careful if this is your first time! Be careful every time!**

Bottling is an important time to be on your game. You will only get one chance to do this properly so it must be taken very seriously.

Why?
A winemaker interested in bottling a wine sterile will want to eliminate all bacteria and yeast from the bottling line and final filter prior to bottling day. Even winemakers who are bottling unfined and unfiltered wines sterilize their bottling lines before bottling for extra security. Steam, when used properly, is lethal to all living organisms.

How?
Live steam is the key to doing this procedure properly. If an orifice or filler spout is expected to be sterile, a flow of live steam must be coming from that area. Below are instructions for steaming a bottling line and final filter prior to bottling.

Before starting the below procedure, it is recommended to rinse out all the areas that will come into contact with the steam. Do a visual inspection for dust or foreign matter in areas that will come into contact with the wine – filler bowl, filter housing, filler spouts, etc.

1. Secure a source of steam that will be abundant enough to handle the set-up. A six-spout gravity flow filter may require less output from the steamer than a 12-spout filler.
2. Secure a wine hose that will allow steam to flow through it safely and use it for all the connections downstream of the steam. This hose should also be approved for wine.
3. Have the steamer as close to the bottle filler as possible while also allowing for a filter to be placed prior to the filler. (The closer the items are – the shorter the runs – leading to less error and faster ramp-up in the steaming cycle.) Some bottling units are nice, since a cartridge filter may be attached directly to the bottom of the unit minimizing hose length.
4. Place a stainless steel T with two valves prior to the cartridge filter and a T with two valves after the cartridge. Accurate pressure gauges should be

placed on the Ts so the winemaker can monitor the pressure during steaming and during bottling. (Remember to use pressure gauges that are designed to be steamed!)

5. Attach the hoses as if pumping wine through the filter and to the bottling line.
6. Open all valves in the beginning. (If using a mono block – make sure the automated solenoid valve is in the open position to allow steam into the filler bowl.) (If a safety blow off valve is not on your steamer – please install one yourself or take it to a qualified mechanic to install one.)
7. Turn on the steamer to initiate the creation of steam leaving all the valves downstream of the steamer open.
8. During the ramping up of the steam, condensed water may flow out of these open valves. Allow this to happen until it turns to steam. Then turn the valve toward the closed position but do not fully close. Leave the valve cracked open to allow a small amount of live steam to flow from the valve. This insures that the valve(s) has come up to the steam temperature and that it will remain at that temperature as long as live steam is flowing from it.
9. 'Chase' the steam through the complete setup and throttling back valves as steam appears.
10. When a full set of steam has reached the final destination of the end of the run (this may be the ends of every filler spout or the leveling mechanism in some mono blocks, etc.) – one can start timing the operation. Double check that all the filler spouts are open and that steam is flowing.
11. Places to look in the filler bowl may be a drain valve. Make sure that it is cracked open slightly to allow for a free flow of steam.
12. Step back and look at the operation asking yourself, 'Is steam getting everywhere and on all the surfaces that may come into contact with the wine?' If the answer is – 'YES', proceed to timing the operation.
13. Most winemakers steam for a minimum of 18 minutes and up to 25 or more should do little or no harm.
14. During the steaming operation – one may take steaming temperature crayons around to double-check they have achieved the desired temperature at a specific location on the bottling line. Steaming crayons are pencil-like devices made of materials that melt at certain temperatures. Caution is necessary when using these crayons on parts that may contact the wine. If wishing to test an area coming into contact with the wine – run a trial steaming operation – use the crayons – allow the line to cool and then thoroughly clean the area the crayons have contacted.
15. During the steaming operation – continue to check on the operation to make sure the function is continuing as planned and for safety reasons. Check that filler spouts have not 'jiggled' into the closed position.
16. After the steaming operation time limit has been met, the operator may once again check to make sure all the orifices are steaming as needed. Then turn off the steam source and allow for cooling of the lines, spouts and cartridge(s), etc.

17. Allow the system to come to room temperature and perform an integrity check (procedure to be covered in a future article) on the membrane filter to ensure the unit will perform properly at the rated micron level. Make sure the parts used to do this are free from microbes and that they have been cleaned with a 70% ethanol solution or equivalent protocol.
18. Do not disconnect any of the lines at this juncture. Use a spray bottle of the ethanol referenced above on all areas that have a possibility of compromised sterility. When in doubt – spray it! Filler spouts – too! The winemaker has a totally sterile system from the final filter downstream to the filler spout!
19. Aseptically close the filler spouts so they will retain wine when wine flows to them.
20. Attach the upstream line that was connected to the steamer to a source of pre-filtered wine.
21. Start the flow of wine slowly through the system once again 'chasing' the wine through the cracked valves. Some water may be allowed to drain off before wine reaches its destination.
22. Once the wine is completely through the system – complete the cycle by running several sets of bottles through the filler and returning that wine to the wine tank being bottled. This will ensure the first bottle of wine off the line will be exactly the same composition as the last bottle of wine.
23. Resume the normal bottling operation.
24. Once bottling is complete, make sure to rinse all the areas of the filler and final filter to remove any residue of wine. Some winemakers will actually rinse and then resteam the line without the final filter. This is a great idea especially after running a cuvee for sparkling wine on your everyday bottling line. Remember one will still want to steam prior to the next bottling too!
25. Send a bottle off to a certified lab or test in house under a microscope to make sure the wine is indeed clean and refermentation or a malolactic is not a concern. Taking samples at different times of the days bottling run can be a great idea to help identify any problem areas if they should occur later during the bottling day.

Some other helpful hints:

- Use water that is clean and free of minerals to extend the life of the steamer. Also, some water issues may clog the filters prematurely or during the steaming cycle.
- If the bottling line has ball valves on the filler or other areas – make sure these are physically clean and sterilized properly with steam. This is an area that can create cross-contamination issues with bottling.
- Contact your final filter supplier to make sure the procedures about to be incorporated are in line with their recommendations and to see if they recommend other helpful advice about their product and the specifications of their product.

- Contact your equipment dealer to make sure the equipment will hold up to the procedure and to hear potential areas of concerns. They may be familiar with other wineries that have gone through these procedures so they may be able to give helpful tips and suggestions.

If looking to sterile bottle your wines for the first time – take the above steps and recommendations and implement them to cater for your specific bottling line setup. Every line is different with a new set of places on which to focus or of which to be aware. Open the lines of communication with your suppliers, winemaker and bottling crew to make sure the above can be implemented successfully.

7.4 What does sterile bottling involve?

T. E. Steiner

Introduction

The purpose of sterile bottling is to remove all wine-tolerant microbes and to ensure that there is no post bottling instability threat or concerns in the finished product. Successful sterile bottling involves several important aspects; sterilization of the bottling line and filter, sterile filtration, the bottling process and bottling room (environment). This section will primarily discuss 'cold sterile bottling' (18–20°C) and will not cover techniques used in hot bottling or the addition of chemical sterilants. Chemical preservatives and sterilants will be discussed later in this chapter.

Since sterilization of the filtration unit and bottling line has been discussed earlier, the focus of this section will be on the importance of filtration and the bottling room environment in achieving successful sterile bottling practices.

A major process of sterile bottling comes through filtration of the wine. Assuming that the wine has had time to clarify in the cellar and has gone through proper filtration prior to bottling, sterile filtration at the bottling line is critical in assuring wine stability in the bottle. The ideal process of sterile filtration prior to bottling is to filter the wine through a depth pad, cartridge, or fine grade of diatomaceous earth prior to being filtered through a sterile membrane cartridge filter.

Depth filters

Depth filters depend on trapping microbes and particles in an angular path that weaves in and around a matrix of silicates and fibers. Depth filters are given a micron (μm) designation relative to a nominal rating. The nominal rating is not a true micron rating in determining the size of microbe trapped in the filter, but relates to the average or nominal tightness of the filter pad, general retention or degrees of tightness. This can also be expressed through the terms of a rough, polish and sterile pad filtration. Although, the use of depth filters can theoretically be used in sterile filtration, these filters are best employed as a 'pre-filter' in helping reduce suspended matter and microbes such as yeast, thus extending the life and efficiency of the final membrane filter. Popular pre-filter sizes of 1.2, 0.80 or 0.65 μm are nominal ratings and commonly used.

Membrane filters

As mentioned above, membrane filters are considered best for obtaining wine sterility. Membrane filters are mainly composed of synthetic polymers (including cellulose esters). They are commonly supplied as cartridges but can also be in the form of stacked disk filters. The rating of membrane filters is based on the membrane pore size (in microns). Since membranes lack depth, their ability to retain particulate matter, without plugging, is relatively low.

Wine packaging and storage   161

Therefore, it advisable to utilize a sterile membrane filter in-line after a depth or 'pre-filter'. A membrane rating of 0.45 μm is recommended for removing all wine microbes in providing a stable wine.

Integrity test

Although effective, a membrane filter has limitations based on length of use, the wine matrix and volume of wine. Therefore, it is critical to perform an integrity test before and after each bottling run to ensure filtration performance. Filter failure due to a rip, tear, and defect or plugging will be detected by these tests. Arguably, the most common integrity test being performed is known as the 'bubble point' test. This method checks membrane integrity by determining the pressure required to force gas bubbles (mainly N_2) through the filter's moistened surface into a vessel containing water at the outlet end. The bubble point test can be performed as follows (Fig. 1):

1. Run sterile water through the membrane filter.
2. Shut off valve at location (#1).
3. Connect the gas cylinder line (N_2) to the inlet (#2) of the filter housing.
4. Place the outlet hosing (#3) into a vessel containing water (#4) and open the outlet vent on the filter housing.
5. Open the valve on the gas cylinder (#5) and regulate a light flow around 5 psi.
6. Observe and record the pressure required to force a small continuous stream of bubbles into the water containing vessel on the outlet side.
7. After completion of the test, close the outlet vent and open the inlet vent valve to depressurize filter housing.
8. If membrane passes the bubble point test, the filter is ready to use.

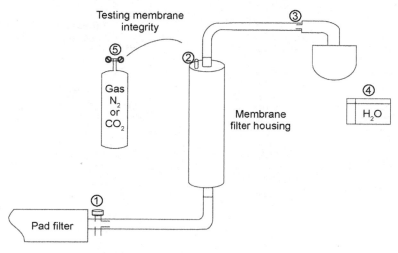

Fig. 1   Integrity testing of membrane filter. Source: Stamp (1987)

162 Winemaking problems solved

To check filter integrity, compare the observed pressure reading ('bubble point') with that defined by the manufacturer. Each pore size will have a corresponding specific gas pressure provided. The smaller the pore size, the more pressure required to force bubbles through the membrane unit. The filter should be considered faulty if the pressure reading is below the manufacturer's recommendation.

Important factors
It is also important to locate/check valves for strategic sampling for microbiology analysis. Sample ports can be placed between the pre and final filter in addition to immediately downstream of the final filter.

The location of pressure gauges before and after the pre and final filter housing is recommended in monitoring any pressure changes. Fluctuations in pressure may indicate a possible problem with either the depth or membrane filtration unit. In addition, any pressure fluctuations in the system may cause the filter to fail, releasing microbial contamination downstream.

It is also recommended to locate the final filter immediately after the pre-filter in preventing any further protective colloids and colloidal aggregates from forming after pre-filtration in clogging up the sterile membrane filter and possibly causing a colloidal haze in the bottled product (Meier 1994). A simple sterile bottling system and filtration schematic is supplied in Fig. 2.

Bottling room and environmental factors
The ideal bottling room is a facility containing only the bottle sterilizer, filler, corker and various conveyors in moving the glass. All other operations such as case dumping, labeling and finishing operations should be conducted outside of

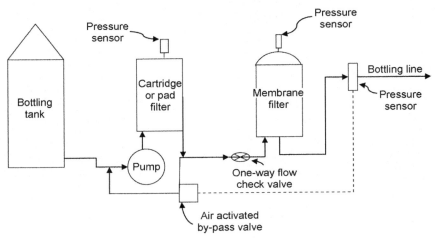

Fig. 2 Sterile bottling system. Adapted from Stamp (1987).

the bottling room. In addition, the bottling room should be located away from the fermentation area and isolated from general traffic areas for the purpose of preventing the risk of a large influx of *Saccharomyces* yeast contaminating the bottling room and product.

It is recommended that the bottling room have a supply of positive air pressure. The air should also be filtered for the purpose of preventing airborne microbial contamination. According to Donnelly (1977), the use of air filters reduced the airborne organism count by approximately 89%.

The bottles should enter and exit the bottling room through small portals with all entrances and exits being supplied with air curtains to help reduce airborne microbial contamination from outside the bottling room.

Another important reason for an enclosed bottling line is to limit excess worker traffic as a possible source of contamination. All workers should wear clean clothes or laboratory coats while in the bottling room. It is advisable for workers to wear non-slip boots in addition to having boot washing stations at the entrance. Workers' hands should be washed routinely in reducing the threat of cross contamination to any part of the bottling line or bottles.

When bottles are produced, they are sterile and generally remain in that condition if they are shrink wrapped and not stored for long periods of time in the winery. If stored for longer periods in the winery, there is a chance of environmental microbial contamination occurring. In addition, the bottles may contain dust or small particles of cardboard that need to be flushed out and sanitized with filtered air, paracetate solutions, ozone or a stream of sterile filtered water. Whether washing manually or automated, the bottles should be inverted for proper draining. In some cases, wineries simply use a jet of sterile filtered air for removal of dust or cardboard particles. When handling bottles, it is extremely important to handle the bottles below the neck.

Although time and effort are needed in achieving proper sterile bottling techniques, understanding and performing essential procedures will help reduce the threat of microbial instability in the bottled wine and ultimately increase the quality and aging potential of your wine.

References

DONNELLY, D. M. 1977. 'Airborne microbial contamination in a winery bottling room.' *Am. J. Enol. Vitic.* 28 (3): 176–181.

MEIER, P. M. 1994. 'Influence of wine composition on filtration achieving microbiological stability through membrane filtration.' Proc. of The Ohio Grape-Wine Short Course, Wooster, OH, pp. 128–141.

STAMP, C. 1987. 'The principals and practices of cold sterile bottling.' Proc. of The Ohio Grape-Wine Short Course. HCS Department Series 578. Wooster, OH. pp. 61–67.

7.5 What chemical additives can I utilize as an additional source of security in helping prevent the threat of re-fermentation or microbiological instability in the bottle?

T. E. Steiner

Introduction

It is important to realize that there is no absolute guarantee of achieving a 100% microbial-free environment in bottled wine. Bottling sterility is only as good as the sanitation procedures and sterile bottling protocols performed. As stated earlier in this chapter, the best bottling procedure to limit unwanted microbial instability in the bottle is sterile bottling. Therefore, the use of chemicals should not be viewed as a substitute for sterile bottling but may be used in addition to sterile bottling if determined as being necessary. In a situation where you are unsure of your sterilization efficiency and sterile bottling procedures, chemical additives may provide an additional line of safety in preventing microbial instability in the bottle from yeast or bacterial contamination.

Sulfur dioxide (SO_2)

Sulfur dioxide has been used for centuries. It is not used solely for wine production but also extensively in food processing. The proper use and monitoring of SO_2 is essential in the production of high quality wines. Therefore, it is not used only at bottling but is essential throughout the winemaking process.

Sulfur dioxide is unique, because it has both antioxidant and antimicrobial properties. As an antioxidant, SO_2 protects wines from browning by inhibiting enzymic and nonenzymic oxidation. For its antiseptic activity, SO_2 prevents spoilage microorganisms such as acetic acid bacteria, most lactic acid bacteria, molds and certain wild yeasts in wine. Sulfur dioxide is more effective on bacteria and molds than on yeast. Although, SO_2 exhibits a germicidal effect on yeast, some are more tolerant to molecular SO_2 than others. Generally, strongly fermentative yeasts are more tolerant to SO_2 than slower fermenting species (Dharmadhikari 1990). Therefore, at bottling it is essential that SO_2 be used in conjunction with sterile bottling techniques or chemical additives such as potassium sorbate.

Disadvantages of SO_2

Excess additions of SO_2 to must and wine may lead to bleaching of color in addition to an objectionable pungent aroma. High levels of SO_2 noticeable upon sensory evaluation are considered a wine flaw.

Adjusting SO_2 at bottling

Bottling provides the last opportunity to adjust free SO_2 to desired levels. Winemakers should adjust free SO_2 levels based upon wine pH. Often, a molecular SO_2 level of 0.8 ppm has been suggested as an acceptable concen-

Table 1 Free SO_2 levels required to obtain 0.8 ppm molecular SO_2 at various pH levels

PH	Free SO_2 to obtain 0.8 ppm molecular SO_2	PH	Free SO_2 to obtain 0.8 ppm molecular SO_2
2.9	11	3.5	40
3.0	13	3.6	50
3.1	16	3.7	63
3.2	21	3.8	79
3.3	26	3.9	99
3.4	32	4.0	125

Source: C. Smith, Enology Briefs, Feb/March, 1982, Univ. of CA, Davis.

tration for most wines. The data from Table 1 show the free SO_2 levels required to obtain 0.8 ppm molecular SO_2 at various pH concentrations. Generally speaking, these levels will inhibit the growth of spoilage microorganisms and prevent oxidation of wines. A word of caution should be expressed when dealing with wines (mostly red) that have a pH greater than 3.7 in achieving the desired level of 0.8 ppm SO_2. This situation may not be totally feasible due to adding excess SO_2 in both bleaching of color and possible negative sensory attributes. However, most premium reds are inoculated and monitored in going through a malolactic fermentation thus making it less favorable for microbiological contamination due to the absence of malic acid. Therefore, I would recommend monitoring and maintaining free SO_2 levels from 30 to 40 ppm in this case.

Since oxygen uptake can be a problem during the bottling process, the addition of SO_2 should also take this into consideration. If the bottling line is set up properly and specific procedures are followed, a maximum of 1 mg/L of dissolved oxygen should be a desired goal for bottling (Ough 1992). According to Boulton et al. (1999), 4 mg/L SO_2 will react with 1 mg/L of oxygen. If we assume an uptake of 1–2 mg/L of oxygen into the wine at bottling, we would need to add approximately 4–8 ppm extra SO_2 to account for oxygen uptake at bottling. This amount of SO_2 is in addition to the adjustment of SO_2 based on wine pH at a molecular level of 0.8 ppm described above.

The length of bottle storage or aging potential will depend mainly on variety, vintage and maturity. This consideration is more solely based on white wine varieties and styles in relation to additional SO_2 concentrations prior to bottling. A wine meant for consumption at six months may not require more SO_2 addition than mentioned previously. However, wines intended for longer bottle aging may benefit with the addition of 10 to 15 ppm more SO_2 than the additions already mentioned.

Illustrating the need to add additional SO_2 levels based on intended length of bottle aging, winemakers should be interested in knowing the amount of free SO_2 retained during bottle storage. Gallander (1987), illustrates the loss of free SO_2 in sweet table wines (Catawba and Vidal) at six months' storage at 18°C.

Results indicated that the level of free SO_2 was reduced during bottle storage. The average percentage loss in free SO_2 was approximately 20% for the sweet table wines and 17% for the control (dry table wines). Similar results were obtained by Ough *et al.* (1960) with dry table wines. They also reported that low storage temperatures generally resulted in higher wine quality and usually slowed the decline in free SO_2.

Sorbic acid
Sorbic acid is widely used in many food products to control yeast and mold spoilage. This agent has a very low order of toxicity, and findings have indicated that sorbic acid is metabolized like other fatty acids to CO_2 and water. Sorbic acid is an unsaturated, six carbon fatty acid, which is similar to those found in edible fats and oils.

Sorbic acid and its salt, potassium (K) sorbate, have broad-spectrum activity, mainly against yeast and molds. It is used at bottling in prevention of refermentation in sweetened wine by fermentative species of S*accharomyces cerevisiae*. Sorbic acid is much less active against bacteria, and should not be expected to protect wines from bacteria spoilage or malolactic fermentation. In addition, K-sorbate does not have much effect on *Zygosaccharomyces, Brettanomyces* and *Dekkera*. Wine pH values are usually within the range of 3.0 to 4.0, which is highly desirable for sorbic acid activity.

Potassium sorbate is often used to treat wines, because sorbic acid has lower water- solubility. The solubility of sorbic acid in water is 0.15%, by weight at 20°C as compared to K-sorbate at 58.2% (20°C). This salt is a white, crystalline powder and should be stored below 38°C and should not be exposed to moisture, heat, or light.

Disadvantages of sorbic acid
There are some disadvantages in using potassium sorbate. Sorbic acid does not control the growth of bacteria. When SO_2 is too low in preventing bacterial growth, certain malolactic bacteria can metabolize sorbic acid to yield a disagreeable (geranium-like) odor in wines. Also, it should be noted that sorbic acid may be detected at low levels. Using a trained taste panel, Ough and Ingraham (1960) reported that the threshold for detecting sorbic acid in a Pinot noir wine was 135 mg/L. However, the panel was trained to detect sorbic acid in this study, and one panelist was very sensitive to sorbic acid. Before training the panel, the individuals had difficulty detecting the odor or taste of sorbic acid in the wines. Ough (1987) reported that wines treated with sorbic acid develop ethyl sorbate, which most people cannot detect at levels normally present in wines.

The effectiveness of sorbic acid is related to the number of microorganisms found in wines. Yeast and bacteria are capable of metabolizing sorbic acid; thus, reducing its ability to control spoilage. Therefore, it is important for the winemaker to use some common measures in controlling high cell counts such as, using clean and sound fruit, juice clarification, and wine filtration.

Another disadvantage of using sorbic acid in the form of K-sorbate is that this salt contains a small amount of potassium, approximately 25% by weight. If wines are not adequately stabilized against potassium bitartrate precipitation, an additional amount of potassium may cause the formation of tartrate crystals in the bottle. The winemaker needs to make sure that the wine is stable for potassium bitartrate precipitation prior to bottling.

Concentration and rates of addition
Ough and Ingraham (1960) conducted a study to test the yeast-inhibiting properties of K-sorbate in sweetened wine in the presence of adequate levels of SO_2. Using a white wine, they reported that inhibition of yeast growth and fermentation was found at relatively low levels (80 ppm) of sorbic acid. These results are slightly lower than those used in most commercial wines. The discrepancy might be explained by differences in wine pH, alcohol concentration, level of microbial contamination, and yeast strains. Therefore, the recommended levels of sorbic acid in most table wines are generally within the range of 100–200 ppm. With these levels, it is important to use an adequate level of SO_2 and proper vinification procedures (i.e., filtration) to reduce yeast populations.

Since K-sorbate does not contain 100% sorbic acid, it is essential to know the percentage of sorbic acid in this salt. On an equivalent weight basis, K-sorbate contains about 75% sorbic acid. Therefore, higher amounts of K-sorbate are required to obtain the same activity of sorbic acid. The following equation may be used to calculate the amount of K-sorbate for treating wines:

$$\text{Wt. of K-sorbate} = \frac{Y \times 3.8 \times 1.34 \times Z}{1000} \qquad 7.1$$

where: Y = Volume of wine in gallons
3.8 = Conversion factor to change gallons to liters (L/gal)
1.34 = Conversion factor to change sorbic acid to K-sorbate
Z = Desired ppm (mg/L) of sorbic acid to be added
1000 = Convert mg/L to gms/L

Example: For a 125 ppm sorbic acid treatment, calculate the weight of K-sorbate to be added to 500 gallons of wine.

$$\text{Wt. of K-sorbate} = \frac{500 \text{ gal.} \times 3.8 \text{ L/gal.} \times 1.34 \times 125 \text{ mg/L}}{1000}$$
$$= 318 \text{ gms or } 70 \text{ lbs or } 11.2 \text{ ozs.} \qquad 7.2$$

TTB rulings indicate that finished wines shall contain no more than 300 mg/L of sorbic acid in wine.

Dimethyldicarbonate
Dimethyldicarbonate (DMDC) is another option to use for yeast inhibition prior to bottling. DMDC is the methyl analog of diethyldicarbonate (DEDC). Diethyldicarbonate was banned due to the carcinogen ethyl carbamate being

formed in wine. According to Ough (1992) DEDC was a very effective yeast sterilant and had mostly replaced sorbic acid as a fungicide. Since DMDC has the benefit of not forming ethyl carbamate, it is not considered a carcinogen and has been approved for use in wines. Although, DMDC produces low levels of methanol, the amount typically used in wine does not approach legal levels relating to health concerns.

Dimethyldicarbonate is used at bottling and sold under the trade name of Velcorin®. According to Porter and Ough (1982), the inhibitory process starts with the hydrolysis of DMDC followed by denaturation of the fermentative pathway enzymes. Fugelsang (1997) reports that this product is lethal towards yeast and most lactic acid bacteria at the maximum level of 200 mg/L allowed in the United States.

However, effectiveness of DMDC is related to the alcohol concentration and temperature of the wine (Porter and Ough, 1982). They reported using 100 mg/L in wine of 10% alcohol and 2% residual sugar in reducing the yeast count from 380 cells/ml to zero in 10 minutes held at 30°C. In showing the effect of temperature, they reported that viable cells were observed in the same wine held at 20°C after 10 minutes. Therefore, there appears to be a synergistic effect between the higher alcohol concentration and higher temperatures used in this study.

The addition of DMDC for larger wineries, according to Ough (1992), is typically accomplished by atomizing it into the bottling line using a proportional pump. Smaller sized wineries will dissolve the appropriate amount of DMDC to be used in a small portion of absolute ethanol then add it directly back into the wine to be bottled.

Several limitations have been reported with the use of DMDC. Dimethyldicarbonate is very effective on most wine and spoilage yeast; however, its killing properties are less effective on bacteria normally found in wine (Boulton et al. 1999). In addition, since DMDC is hydrolyzed rapidly with no residual activity, secondary contamination can be a problem making sterilization procedures of the bottling line critical in keeping the microbial load down in the bottled wine.

Also, DMDC needs to be handled properly since it can burn the skin and pose a health threat if ingested.

Table 2, according to Kunkee (2005), provides information regarding the possible risk of re-fermentation in the bottle based on the concentration of residual sugar determined by the clinitest® method. Although, the clinitest® method may not provide the accuracy as other titrametric laboratory methods such as the Rebelein (Gold Coast) or Lane-Eynon procedures, it provides a quick and fairly accurate determination of the residual sugar content and possible threat of yeast activity. Therefore, the use of this chart may help decide the potential threat of re-fermentation in the bottle and aid in the choice of using a chemical additive as an additional measure of quality control in the bottle. This holds especially true if sterile bottling practices are questionable.

Table 2 Percent residual sugar by Clinitest™ relating to possible refermentation in the bottle

Residual sugar (%)	Comments and possibility of yeast growth
0.2	no fermentable sugars, only pentoses – can't taste it, no chance for yeast growth
0.4	pentoses 0.2%; glucose 0%; fructose 0.2% – can't taste it, a small chance of yeast growth possibly on fructose
0.6	pentoses, glucose and fructose each 0.2% – probably can taste it, a good chance of yeast growth

Adapted from Kunkee (2005).

References

BOULTON, R. B., V. L. SINGLETON, L. F. BISSON and R. E. KUNKEE. 1999. *Principles and Practices of Winemaking*. New York: Springer Science+Business Media, LLC.

DHARMADHIKARI, M. R. 1990. 'Sulfur dioxide in wines.' Proc. of The Fifth Annual Midwest Regional Grape and Wine Conference. Osage Beach, Missouri, pp. 65–74.

FUGELSANG, K. C. 1997. *Wine Microbiology*. New York: Chapman and Hall.

GALLANDER, J. F. 1987. 'The reduction of free sulfur dioxide in Catawba and Vidal blanc table wines.' Unpublished data.

KUNKEE, R. E. 2005. 'Alcoholic and malolactic fermentations for quality table wine.' Vineyard and Winery Establishment Workshops (CD). The Ohio State University/OARDC Wooster, OH.

OUGH, C. S. 1987. 'Chemicals used in making wine.' *C. & E.N.*: Jan. 19–28.

OUGH, C. S. 1992. *Winemaking Basics*. New York: Food Products Press.

OUGH, C. S. and L. J. INGRAHAM. 1960. 'Use of sorbic acid and sulfur dioxide in sweet table wines.' *Am. J. Enol. Vitic.* 11: 117–122.

OUGH, C. S., E. B. ROESSLER and M. A. AMERINE. 1960. 'Effects of sulfur dioxide, temperature, time and closures on the quality of bottled dry white table wines.' *Food Tech.* 14: 1–4.

PORTER, L. J. and C. S. OUGH. 1982. 'The effects of ethanol, temperature, and dimethyl dicarbonate on viability of *Saccharomyces cerevisiae* Montrochet No. 522 in wine.' *Am. J. Enol. Vitic.* 33 (4): 222–225.

SMITH, C. 1982. 'Review of basics on sulfur dioxide.' *Enology Briefs*, February/March, University of California, Davis.

7.6 How significant is oxygen pick up during bottling?

C. Butzke

Wine is usually bottled at room temperature around 20°C (68°F) to reduce oxygen solubility and to anticipate the most likely storage conditions and internal bottle pressure. Before delving into the significance of oxygen pick-up during the bottling process, a quick review of sources and quantities of oxygen exposure during wine aging is pertinent.

What are the basics of air and oxygen contact of wine?

Before and after bottling, the wine has to meet the free SO_2 requirements that correspond to its pH to prevent MLF in the bottle or premature oxidation and browning.

The recommended concentration of free SO_2 in wine equals $0.85 \cdot (1 + 10^{pH-1.83})$. While some enologists have argued a factor of 0.6 instead of 0.85 may suffice, the dose recommended above appears safe and takes the average winery's proficiency for measuring sulfites and pH into account (Table 1).

How much oxygen is added up during bottling under different equipment options?
How much free SO_2 will be consumed by the oxygen picked up during bottling?

Each molecule of oxygen can eventually cause the binding of one molecule of sulfur dioxide, i.e., 8 mg(/L) of O_2 correspond to a reduction of 32 mg(/L) of

Table 1 Dissolved oxygen solubility and uptake during wine processing and aging

	mg/L
Air-O_2 solubility (68°F)	8
Air-O_2 solubility (32°F)	14
Uptake per racking	20
Uptake via topping (per year)	20
Uptake at bottling	0.17–8
Uptake via bark cork (per year)	0.1–560
Uptake via technical cork (per year)	0.4–0.8
Uptake via synthetic cork (per year)	2.8–5.6
Uptake via screw cap (per year)	0.1–0.4
Young *red* wine	
Total O_2 capacity	1000–4000
Optimum O_2 uptake	60–130
Young *white* wine	
Total O_2 capacity = *Oxidized*	36–107
Optimum O_2 uptake	0–16

Table 2 Dissolved oxygen uptake into wine at bottling

Bottling technique	O_2 (mg/L) added			
	At filler	Via headspace	Total	%
Gravity flow	6.60	1.40	8.00	100
	O_2 (mg/L) removed			
	At filler	From headspace	Total	%
1. Vacuum pulled at filler	6.00	–	6.00	75
2. Bottle sparged with N_2	0.51	0.02	0.53	7
3. Vacuum pulled at corker	–	0.15	0.15	2
4. Headspace sparged with N_2	–	1.15	1.15	14
Steps 1–4 implemented	6.51	1.32	7.83	98

SO_2. Practically, and depending on the sophistication of the available bottling line, an extra 10 to 50 mg/L SO_2 should be mixed into the wine to account for SO_2 losses due to the bottling process, including filtration and filling and closing. When using a simple gravity flow filler without sparging the bottle and its headspace with nitrogen, an extra 50 mg/L may be advisable. Be aware though, that an addition of *potassium* (metabisulfite) may compromise the (tartrate) cold stability of the wine (Table 2).

Further reading

BOULTON, R. B., V. L. SINGLETON, L. F. BISSON and R. E. KUNKEE. 1996. *Principles and Practices of Winemaking*. Springer, pp. 420–412.

GAI Macchine Imbottigliatrici s.r.l. Minor ossidazione all'imbottigliamento. http://www.gai-it.com/index.php?method=section&id=354

Scorpex Inc. Variability of Permeability of Corks and Closures. http://www.scorpex.net/ASEVClosures2005RGibson.pdf

7.7 How can I control oxygen uptake at bottling?

T. E. Steiner

Introduction

The benefits and drawbacks of dissolved oxygen in wine can be discussed at great length. However, to extend the aging potential and prevent undesirable changes in the wine due to oxidation, a winemaker must recognize that in most cases oxygen is considered to be detrimental in the production of a high quality product.

Benefits of limited oxygen

In some cases, oxygen exposure in the must/juice otherwise known as hyper-oxidation has been associated with stabilizing white wines from further browning oxidation during the vinification process. This enzymatic oxidation occurs in must/juice absent of sulfur dioxide (SO_2). During enzymatic oxidation, certain phenol groups react with oxygen to produce yellow quinones. These compounds in turn react with more oxygen to yield brown colored products. This process stabilizes further browning reactions in wine from this source (Ough 1992). Although, the author of this text considers grape juice oxidation as being detrimental to producing wines of high quality, this oxidative process is not implicated for the most part in oxidative reactions occurring in wine.

Oxygen is also essential during the initial stages of alcoholic fermentation for healthy yeast propagation and fermentation. Residual oxygen is then completely removed by the increased production of carbon dioxide (CO_2) during the fermentation process.

Some controlled oxygen exposure may be beneficial in red wines during barrel aging. This increases phenol polymerization and improves color stability and softening of the palate in red wines (Zoecklein *et al.* 1995). A cellar procedure for controlled oxygen addition accomplished in red table wines known as micro-oxygenation is reported to reduce harshness and softens the palate. It is important to understand that micro-oxygenation is intended to avoid excessive accumulation of dissolved molecular oxygen in the must or wine that causes oxidation (Smith 2002). However, the advantages of micro-oxygenation needs further research performed and should be performed by trained personnel only in recommending this technique.

Oxygen elimination prior to bottling

Generally, oxygen is detrimental to wine quality especially from the end of fermentation through wine storage and bottling. The presence of oxygen during the latter stages of wine production can increase browning reactions, chemical and microbiological instability and the production of off aromas such as acetaldehyde.

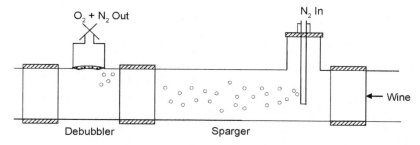

Fig. 1 Schematic diagram of an inline oxygen sparger. Source: Ough (1992).

Attention must be given during the vinification process to avoid those potential sources for oxygen pickup and prevent excess oxygen from dissolving into the wine. Key sources for oxygen pickup include: racking, excess headspace, pumping, filtration and bottling. Depending on temperature, dissolved oxygen levels can range from 6 to 9 mg/L in wine. Higher levels are expected at lower temperatures (Boulton et al. 1999). Since the rate of oxidation increases with temperature, it is critical to add the appropriate amount of SO_2 based on wine pH. Furthermore, when kept at low temperatures, such as during cold stabilization, protecting the wine from air and keeping tanks full are essential to minimize oxygen absorption in wines (Gallander 1991). Other practices such as filling tanks from the bottom, inspecting for leaky pump seals and securing any loose hose connections on the inlet side are necessary to lowering oxygen pickup. Prior to bottling, excess oxygen in wines can be removed by using an inline sparger. This introduces an inert gas like nitrogen (N_2) or CO_2 through a porous stainless steel cylinder suspended in the wine. As the wine passes around the sparger, gas bubbles enter the product and displace the dissolved oxygen. The bubbles will rise to the top of the tank releasing the inert gas and oxygen (Fig. 1). For this procedure, the use of CO_2 as an inert gas is less effective and may excessively carbonate (saturate) the wine prior to bottling; therefore, N_2 is preferred (Ough 1992).

Oxygen elimination at bottling
Bottling is the last process where added dissolved oxygen can have a significant negative impact on the aging potential and quality of the wine being released directly to the consumer. Thus, extreme care must be employed in minimizing the amount of oxygen entry at bottling.

Oxygen has the potential to dissolve into the wine at every stage of the bottling process. A recommended level for total dissolved oxygen in bottled red wines should be below 1.25 mg/L and 0.6 mg/L for white, blush and rosé wines (Fugelsang 2009). Major sources of oxygen diffusion into wine at bottling occur during wine transfer, filtration, filling and headspace levels of the bottling tank, filler and bottle. Each process will be described in further detail below.

When transferring wine to the bottling tank, it is advisable to purge the tank and transfer lines with N_2 or CO_2 prior to filling. If any headspace is present after filling, it is important to use an inert gas on the surface to prevent oxygen from dissolving into the wine. Often, a mixture of N_2 and CO_2 can be beneficial especially for white wines. Maintaining a slight but constant pressure over the headspace is recommended. Although CO_2 levels ranging from 300 to 600 mg/L can enhance a young white or light red wine (Peynaud 1984), caution must be exercised that excessive pressure may cause too much CO_2 absorption providing a noticeable tactile sensory perception and possible bubble formation. In addition, excessive CO_2 levels can cause an increase in pressure possibly pushing the cork out after bottling. Therefore, the use and monitoring of CO_2 in the wine prior to bottling by carbodoser are beneficial in adjusting concentrations up or down accordingly for these purposes. The carbodoser is a relatively simple technique involving a glass tube measuring the amount of CO_2 out-gassed from a fixed volume of wine. Comparing actual results with a calibration curve provides the concentration of CO_2 in mg/L of wine.

Wine filtration prior to bottling is another source for oxygen pick up. During filtration, it is important to operate the filtration unit according to the manufacturer's directions, making sure all connections and pads are tight to prevent oxygen entry. Purging of air from the filter pads and transfer lines is also a recommended practice.

Wine entering the filler bowl is typically one of the most problematic sources for oxygen pick up. The filler bowl should also be covered with an inert gas to reduce oxygen pick up. Depending on the type of filler used, filling of wine into bottles can increase the levels of dissolved oxygen by 0.5 to 2.0 mg/L (Peynaud 1984). The length of the fill spouts as well as the type and force of the jet may influence the amount of dissolved oxygen. Therefore, it is advisable that filling tubes be as long as possible depending on the bottle. Providing vacuum prior to filling and flushing with 2 to 3 volumes of N_2 has been reported to lower oxygen absorption at bottling (Boulton et al. 1999).

After filling, bottle headspace is another source of oxygen absorption. This is due, in part, to the variability of the bottle headspace, which is influenced by such factors as, wine temperature, solubility of gases in the wine, bottle size and shape. To help reduce oxygen ingress at this stage, the injection of an inert gas such as N_2 or CO_2 can reduce the amount of oxygen in the headspace. According to Peynaud (1984), a small amount of CO_2 supplied to the bottle headspace will help replace the oxygen and diffuse into the wine causing a depression which also helps prevent the problem of wine leakage due to expansion. In addition, a bottling line supplied with a vacuum filler is also effective in reducing the amount of oxygen in the headspace. Similarly, a controlled dosage of liquid N_2 into the wine after filling is another good option in flushing oxygen from bottle headspace for screw-cap operations (Crochiere 2007).

The corking machine may vary on whether it supplies a vacuum or not prior to cork insertion. According to Crochiere (2007), if set up properly supplying a vacuum at corking can help reduce the amount of oxygen absorption into the wine.

Whether using inert gas sparging, pulling a vacuum, liquid N_2 dosing or a combination of these procedures, it is advisable to keep the time and distance from the filler to the corking machine as short as possible. In addition, if there is an interruption in the bottling line process, down time may cause the inert gas to escape allowing oxygen to concentrate back into the headspace of the bottle. Therefore, if a bottling line stoppage has occurred, it is advisable to remove all bottles in question and dose them again or discard them from the bottling line.

The last important item of the bottling process that influences oxygen absorption in wine and ultimately affects aging potential is the closure. Today, there are many wine closures available each having different properties. Two major functions affecting oxygen pickup in bottled wine include closure recovery time from compression and the rate of oxygen permeation. Lopes *et al.* (2007) indicated that the level of oxygen permeation is lowest for screw caps and 'technical' corks, intermediate for conventional natural cork stoppers, and highest for synthetic closures. Further, they showed that differences in oxygen pickup varied among grades of each closure. This variability could then provide an explanation for bottle to bottle variation. This finding was in agreement with the results reported by Crochiere (2007). Both studies reported the need to be more consistent in production standards of each type of closure as it relates to compression recovery and oxygen ingress rates.

In conclusion, oxygen incursion at bottling can have a significant negative impact on wine quality and aging potential. Therefore, the recognition and knowledge of how one can control or limit the amount of oxygen entry at bottling are critical.

References

BOULTON, R. B., V. L. SINGLETON, L. F. BISSON and R. E. KUNKEE. 1999. *Principles and Practices of Winemaking*. New York: Springer Science+Business Media, LLC.

CROCHIERE, G. K. 2007. 'Measuring oxygen ingress during bottling/storage.' *Practical Winery and Vineyard*. (January/February): 74–84.

FUGELSANG, K. C. 2009. Personal communication

GALLANDER, J. F. (1991). 'Wine oxygen level: how much is too much?' Proc. of The Ohio Grape-Wine Short Course. HCS Department Series 621. Wooster, OH, pp. 48–51.

LOPES, P., C. SAUCIER, P. TEISSEDRE and Y. GLORIES. 2007. 'Oxygen transmission through different closures into bottled wine.' *Practical Winery and Vineyard*. (January/February): 65–71.

OUGH, C. S. 1992. *Winemaking Basics*. New York: Food Products Press.

PEYNAUD, E. 1984. *Knowing and Making Wine*. New York: John Wiley and Sons.

SMITH, C. 2002. 'Micro-oxygenation in extended maceration and early stages: immediate effects and early stages.' Proc. of The Ohio Grape-Wine Short Course. HCS Department Series 726. Wooster, OH, 25, pp. 54–63.

ZOECKLEIN, B. W., K. C. FUGELSANG, B. H. GUMP and F. S. NURY. 1995. *Wine Analysis and Production*. New York: Chapman & Hall.

7.8 How much oxygen can I have in my bottles at filling?

R. Gibson

In my opinion, oxygen in the bottle (or any other package) at filling should be kept to an absolute minimum. For many wines, the oxidation clock starts ticking as soon as the pack is sealed, due to diffusive oxygen ingress into the package. Any oxygen in the pack simply shortens the time that the wine will remain in prime condition for the consumer.

Oxygen in the pack can come from three sources. Oxygen can be entrained in the wine. This oxygen might already be present in the wine in the bottling tank, or it may be taken up the wine during the filling process. Oxygen in the wine in the bottling tank can be removed by sparging with inert gas, but control of oxygen uptake during filling requires close attention to prevention of air contact and turbulence as the wine is delivered from the bottling tank to the package.

Oxygen in the pack headspace can also make a significant impact. Remember that vacuum can only remove a proportion of air. Inert gas dosing prior to vacuum is a good technique when using inserted closures. Aim for a final internal pressure in equilibrium with the atmosphere, with all of the gas in the headspace composed of nitrogen or carbon dioxide – that is, free of oxygen! Do not attempt to minimise headspace oxygen content by reducing the headspace volume to very low levels when using inserted closures. This can lead to significant leakage or closure movement issues if the bottles are warmed up at a later stage. Without headspace, hydraulic pressure is applied to the closure as soon as the wine volume expands.

If you are using screw caps, use an effective inert gas dosing system (preferably dosing liquid nitrogen or solid carbon dioxide) to remove air from the bottle headspace. Some pack types, such as bag in box and cans, can have significant headspace volume relative to the volume of product they contain. Oxygen management of the headspace in these packs is critical.

Oxygen can also be entrained in the packaging materials into which wine is filled. Bottles and bags can contain oxygen if they are not treated to remove air prior to fill. Closures can emit air when they are compressed, contributing to the overall oxygen load in the pack.

Why be concerned with pack oxygen, some might say – oxygen reacts with antioxidants such as sulphur dioxide and does not have much impact on wine quality. It seems, however, that exposure to large amounts of oxygen at one time can cause ongoing impacts on wine quality over time. The reasons for this are not clear, but may be associated with the inability of the reaction between SO_2 and oxygen to proceed at sufficient pace to stop the reaction between oxygen and other wine components. In red wines, the oxygen may be absorbed by the phenolic components that are present without much quality impact. In whites, the effects may be more noticeable.

In my experience, ascorbic acid additions to whites certainly help to manage the impact of oxygen that is included in the bottle during fill. Ascorbic acid acts as an intermediary in the reaction between oxygen and sulphur dioxide, speeding up the process. Use of ascorbic acid results in less depletion of free SO_2 when wine is exposed to the same amount of oxygen as some oxygen is included in the inert dehydro-ascorbic acid by-product of the reaction between oxygen and ascorbic acid. Studies have shown that white wines with ascorbic acid stay fresher for longer under oxidative conditions, and develop less total color (although the absorbance at some wavelengths may be slightly increased).

There are concerns that residual ascorbic acid in wine with no free SO_2 will cause rapid oxidation due to the formation of hydrogen peroxide. Reaction stoichiometry shows that 1 mg of oxygen will directly react with 4 mg of SO_2. When ascorbic acid is present, 1 mg of oxygen reacts with 5.5 mg of ascorbic acid. The active by-product of this reaction, hydrogen peroxide, will then react with 2 mg of SO_2. Use of up to 100 mg/L of ascorbic acid should present no problems, as the free SO_2 content of wines prepared for filling into oxidative packs should always well exceed the 18 mg/L that this level of ascorbic acid can react with.

There are concerns about the formation of reduced characters in some packs that do not allow oxygen entry, such as screw caps incorporating a liner with a layer of tin. Incorporation of oxygen in these packs at bottling may delay the onset of reduced characters, but this practice will not prevent their formation in the longer term as the compounds responsible for their formation are in equilibrium reactions. They are not permanently altered by exposure to oxygen.

In summary, oxygen in the pack at filling does not give any benefit and may cause quality loss. Oxygen levels should be kept as low as possible, and steps such as the use of ascorbic acid should be taken to offset the impacts of any oxygen unavoidably included in the pack during the filling process.

7.9 How many corks out of a 5000-cork bale would I need to sample to assure a taint rate of less than one bottle out of five cases?

C. Butzke

The rate of *cork taint*, the musty, moldy off-odor in wine, stemming by and large from the growing, processing and shipping of natural bark corks, has been significantly reduced in recent years due to pressure from winemakers and subsequent improvements in the production process and quality control at cork manufacturers, distributors, and associated wine laboratories. In the mid-1990s, the taint rate was estimated between 3 and 5% of bottles, about one in every other case. As a consequence, by 2009, at one of the big international wine competitions, the use of bark corks as bottle closures had dropped to less than two-thirds of all entries. The problem of *TCA* (2,4,6 trichloroanisole), the main impact component for cork taint, has not completely disappeared, much to the dismay of winemakers and consumers alike, so if your winery continues to have problems with tainted closures, below are the proper sampling procedures to keep the taint rate below 1.5%. This translates into one out of 67 bottles, an occurrence that even for a super core wine drinker will be rare enough to be arguably acceptable.

Simply and logically put, if one wanted to assure that only one in 67 corks is bad, one would have to sample at least one out of 67 corks. In larger lot sizes, the actual sampling size increases slightly to decrease the statistical probability of accepting a large lot of tainted corks. For each 5000-cork bale and when using the simplest fixed-sample-size plan applicable to wine closures (ANSI/ASQC Z 1.4-1993 sampling plan for normal inspection, general inspection level I; supplier with good track record), the winemaker has to sample, soak and sniff 80 corks. If 4 or more of those 80 corks were tainted, the lot should be rejected and sent back to the supplier.

Corks are a natural and renewal product from the bark of the cork oak tree. Therefore, the sampling of such a naturally variable part of wine packaging cannot be compared to, e.g., ammunition, mass-produced by precision machinery. The use of *reduced* sampling protocols such as the old military specs MIL-STD-105E, that may be applicable to M16 bullets, cannot be used for wine corks.

No winemaker should have the time to sample and test 80 corks per each 400 cases bottled. Any sampling shortcuts, however, are a complete waste of the enologist's time. If the winery lab does not have the time or resources to conduct a full sampling and sensory evaluation test, then the burden must be passed on to the manufacturer or distributor of the closure. Both trust and verification are essential to a lasting relationship between winemaker and closure manufacturer. The perfect and precious closure is what makes wine different from all other beverage, complements its special value-added image, and remains a conversation topic in itself among educated consumers. However, it should never compromise the great work that the winegrower and winemaker have created.

More flexible but slightly more involved quality control plans include the recommended Fraction Defective Sampling Plan, and the Sequential Probability Ratio Test. All three plans are valid as long as the protocols are applied correctly. None of the plans can circumvent the principles of applied statistics.

Finally, it is crucial that the tasting room staff is trained to recognize TCA immediately when a new bottle is opened as it could affect up to 25 potential customers who taste their 30 mL sampling of a tainted wine.

Simple cork soaking procedure:

- Place ONE cork in a 50 mL disposable centrifuge tube.
- Fill with vodka-water mix (dilute 40% (80 proof) generic vodka to 10% alcohol by volume with distilled water (1 part vodka plus 3 parts water)); close.
- Soak for 24 hours at room temperature.
- Smell opened tube for musty, moldy TCA off-odor.
- Discard tube, cork and testing solution.

Detailed sampling and training options can be found in Butzke and Suprenant (1998).

Reference

BUTZKE, C. and SUPRENANT, A. 1998. *Cork Sensory Quality Control Manual*. University of California ANR Publication Number 21571. http://books.google.com/books?id=uTQEEAWTKFkC&printsec=frontcover

7.10 Will screw caps make my wine better?

R. Gibson

The time between bottling and consumption is critical for wine quality and style. Many changes can take place in the bottle before the consumer removes the closure and samples the contents. The risks that can occur in the period between bottling and opening include taint, oxidation and reduction.

A good cork is a great seal for wine bottles. However, bad corks cause problems and it isn't possible to tell which corks are good before they go into the bottle. Issues with traditional cork closures are well documented. Taint and variable oxygen transfer rates have led to increasing adoption of alternative bottle closures to cork for wine, including screw caps.

Screw caps are generally free of taint, and remove this risk for wine quality. However, the introduction of a new closure type can introduce new risks. If these risks are not managed well, wine quality can suffer and your wine will be no better and may be worse than it was when sealed with a bad cork! The major risks that can occur with screw caps are oxidation and reduction.

The oxygen barrier properties of screw caps are strongly influenced by the liner (also known as the gasket or wad) in the top of the cap. High barrier liners incorporate a layer of metal (usually tin) in their structure. Liners made up of polymers only will allow more oxygen transfer.

The use of high oxygen barrier liners allows wines to develop in the bottle without the influence of oxidation or taint. However, screw caps with high barrier liners can promote the formation of reduced characters in the bottle. Wines can appear perfectly clean when filled, but the low oxygen conditions generated by high barrier caps can induce the formation of sulphide aromas from benign precursor compounds. The prediction and control of the formation of these aromas are difficult. Care is required to ensure that preventative steps are taken to prevent the formation of sulphide compounds of any sort during the winemaking process, including careful attention to yeast nutrition during fermentation.

Low barrier liners may allow oxidation to occur in the time between bottling and consumption of the wine. If low barrier liners are used, care must be taken to manage the length of time that product spends in the distribution system to ensure that the consumer has opened and enjoyed the bottle before oxidised characters arise in the wine. If low barrier liners are used, free sulphur dioxide levels should be at the higher end of the normal range and ascorbic acid use should be considered for white wines.

The application of screw caps of any sort has an influence on their ability to seal. The downward force used during cap application determines the compression of the liner and the thread roller settings influence the retention of the compressive force of the liner against the bottle rim. Torque on caps must also be applied very carefully. The quality of the bottle is also important in ensuring that good seal is made – the rim and thread must be free of flaws and the bottle must not tilt. If these factors are not right, rapid air entry and oxidation can result.

Screw caps can be susceptible to damage after application, resulting in loss of seal integrity. The BVS type bottle finish combined with cap redraw provides protection against loss of seal integrity.

The headspace that is left in screw cap bottles is generally greater than that in cork sealed bottles. Treatment of this headspace should be carried out at bottling to remove air, preferably using effective inert gas displacement.

If these risks are managed, screw caps can do a very good job of maintaining wine quality between bottling and the consumer. Your wine may well be better than it is under other seals!

7.11 Can synthetic closures take the place of corks?

R. Gibson

Synthetic closures made of foamed polymers have become a prominent closure choice for wine. These closures are inserted using conventional corking equipment. Synthetic closures have several technical attributes that make them an attractive alternative to corks. However, they also introduce risks that must be managed if they are to provide a viable wine closure option.

Synthetic closures made by competent producers from assured raw materials approved for food contact do not have the capacity to taint wine. However, it should be understood that the polymers used for synthetics can absorb environmental contaminant compounds. If these compounds are transmitted to wine after bottling in concentrations exceeding their sensory thresholds, taint may occur. Synthetic closure producers must be vigilant to ensure that post-production contamination does not occur from sources such as shipping containers, wooden pallets and cardboard packaging.

Synthetic closures give a tight liquid seal against the bottle neck. However, the surface of many synthetic closures appears to be more easily affected by corker jaw damage than cork closures. When leakage at points on the closure circumference that are separated by 90° ('four points of the compass') is observed, it can be assumed that corker jaw damage has occurred. In addition, the rapid recovery of synthetic closures after compression can lead to the retention of high positive pressures in the bottle if vacuum or carbon dioxide corking is not used. This may lead to leakage if bottles are laid down.

It is known that synthetic closures do allow oxygen diffusion from the atmosphere into the bottle. This leads to gradual oxidation of the wine. The diffusion rate depends on the polymers that are used, the density and length of the closure and the size of the cells in the structure. Synthetic closures cannot be considered for the long-term maturation of wine in the bottle. There may be some short-term benefit of oxygen exposure in the bottle for younger red styles, but ultimately, the bottle contents will become oxidised.

This may be seen as a drawback, but in effect the shelf life that can be achieved with good synthetics is more than adequate for commercial wines that are turned over quickly and not cellared by the consumer. In addition, oxygen transfer is consistent and predictable from closure to closure. Synthetic closures provide the same oxygen barrier upright and laid down, unlike some batches of cork that can cause extreme variation in oxidative status when bottles are stored upright.

The techniques that can be used to offset the impacts of oxygen diffusion are similar for wine packed with synthetic closures and other oxidative packs, maintain low oxygen levels in the wine and headspace at bottling, maintain high levels of antioxidants, manage product storage temperatures and manage stock age.

Scalping (adsorption of aroma compounds by packaging materials) is a phenomenon that is well understood in food packaging. It has been shown that synthetic closures have a higher scalping potential than corks for a number of wine aroma compounds. Some compounds, such as oak aroma components, are not affected and compounds that are scalped may have limited impact in some wine styles. However, it is prudent to remember that scalping could have an effect on wine quality after bottling when synthetic closures are used.

Inserted closures must be easy to remove from the bottle neck, and synthetic closures have gained something of a reputation for extraction difficulties. The art of surface lubrication is long-established for cork closures: materials, quantities and application techniques have been honed empirically since surface treatment was introduced many years ago. Synthetic closure producers have had less time to get it right. Reputable producers should be able to provide evidence of controlled lubricant application, and extraction results that demonstrate that it works. Control of product density also contributes to consistent extraction forces.

As discussed earlier, synthetic closures make a tight seal against the bottle neck. This can make them more susceptible to movement due to pressure in the bottle than corks, which tend to vent pressure changes. Pressure can be generated by excessive vacuum at closure insertion (atmospheric pressure may be able to push the closure into the neck) or warming of the wine (pressure generated in the bottle may push the closure out of the neck).

In summary, synthetic closures can provide an alternative to cork for some wines, but their properties introduce risks that must be managed to optimise wine quality.

7.12 What precautions do I need to take when using 'bag in box' packaging?

R. Gibson

Bag in box (BiB) provides a number of benefits over conventional packaging for the consumer. The convenience of drawing off a few glasses at a time when needed, combined with other benefits such as light package weight, fuel the increasing demand for this packaging configuration.

Packaging wine in BiB presents a number of technical risks and challenges. If these risks are not recognised and managed, BiB packaging will not deliver wine to the consumer in good condition.

First and foremost, BiB creates an oxidative package environment for wine. The walls and tap of all bags allow oxygen ingress by diffusion. Some bag types allow more oxygen ingress after they have been transported. Wine in smaller packs, with a larger surface area to volume ratio, is more susceptible to oxidative change than wine in large packs. From the moment the bag is filled, the oxidation clock is ticking. If wine is to be presented to the consumer in the best possible condition, a number of steps must be taken to ensure that oxidative degradation is minimised.

Firstly, make sure that the antioxidant content of wine at filling is at the higher end of the scale. Sulphur dioxide content should approach the 50 mg/L free mark. Forget about molecular SO_2; free SO_2 levels that give normal molecular SO_2 targets are simply not high enough to provide adequate protection of BiB wine quality over time. Consider using ascorbic acid in white wines. There is no doubt that ascorbic acid provides added protection against oxidation and reduces the consumption of sulphur dioxide.

Next, make sure that the oxygen load in the pack at filling is minimised. Any dissolved oxygen in the wine and oxygen in the headspace simply erodes the ability of the wine to resist the inevitable oxidative change that will take place over time.

Planning and logistics have a large impact on getting BiB wine to the consumer in optimum condition. Plan to pack BiB wine only as the market requires it. If you are selling BiB wine offshore, consider bulk shipment of the wine to the market with filling of the pack in the country of sale. Don't build large stock holdings that will be sold over 6 or 12 months. Practise strict warehouse stock rotation. Monitor and control warehouse and distribution temperatures – warm temperatures accelerate oxidative change. Consider printing a 'best before' date on the package. Monitor stock age in the retail system; don't let consumers tell you when there is a problem through product returns.

Oxidation is not the only risk with BiB products. The BiB pack can swell if it contains excessive amounts of gas, especially in warm weather. Always monitor the carbon dioxide content of wine that is filled into BiB, and maintain the level below 0.8 g/L. If the CO_2 content is too high, sparge with nitrogen. When handling BiB wines, do not use CO_2 sparging or blanketing to remove oxygen or reduce air exposure – use nitrogen, which is virtually insoluble in wine, instead.

Correct operation of the BiB tap is essential. Tartrate crystals can jam taps open, leading to considerable consumer dissatisfaction! BiB packs are often left in refrigerators for extended periods. Wine packed in BiBs must be completely stable to tartrate deposition. Ensure that rigorous stabilising and testing procedures are carried out.

Secondary fermentation in any retail package is undesirable, but in BiB this problem can lead to extreme difficulties as packs can swell and burst throughout the distribution and retail system, leading to considerable damage to other stock and premises. Make absolutely certain that sterility is maintained (high sulphur dioxide levels help) and that microbiological monitoring is carried out. In addition, ensure that stock tracing and product withdrawal procedures are in place and ready to be activated if necessary.

Under certain circumstances, some BiB packs can allow the passage of environmental taints to wine after filling has taken place. For example, taint compounds derived from the interior of shipping containers can pass through the bag wall and enter wine when BiB wine is shipped offshore.

Finally, be aware that the stacking strength of BiB packages is much lower than that of conventional wine packages. To ensure best presentation to the consumer, limit the height of stock storage in the warehouse.

If these risks are managed, BiB packaging allows wine to be presented to the consumer in a convenient pack – and in good condition.

7.13 What's the best way to store a wine bottle: sideways, upside down or closure up?

C. Butzke

Empirically, for the past few hundred years or so, wines are best 'laid' down on their sides during aging in all traditional cellars at temperatures around 13°C (55°F). Without shelves one can stack bottles in layers 20 high, saving space and keeping the cork in direct contact with the wine.

In modern grocery stores, on the other hand, wines are displayed upright on shelves so that consumer can read the fancy labels, today sadly the main guide for many consumers to make a decision about which bottle to buy. Depending on the popularity of the wine, the bottle may stand in the store for a year or more, not just catching dust but aging prematurely based both on temperature and bottle position.

According to wine folklore, if a bottle is finished with a bark or synthetic cork, a certain amount of outside air is assumed to pass through it over time, allowing the wine to 'breathe' and develop. This assumption is both right and wrong. Gas does pass through the closure but it is both unpredictable and highly variable. The porosity (fraction of volume which is occupied by gas) can vary greatly between individual natural corks.

It should not be a requirement for proper bottle aging of wine as most aging reactions are actual independent of the influx of oxygen. Nonetheless, excessive air (21% oxygen) influx can oxidize a wine, while a hermetical seal such as a screw cap may bring out reductive winemaking flaws (see **11.9**).

What relevant physical phenomena are involved when bottled wine is stored?

In a bottle resting sideways, the wine (liquid) will be in direct contact with the cork (solid) which is in contact with air (gas), thus two phenomena are at work:

1. *Concentration*-driven *diffusion* (described by Fick's Law) as oxygen-rich air passes from the outside through the cork into the wine.
2. *Pressure*-driven *permeation* (described by Henry's Law and Darcy's Law) which depends on the gas pressure inside the bottle and on the outside, and increases exponentially with the wine's temperature. Daily temperature fluctuations, e.g. during unprotected shipments of wine, will exacerbate this effect.

In a bottle standing upright, the wine is separated from the cork by a gas-filled headspace ('ullage'), usually about 3 to 6 mL. The initial ullage at bottling can be dialed in by the winemaker, and serves as a cushion to mediate variations in pressure as the wine liquid expands and contracts during shipment and storage under varying temperatures. In this situation, one has the outside air separated by the cork and another layer of gas. Because of the much smaller viscosity of the headspace gas, a mix of wine volatiles and nitrogen, the gas exchange between

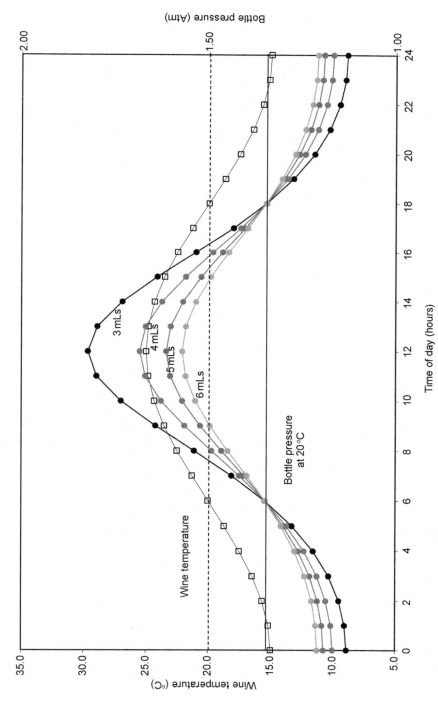

Fig. 1 Changes in internal bottle pressure under a 5 degree rise and fall from 20°C and bottle headspaces from 3 to 6 mL. Courtesy of Dr. Roger Boulton, Stephen Sinclair Scott Professor of Enology and Chemical Engineering, Department of Viticulture and Enology, University of California.

the inside and outside of the bottle is theoretically increased by a factor of 100. Depending on how perfect, i.e. permeation-resistant, the cork is, this can completely and prematurely oxidize a delicate Riesling or Sauvignon Blanc, it might have little or no effect on a heavily (*over-*)extracted, radical-quenching red.

When comparing closures or conducting closure studies, it must be noted that the loss of SO_2 through an imperfect closure via diffusion should not be confused with the added loss due to oxygen radical reactions within the wine.

Wine is usually bottled at room temperature around 20°C (68°F) to minimize oxygen solubility and account for the most likely storage conditions it will see between the winery's warehouse, distributor's warehouse, truck trailer, grocery store shelves and consumer's living room. If the ullage is small (3 mL), the internal pressure may rise by 40% from 1 atmosphere (atm) to 1.4 (15 to 21 psi) if the wine warms up from 20°C to 25°C (77°F), which leads to permeation loss of SO_2. Similarly, if the temperature drops to 15°C (59°F), the pressure inside the bottle will drop to a light vacuum at about 0.8 atm = 12 psi, which leads to more oxygen permeation from the outside. Alternatively, the pressure on the cork in a wine bottled with a nitrogen cushion of 6 mL, will increase less than 20%, to 1.18 atm (18 psi) at 25°C (Fig. 1). Depending on the closure, and its coatings and the consistency of the bottle neck diameter, this may mean the difference between corks pushing or not.

Note that many high-end red wines often have very long (50 mm) and physically perfect corks, an indication that significant permeation and/or diffusion by air is not a desirable aspect of their bottle aging process. A perfect cork may let 0.075 mg of oxygen per year into a standard bottle, so even a 50-year old bottle of Bordeaux would only see about half a saturation with oxygen (6.5 mg/750 mL bottle at 20°C). In comparision, racking and topping of barrels during barrel aging each contribute the equivalent of 15 mg O_2/bottle per year.

Another issue that sometimes comes up is the practice to store bottles upside down right after bottling so that the cork can soak up wine and improve the sealing capacity. While this may have been an appropriate practice in days gone by, modern cork coatings of paraffin and silicon will provide such a tight and immediate seal between cork and glass bottle neck that it does not matter which way is up. However, wine cases are commonly stored on pallets during storage and shipments with the wine bottles either right side up or upside down, not sideways. Therefore it is preferred that bottles go into boxes and onto pallets upside down to minimize the above mentioned permeation effects.

Further reading

BOULTON, R. B., V. L. SINGLETON, L. F. BISSON and R. E. KUNKEE. 1996. *Principles and Practices of Winemaking*. Springer, p. 422.

BOULTON, R. B. 2005. The Physics of Wine Bottle Closures. Presented at the 56th Annual Meeting of American Society for Enology and Viticulture – Science of Closures Seminar, Seattle, June.

7.14 What effects do post-bottling storage conditions have on package performance?

R. Gibson

The package has been sealed; the closure has gone on or into the bottle and the bottling process has finished. All is now sweet until the product reaches the consumer, right? Wrong! The conditions of post-bottling storage can have an impact on package performance and wine quality.

First of all, consider the impact of storage temperature. Higher temperatures promote faster chemical reactions, including the rate at which oxygen reacts with wine components. The rate of oxygen entry through packages that allow oxygen diffusion (such as synthetic closures and bag in box packs) depends in part on the oxygen concentration difference between the outside and inside of the pack. More rapid depletion of oxygen inside the pack (through more rapid reaction of oxygen with wine components) drives more rapid diffusion, as a larger concentration gradient is maintained. Oxidation of the wine is more rapid, decreasing shelf life and reducing quality.

Higher temperatures also cause wine expansion. This increases the pressure in the package. The pressure may be sufficient to cause leakage or closure movement in the bottle neck, compromising pack integrity. This issue is exacerbated by the use of small headspace volumes. Very low temperatures can also be a problem – wine expands when it freezes. Frozen bottles of wine (caused by low ambient temperatures or inadvertent use of refrigeration) generally show catastrophic failure. Stock that has been frozen is generally unsaleable due to leakage, closure movement, broken bottles and deformation of screw caps.

Fluctuating storage temperatures can also be an issue, especially for bottles sealed with natural corks that are stored in an upright position. Some corks seem to vent gas pressure – they allow bulk transfer of gas to compensate for pressure differences between the interior and exterior of the bottle. When wine warms up, pressure in the bottle is created as the wine volume expands. Some of the headspace gas may leave the bottle through the cork. When the bottle cools down, the volume of wine contracts and air may move through the cork into the headspace to equalise the pressure in the bottle with the atmosphere. Each time this cycle happens, a small amount of oxygen is introduced into the bottle. This oxygen can then react with wine components. Over time, the wine becomes oxidised – more oxidised than it would be if the temperature was constant. This effect does not seem to happen when bottles sealed with corks are laid down. Bottles sealed with closures that do not vent, such as synthetic closures, do not show these issues when they are upright and exposed to fluctuating temperatures.

It has been suggested that upright storage of cork-sealed bottles may prevent the transfer of taint compounds to wine from the cork. In my experience, upright storage provides no protection against taint transfer – taint compounds migrate very quickly in the vapour phase in the headspace – and significant oxidation risks are introduced.

The integrity of screw cap seals can be affected by post-bottling storage conditions. Damage to the corner of the cap by knocking against a hard surface can reduce liner compression against the bottle rim and facilitate gas entry or product leakage. Stock stacking must also be considered for upright stored bottles with screw caps. Excessive weight on the cap can compromise the seal. Post-packaging performance is improved by use of the BVS bottle finish and redraw of the cap to improve contact of the liner with the bottle rim.

Flexible packages such as bag in box can be susceptible to transport damage after filling. Repetitive movement, such as that caused by road transport, can lead to flex cracking of the package structure. Breakdown of the structure can lead to rapid oxygen ingress, and in extreme cases, leakage of the bag contents.

Storage time is a factor that must be considered for packages that allow oxygen ingress and provide a limited shelf life for wine. Stock age must be managed to ensure that wine quality for the consumer is optimised. Use first in-first out (FIFO) warehousing principles and plan filling to match demand, avoiding large stock build-up.

Some packages, such as polymer bag in box structures, can allow the passage of contaminant compounds from the environment in which they are stored into the wine they contain. If the concentration of these contaminants exceeds their sensory threshold, wine taint can occur. Ensure that stock storage and transport environments are free of contamination sources.

With attention these stock storage factors, the chances of delivering wine to the consumer in the best possible condition are enhanced.

7.15 What are the optimum environmental parameters for bottle storage and what effect do these parameters have on wine quality?

T. E. Steiner

Introduction

After a great vintage and paying special attention to sound winemaking practices, one anticipates the release of a high quality wine. However, some winemakers neglect the importance of bottle storage conditions. All of the hard work that has accumulated from growing the grapes through bottling the wine may be at risk if one does not store the bottled wine under appropriate conditions. Wine quality is often affected by light, humidity, time and temperature during bottle storage. Bottle storage conditions can help further develop an excellent wine or destroy the quality in a short period of time.

Effect of temperature on wine quality during bottle storage

In the author's viewpoint, the most important factor in controlling wine quality during bottle aging is temperature. Bottled wines exposed to excess heat and cold can decrease wine quality to an unacceptable level. In addition, damage can also occur in bottled wine from variations in temperature.

Generally, chemical reactions in wine occur more slowly at cooler temperatures than higher temperatures. In bottle storage, extreme cold temperatures cause precipitate formation (potassium bitartrate) and often wine freezing may result in bottle breakage or cork displacement. Precipitate formation at cold temperatures usually happens in wines that are not properly stabilized during the winemaking process. Although, precipitate formation at cold temperatures is mainly associated with the formation of potassium bitartrate crystals, it does not exclude the possibility of other chemical or microbial substances from forming. The freezing point of a wine is typically below $-3°C$ (Ough 1992) and depends on the alcohol content. Freezing wine for a significant length of time may result in bottle breakage. Breakage typically occurs at the bottom or wall of the bottle depending on bottle uniformity, design and thickness (Boulton *et al.* 1999). Also, a wine subjected to partial freezing can cause cork displacement from ice crystals forming in the bottle neck. This causes an increase in pressure due to wine expansion which decreases the headspace (Boulton *et al.* 1999).

Since most microbial growth and chemical reactions increase with higher temperatures, a major concern during bottle storage is extreme high temperatures which prompt wine quality deterioration. Table 1 shows the relative rates of selective reactions that can occur in wines it relates to an increase in temperature relative to $10°C$. For example, in looking at the browning reaction of wine from this table, it is clear to see the negative effect of increasing temperature with rates being reported nearly three times faster at $20°C$ than at $10°C$ and continue to increase with increasing temperatures.

Table 1 The relative rates of selected reactions in wines

Reaction	Relative rate[a] at temperature (°C)						
	5	15	20	25	30	35	40
Oxygen uptake (Ribéreau-Gayon 1933)	0.35	2.76	7.35	18.9	47.3	114	270
Browning (Berg and Akiyoshi 1956)	0.60	1.63	2.62	4.13	6.42	9.84	14.9
Browning (Ough 1985)	0.57	1.73	2.94	4.91	8.07	13.0	20.7
Total SO_2 decline:							
Red wine (Ough 1985)	0.76	1.30	1.67	2.15	2.72	3.43	4.28
White wine (Ough 1985)	0.90	1.10	1.21	1.33	1.46	1.59	1.73

[a] Relative to the rate at 10°C.
From Boulton et al. (1999).

Ough (1985) found that as the length and temperature of storage increased there was a corresponding increase in absorbency and browning. Similar results were reported by Sims and Morris (1984) with red muscadine wines. Ough (1985) also showed that sensory results for both red and white wines in this study decreased in score with a corresponding increase in storage temperature. It is important to note that most sensory attribute scores improved with higher amounts of SO_2 present in both white and red wines (Fig. 1). Therefore, it is important to make sure adequate SO_2 levels are monitored and maintained prior to bottling in helping keep the wine sound over the required storage period.

From the information presented above, it is the author's suggestion that optimum temperatures for bottle aging or case storage be between 13 and 20°C

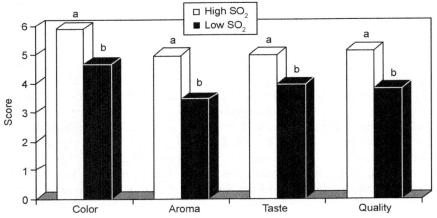

[1]Ratings from 1–10 with 1 the lowest and 10 the highest.
[2]Results presented as composite values of temperature and SO_2 effects.
[a,b]Ratings with different letters within same category indicate significant differences ($P < 0.01$).

Fig. 1 White wine sensory evaluations[1] of color, aroma, and quality in relation to SO_2 concentration.[2] Adapted from Ough (1985).

for both white and red wine varieties. I strongly believe 'the colder the better' in keeping a wine fresh and young during bottle storage and increasing shelf life potential. Granted, some bold, heavy bodied, tannic red wines may benefit with temperatures approaching 65 to 70°F during bottle aging for a period of time in helping develop the wine in both aroma and taste pertaining to an earlier release date. Therefore, having two bottle aging facilities with one being designated for white wines and one for red wines may be a good option if financially feasible. Temperatures for white wine bottle storage could occur between 10 and 17°C while red wine bottle storage could occur between 17 and 20°C.

Effect of temperature fluctuation on wine quality during bottle storage
Relating to temperature and its effect on wine quality during bottle storage, variation or fluctuations in temperatures are also critical to prevent at this stage. Temperature fluctuations during bottle storage are a concern through thermal expansion and contraction of both the cork and bottle by increasing the vapor pressure. This situation can lead to wine seepage and oxygen re-entry into the bottled wine.

Fluctuations in temperature can also trigger both chemical and microbial instabilities. As mentioned previously, potassium bitartrate precipitation can occur as wine storage temperatures approach freezing. In addition, protein precipitation (casse) can occur more readily from elevated temperatures during storage. Pocock and Rankine (1973) reported that 40% of wine protein became insoluble at 40°C held for a 24-hour period. Chemical instabilities such as ferric and cupric casse can also cause sediments due to temperature fluctuations. However, these instability problems are usually prevented through chemical analysis and proper stabilization methods prior to bottling.

Effect of light on wine quality during bottle storage
Light can also play a role in proper storage conditions. Interrelated with protein instability, cupric casse can form in white or rosé wines under reductive conditions within several days when exposed to indirect sunlight and may occur after several months when stored in cases or dark cellars (Peynaud 1984). Since light catalyzes the consumption of oxygen, even a few minutes under direct sunlight can have a negative impact on wine quality and may produce off flavors (Boulton et al. 1999). The use of colored bottles can help prevent these chemical reactions from occurring by excluding ultraviolet, violet and blue wavelengths from degrading the wine. Therefore, the best practice is to store bottles in a dark cellar.

Effect of humidity on wine quality during bottle storage
For proper humidity during bottle storage, there are several rules of thumb for keeping the cork moist and swollen. Since it is recommended to keep new corks

in an environment providing 50 to 70% humidity to control cork moisture, many people follow this practice for bottle storage conditions. However, since the bottles are full of wine, this provides an environment of 100% relative humidity in the headspace, thereby keeping the cork moist and swollen. In many cases, following a 24-hour period after bottling, the bottles are inverted to keep the wine in contact with the cork irrespective of the humidity.

In conclusion, ideal bottle or case storage conditions for both white and red varieties should occur from 13 to 20°C with an emphasis being placed on keeping the temperature constant. In addition, keeping stored wines out of direct and indirect sunlight can also be helpful in preventing the possibility of a metallic casse from forming and further chemical reactions causing deterioration in wine quality. Therefore, insulating the bottle storage area well is extremely important if not provided underground. It is also important during shipping and transit of wine to provide proper conditions as described above for the purpose of maintaining wine quality.

References and further reading

BERG, H. W. and M. AKIYOSHI. 1956. 'Some factors involved in browning of white wines.' *Am. J. Enol. Vitic.* 7: 1–7.

BOULTON, R. B., V. L. SINGLETON, L. F. BISSON and R. E. KUNKEE. 1999. *Principles and Practices of Winemaking*. New York: Springer Science+Business Media, LLC.

OUGH, C. S. 1985. 'Some effects of temperature and SO_2 on wine during simulated transport or storage.' *Am. J. Enol. Vitic.* 36 (1): 18–21.

OUGH, C. S. 1992. *Winemaking Basics*. New York: Food Products Press.

PEYNAUD, E. 1984. *Knowing and Making Wine*. New York: John Wiley and Sons.

POCOCK, K. F. and B. C. RANKINE. 1973. 'Heat test for detecting protein instability in wine.' *Australian Wine Brew. and Spirits Rev.* 91 (5): 42–43.

RIBEREAU-GAYON, J. 1933. Contribution des oxidations et réductions dans le vins. Applications à l'étude du vieillissement et des casses. Doc. thesis, Bordeaux, France: University of Bordeaux.

SIMS, C. A. and J. R. MORRIS. 1984. 'Effects of pH, sulfur dioxide, storage time, and temperature on the color and stability of red Muscadine grape wine.' *Am. J. Enol. Vitic.* 35 (1): 34–39.

7.16 What temperatures can my wine be exposed to during national and global shipments and storage once it leaves the sheltered winery?

C. Butzke

Bottled and bulk wines change with time in ways that are greatly influenced by storage temperatures. Traditionally, wine is stored at cellar temperatures between 10 and 16°C (50–60°F) and our empirical expectations of a properly aged wine are based on this temperature range. The hundreds of concurrent aging reactions that contribute to the aroma, flavor, structure, color and healthiness of a wine, all proceed at individual rates. They are exponentially and differently accelerated by elevated temperatures. Because of these multiple and unpredictable changes in the aging of the wine, temperatures higher than the ideal long-term storage range shown above are generally considered undesirable. While most commercial wines are stabilized against aesthetically objectionable instabilities such as protein hazes or crystal precipitates, there are no rapid or standardized tests to assure the sensory quality of a wine without actually tasting it.

Aging reactions in wine have substantially different rates, which explains why 'speed-aging' just by elevated temperature alone will not yield a wine that was aged around a traditional cellar temperature of around 13°C (55°F). Some examples of temperature increases that would double the aging reaction rate are 3.8°C for O_2 uptake, 7.8°C for browning, 16°C for ethyl carbamate formation, and 39°C for SO_2 decline in white wines.

The aroma (by nose) and flavor (by mouth) qualities of a wine are the most important criteria that determine the value of the product that can range across multiple orders of magnitude ($2 to $5000 per bottle). Both the high and the low temperature tolerances of a given wine depend on its original chemical composition, stabilization treatments at the winery, as well as its provenance and previous storage history. On the positive side, wines stored at these temperatures can age and improve with the development of 'bottle bouquet' and the 'softening' of harsh characters for a more supple 'mouthfeel'. This ability to develop and become more complex in the bottle has elevated wine from a mere beverage to a glorious and precious experience.

The greatest storage hazards for wine are associated with elevated temperatures as well as with temperature fluctuations. Any elevated temperatures, above 16°C (60°F), accelerate the maturation process, may change their varietal character or their sense of origin ('terroir'), and can shorten the life expectancy of a wine (especially of whites). Temperatures above 29°C (85°F) greatly and untypically age most wines, leading to undesirable aroma, flavor and color changes. Diurnal (day vs. night) temperature spikes during the commercial shipping of wine are not unusual but should be avoided. Within the entire distribution chain from winery to wine consumer, wines should never see an even short-term exposure to temperatures of 30°C (86°F) or above. It can be argued that wine should be shipped and stored under conditions that are at least equal to much less precious commodities such as milk, ice cream or produce.

What's too hot?
1. Visible protein hazes are occasionally precipitated by elevated temperatures in marginally heat-stabilized wines. These hazes constitute only an aesthetic flaw, but over-stabilizing a wine against their appearance can hide damage to its sensory qualities.
2. Temperatures above 16°C (60°F) may stimulate the growth of dormant microbes, leading to off-flavors, hazes, and excess carbon dioxide.
3. Temperature fluctuations test the integrity and position of the bottle closure, especially corks, and can lead to the introduction of air to wine, with rapid spoilage following.
4. Leakage and seepage of wine past the closure caused by excessive temperatures and resulting headspace pressure will damage the label and other packaging materials and may make the wine unsalable.
5. Storage at elevated temperature may cause excessive extraction of (off-) odors from the bottle closures as well as increased scavenging and permeation-based loss of protective sulfur dioxide or certain wine aromas.
6. Early experience with shipping barrels of wine in excessive heat produced accelerated oxidation. A significant browning of white wines, with aldehydic aromas of Spanish Sherry, caramel and nuttiness, was given the name 'Madeirization'. This is usually a grave defect unless the wine was intended to be an oxidixed style (e.g., Madeira, Sherry, Port) in the first place.
7. Most wines naturally contain traces of a precursor to a probable human carcinogen, ethyl carbamate that can form at accelerated rates under elevated temperatures, especially above 30°C (86°F).
8. High temperatures also test the integrity of sparkling wine bottles by causing the already high pressures (6 atm = 90 psi) to rise dramatically. Consequently, loss of sparkling wine by bottle bursting has been reported as a result of high temperature exposure.

What's too cold?
1. Wines stored at cooler than recommended temperatures – below 10°C (50°F) – may not develop their full potential for aroma by nose and flavor by mouth.
2. Storage at extremely low temperatures – 0°C (32°F) and below – for as little as 1 hour can cause the natural precipitation of potassium bitartrate in the form of visible crystals in white wines and in the form of colored crystals or a sludge-like mixture of crystals in red wines. This material is not re-soluble in the wine. While its presence is considered only a visual defect, it can be confused by consumers with broken glass which may create litigation issues.
3. At temperatures below −5°C (23°F) wines with an alcohol content of 14% by volume and below will start to freeze, causing corks to push out and eventually, bottles to break.

4. Moving bottles from very cold storage to a warmer environment will cause condensation of water on the bottle, labels, and cork depending on the relative humidity of the surrounding air. This can easily lead to mold growth and significant damage to the entire package.

What position?
1. Ideally, all wines are stored in refrigerated warehouses and shipped in refrigerated containers/trucks with the temperature exposure continuously logged to assure that the refrigeration system has indeed been working.
2. Wine bottles closed with corks (both bark or synthetic) should be stored upside down or sideways to minimize the gas exchange between the outside atmosphere and the wine inside to avoid accelerated oxidation. Individual cases of wine should be marked to indicate which position the bottles are in.
3. In very hot climates, including, e.g. shipments across the equator or through the Panama Canal, the ability of refrigerated containers to maintain a steady cool temperature may be compromised.
4. In general, it is recommended to avoid shipping wine unrefrigerated during the hottest months of the year.
5. When wine is storage or moved in containers on truck, railroad cars, or container ships, especially in or through hot climates, one needs to be aware that wine stored in the upper part of the container exposed to direct sunlight will heat up the fastest.
6. If refrigerated trucks are not available, it is preferable to use non-isolated containers in order to avoid the accumulation of hot air in the container space above the pallets of wine during the day and continuous heat dissipation during the night. Instead, the pallets inside the container should be covered with insulation blankets or similar materials.
7. Bulk wine trucked in not completely filled tankers tends to slosh significantly, increasing the loss of sulfur dioxide and the risk of oxidation if the head space in the tank is not sparged with inert gases such as nitrogen or argon.

If storage is provided without excessive (UV) light exposure, and the temperature is constant (± 1–$1.5°C$ or (± 2–$3°F$)) and in the ranges listed, typical white table wines should retain their quality for about three years after bottling. Red table wines, depending on winemaking style, may retain or improve their quality for three to ten years. Note that these are very general guidelines, as some wines are made to be consumed within their first year while others may age gracefully for several decades. However, experience and research have shown that improper storage conditions, during distribution through the entire supply chain, from producer to négociant to shipper to importer to distributor to retailer or restaurant to consumer, are the most common cause of loss of a wine's quality and value. Documentation of both authenticity and provenance, temperature monitoring of shipments and storage conditions, and the traceability of wines during their entire lifespan is crucial to prevent such losses.

Direct, internet-regenerated shipments of wine are a popular way to make winegrowing both sustainable for the winery and regional wines more accessible and affordable to the consumer. However, the major global package shipping companies do not have logistical systems in place that protect wine adequately from damage during shipment and intermediate storage. Obviously especially in the summer time, the non-refrigerated local delivery trucks are loaded in the early hours of the morning at distribute packages until the late afternoon. Within this 13+ hour time window, wines are frequently exposed to temperatures above 26°C (79°F).

Unfortunately, many commercial wines are so over-stabilized that adverse transport conditions that affect the sensory properties and quality of the wine go unnoticed because the wines stay visually intact. No precipitation and haze of heat-instable protein can be observed because the wine was probably treated with a substantial amount of bentonite which by itself has the capacity to strip a wine of aroma compounds.

Global tracking of individual cases or bottles of wine using GPS and RFID technology has become readily available and is used to protect many of the world's most precious wines.

Recommended global wine shipping and storage conditions

Heat exposure

Storage time	Temperature
Never	30°C (86°F) or above
Spikes of 30 min or less	29°C (85°F) or below
1–4 weeks	24°C (75°F) or below
Long-term storage	10–16°C (50–60°F)

Cold exposure

Storage time	Temperature
Never	−5°C (23°F) or below
Spikes of 30 min or less	0°C (32°F) or below
1–4 weeks	10°C (50°F) or below
Long-term storage	10–16°C (50–60°F)

Further reading

BERG, H. W. and AKIYOSH, M. 1956. Some factors involved in browning of white wines. *Am. J. Enol. Vitic.* 7, 1, 1–7.

BUTZKE, C. E. 2008. Wines. In: *Commodity Storage Manual*. World Food Logistics Organization.

OUGH, C. S. 1985. Some effects of temperature and SO_2 on wine during simulated transport or storage. *Am. J. Enol. Vitic.* 36, 1, 18–22.

RIBEREAU-GAYON, J. 1933. Contribution à l'étude des oxydations et réductions dans les vins. Application à l'étude du vieillissement et des casses, Delmas: Bordeaux, France.

http://www.eprovenance.com

8

Winemaking equipment maintenance and troubleshooting

8.1 What are common problems in the stemming-crushing operation and how can these be remedied?

H. Bursen

A successful crush must be completed under critical time constraints. Stems and other undesirables must be separated out, and fruit must be processed before spoiling. The time window can be shorter or longer depending on various factors, such as the ambient temperature and the condition of the fruit at harvest. But crushing is never a leisurely affair; the fruit is at risk until it is deposited in the fermenter.

Therefore it is important to know how the stemmer/crusher works, and to know the causes of common problems in this crucial procedure. First, a short description of the two most common types of stemmer/crusher: centrifugal and the roller-and-wiper type.

Centrifugal stemmer/crusher
Centrifugal machines speedily process large amounts of fruit. Crushing and stemming happen in a single step: the grapes feed into a perforated cylinder fitted with a whirling central shaft. Rods on the shaft slam into the bunches. This ruptures the berries, and whips the tough stem around so fast that the grapes are left behind, falling through the cylinder's holes. The cylindrical cage rotates, too, encouraging the berries to fall into a collecting trough. Meanwhile, the stems won't fit through the holes. The rods are arranged in a screw pattern, so each strike moves the stem further down the helix to the discharge end. This is speedy, but the rough handling yields juice with many tiny pieces of grapeskin

and pulp, causing extraction of bitter components from these solids and a high proportion of muddy sediment, unsuitable for quality wine production.

Roller and wiper type stemmer/crusher

Roller/wiper types offer quality, though at a somewhat slower throughput. Here the stemming and the crushing are two separate operations. In fact, it is possible to use a separate stemmer and crusher. The grapes are crushed by passing between two counter-rotating lobed rollers, set far enough apart to avoid crushing the seeds. Stemming again involves a rotating helix and perforated cage (sometimes rotating). But this time the rods (often with rubber paddles on the ends) rotate slowly, wiping the grape bunches across the holes in the cage. Individual berries fall through into a collecting trough, while the longer and tougher stems are wiped along the helix until ejected at the discharge end. The key here is that this gentle action doesn't pulverize the grape flesh. The result: higher yields of cleaner juice with fewer solids and fewer undesirable components are extracted. This type of crusher/stemmer is recommended.

Common problems

Too many berries ejected with stems
Cause: rods spinning too fast
Cure: lower machine speed

Cause: clogged cage – feed rate too high, or machine speed too low
Cure: increase machine speed, or reduce fruit feed

Cause: clogged cage holes (*botrytis*-infected fruit)
Cure: reduce fruit feed, occasionally clean cage

Cause: stem pile backup
Cure: remove stem pile

Too many stem pieces in the processed grapes
Cause: excessive speed is cutting stems
Cure: lower machine speed

Cause: cage holes too large
Cure: different cage (only available on some models)
Cure: post-stemming sorting table

Bridging of fruit at feed hopper
Cause: fruit feed too rapid
Cure: reduce feed rate

Cause: crushing rollers overloaded
Cure: avoid models lacking infeed auger – auger feed to rollers will space fruit more evenly

Note: crushers work best with steady feeding, neither overloaded nor 'starved'.

Clogging of augered output trough (where fitted)
When the output auger is turning, it pushes fruit into the narrow outlet neck. If for some reason this fruit is not removed, the result will be a tightly packed plug in the outlet fitting. Here are two common causes of a clogged output trough:

Cause: auger turning when pump is shut off
Cure: shut off crusher and reverse the pump – if you're lucky this may blow out the plug, if not, dig out the plug
In the future don't run the crusher for more than a few seconds when the pump is off. Try to match fruit feed rate to pumping rate.

Cause: starved pump loses prime, causing packing as above
Cure: feed fruit steadily, and/or lower the pump speed
Try to match fruit feed rate to pumping rate.

8.2 Why won't my electric motor start?

H. Bursen

The vast majority of winery equipment is run by electric motors. When motor problems arise, the last resort is to call in a specialist. This is sometimes an independent electrician or motor shop, and sometimes a technician who works for the equipment manufacturer or supplier. But even an electrically challenged winemaker can do some preliminary diagnosis. This has several good effects:

1. It is often possible to fix the problem oneself, without much effort.
2. Preliminary troubleshooting can save the expert's time, saving money.
3. Feeling competent, rather than helpless is good for the winemaker's blood pressure, and probably for the wine.

When a machine is brand new or under warranty, it's best to start with a call to the supplier. The supplier should be able to help with simple troubleshooting, and should also let you know whether you may attempt repair without voiding the warranty. This article assumes that the reader has very little familiarity with electric motors. What happens when we turn on the switch and the motor doesn't start?

Safety note: Always unplug the device before working on it!

Symptom: Nothing happens when switch is turned on
Cause: no power at the outlet
Simple check: examine electric panel to see if circuit breaker is tripped
Cure: reset circuit breaker switch
If that doesn't work, try another outlet – preferably in a different part of the building. If the motor runs on a different outlet, the cause is in the building power circuit, not in the motor.

Suppose outlet has power but motor still doesn't run:

Cause: short circuit
Check: Smell the motor. A strong smell of burnt plastic means motor windings are probably burnt out. A classic short circuit. Time for a new motor.

Cause: loose connection
Simple check: **unplug the device** and check for loose connections:

1. connections between power cord and the plug prongs (only practical if the cord and plug are not a sealed unit)
2. connections between power cord and on/off switch
3. connections between on/off switch and the motor itself.

If connections are loose or corroded, correct the situation and try again. If still 'dead', it could be the on/off switch.

Cause: bad power switch (on/off switch)

This requires a meter for testing the **unplugged device**: With meter set to 'ohms', place meter probes on inlet and outlet connections of switch. When switch is 'on', should show near zero resistance (zero ohms). With switch 'off', should show infinite resistance. If infinite resistance when switch is 'on', replace the switch.

Symptom: when power is on, motor hums, but doesn't turn
(*Note*: We assume this is a single-phase motor, 110 or 220 volts. See **8.3** for a 3-phase motor with similar symptoms)
Cause: worn front shaft bearing is binding up.
Simple check: **with motor unplugged**, try to wiggle the front end of the motor shaft. If it has substantial 'play', the bearing needs replacing. *Note*: Sometimes it is more practical to replace the entire motor.

Cause: bad start capacitor.
If there is no substantial play in the shaft, the start capacitor may be damaged. Many single-phase motors require a start capacitor to begin turning. This device is usually a small metal can. Some motors have 2 capacitors, one for starting and one for running.
Simple check: **unplug the motor** and give it a quick spin by hand. While it's still spinning (and with your hand out of harm's way!), plug it in and turn power switch on. If the motor continues running, the start capacitor has a problem.
Cure: **with motor unplugged**, replace the capacitor with another of similar rating. If the motor starts, but doesn't run smoothly, or up to speed, then the second 'run' capacitor (if it has one) may be damaged, too. In that case, **with motor unplugged**, replace run capacitor too.

If that doesn't help, it's time to call in a professional.

8.3 What other problems might I encounter with my electric motor?

H. Bursen

We assume here that the winemaker has at least a little experience with electrical equipment, and can use a multimeter safely. Training material is available online, for example at: http://www.youtube.com/watch?v=yrpmq2FCnqs&feature=related

When a machine is brand new or under warranty, it's best to start with a call to the supplier. The supplier should be able to help with simple troubleshooting, and should also let you know whether you may attempt repair without voiding the warranty.

1. Why does my pump (fan, filling machine, etc.) keep tripping the circuit breaker?

This problem has many causes. Here are some common ones:

Cause: low voltage. This makes motors run hot.
Some causes of low voltage:
1. The machine's power cord is too long
   Cure: shorten power cord
2. The motor is mismatched (example: a 240 volt motor on a 208 volt line)
   Cure: replace motor to match available voltage
3. The power company lowers the voltage due to too much demand (a 'brownout')
   Cure: none – try to reschedule machine operation.

Cause: circuit breaker
1. Circuit breaker is worn out, tripping when load is normal
2. Circuit breaker is mismatched, with too low capacity
   Cure: replace breaker
3. Adjustable circuit breaker is adjusted too low
   Cure: readjust breaker.

Cause: motor is working too hard.
Examples:
1. Pumping a liquid with too high a viscosity (a water pump trying to pump glycol antifreeze, for example)
   Cure: use a different pump
2. Worn motor bearings cause too much drag
   Cure: repair or replace motor
3. Motor running a machine which needs lubrication
   Cure: lubricate machine
4. Motor bearings need lubrication
   Cure: lubricate if possible. If bearings are sealed, replace bearings or replace motor

Cause: small partial short in motor windings draws extra current
Cure: rewind or replace motor

Cause: weak start and/or run capacitor causes motor to draw extra current
Cure: replace capacitor(s)

Cause: undersized motor drawing too much current and overheating
Cure: replace with larger motor

Three-phase power
Much European-made winery equipment runs with 3-phase power. This is a more efficient form of electricity. Three-phase plugs have 4 prongs. Three of the wires are 'hot' and one (usually green) is the ground.* The most common voltage for small motors is 220. This type of power can be supplied by the power company, or else must be manufactured on-site, usually with a rotary converter.

2. Why is my three-phase motor vibrating and barely turning?
Cause: one of the three legs is not providing power, because
1. a 'hot' wire is corroded, or has vibrated loose
   Cure: clean and reattach wire
2. One leg of the rotary converter (if there is one) is malfunctioning
   Cure: repair converter

3. My three-phase motor (only sometimes) fails to start – when it starts, it runs normally
Cause: rotary converter voltage is too high
Check: turn on several other three-phase machines, and then try starting the problem motor – if the motor starts only when other three phase machines are running, the converter is too large
Cure: have electrician install 'buck/boost' transformers across any two legs of the converter

4. Why is my three-phase motor running backwards?
Cause: hot legs are connected incorrectly – three-phase motors are inherently reversible (they can run either direction, depending on how the hot legs are connected)
Cure: simply switch the connections on any *two* of the hot legs

Note: Some more complex machines use 24-volt controls to prevent dangerous shock (examples: pressure leaf filter, press, automatic corker). If the 24-volt transformer fails, the main motor will not run. If such a machine refuses to start, look for a bypass test switch in the control box. If the main motor runs when the bypass is pressed, only the 24-volt circuitry needs attention.

---

* Note: There are other 4-prong plugs which are not three-phase. Your supplier will know if you are purchasing a three-phase machine.

8.4 What different types of pump are there for wine and/or must transfer?

M. R. Dharmadhikari

Selection of a proper pump is important to efficient winery operation. Several factors need to be considered when choosing pump(s) for wine transfer. In a winery, the liquid transfer involves moving must, lees, juice and wine. These liquids vary in terms of viscosity and density (Amerine and Joslyn 1970). Therefore, the choice of pump will depend on pumping requirements (i.e., what is being moved). The movement of wine can be relatively slow in filtration and bottling, or it can be fast. There can also be some variation in the distance and the height where a wine needs to be transferred. Several other factors that also need to be considered include: pump size and capacity, cost of operation, material of construction and ease of cleaning and maintenance.

Many types of pumps are used in a winery for wine transfer. For economic reasons it is preferable to select a pump that can be used for multiple applications, including wine transfer. Boulton *et al.* (1996) classified winery pumps into two major categories as positive displacement and centrifugal pumps, along with several subgroups. In this question and **8.5** we will discuss the pros and cons of major types of pumps used in a winery for wine, lees and must transfer.

Positive displacement pump
The common types of positive displacement pumps used in the winery include: reciprocating piston, progressive cavity, rotary vane, flexible impeller and diaphragm pumps.

Reciprocating piston pump
These are positive displacement, self-priming pumps designed to handle fluids with high suspended solid contents. In a winery they are suitable for transferring must and lees. Some portable models can also be used to transfer wine.

Progressive cavity pump
These pumps are commonly used to transfer must and also good for pumpovers during red fermentation. They are fairly gentle and minimize solid generation during must transfer. They are self-priming and can move liquid in both directions.

Rotary vane or lobe pump
A rotary vane/lobe pump is considered a versatile winery pump. It is self-priming and can handle many winery applications such as transferring must, lees, juice and wine. It can also be used for pumpovers and wine filtration. For must transfer, a sufficiently large-size pump with an inlet and outlet that is at least three inches or larger in diameter is recommended (Curtis 2006). The pump is also suited for pumpovers due to high throughput and gentle handling of must.

It is more commonly used for tank to tank wine transfers. When equipped with remote control option, it can also handle tank to barrel wine transfer. For wine filtration a pump needs to operate against significant back pressure with a good flow rate. A rotary lobe pump fitted with a variable speed control is a good choice for wine filtration operation. Because of its ability to operate against back pressure this pump is also suited for sterile filtration during bottling of wine.

Flexible impeller pump

Flexible impeller pumps are best suited for transferring juice or wine. They are affordable and can be used with a filter. These pumps should not be allowed to run dry as this can cause rubber impellers to be scorched. They are not useful in must transfer, therefore, they are mainly used for juice or wine transfer in a winery.

Diaphragm pump

Diaphragm pumps have many applications in a winery. They are operated by compressed air and have many desirable features. They are very gentle on juice and wine during transfer. They are easy to clean, and can run dry without damage. Portable models are available now. They can also be used for barrel transfers (easy shut off valve), pumpovers and bottling. Since they are not operated by electricity they can be used in wet environmental conditions in winery without the dangers of electrical hazards.

Centrifugal pumps

Centrifugal pumps are very efficient in moving large volumes of wine. They are usually not self-priming and require a flooded suction. When using this type of pump it is usually placed below the level of the tank outlet. While running they can be shut off, or the flow can be restricted at the outlet without busting hoses if the inlet valve on the receiving tank is closed. Because of this feature, the pump is not suited for filtration since filter pumps need to work against back pressure.

References

AMERINE, M. A. and JOSLYN, M. A. 1970. *Table wines: The technology of their production*, Berkley, University of California Press.
BOLTON, R. B., SINGLETON, V. L., BISSON, L. F. and KUNKEE, R. E. 1996. *Principles and practices of winemaking*, Chapman & Hall, New York.
CURTIS, P. 2006. The versatile rotary lobe pump, *Wine Business Monthly*, October.

8.5 What are the strengths and weaknesses of different types of pump?

H. Bursen

One of the principal activities in winemaking is the movement of must, juice and wine from one place to another. For the most part, this involves pumping. But not all pumps are the same. Using an unsuitable pump can be an exercise in frustration; even worse, it can damage the wine. Thus it makes sense to spend some time on the subject. Because there are so many different pumps, this short article cannot include them all. Here is a quick look at some of the most common types, their virtues and deficiencies. Before buying a particular pump, contact customers who already own one.

A word on gravity flow
Some wineries are designed to take advantage of gravity flow, in an effort to avoid unnecessary pumping. Nevertheless, pumps remain a critical winemaking tool in any winery. This article is applicable to every winery, no matter the design.

Why are there so many types of pump?
The answer is that pumps are used for different purposes. For example: transferring wine from tank to tank, filtration, moving must from crusher to fermenter or press, pumping over red must, aerating, etc. No one design is perfect for all of these processes. There is no one 'best pump' for winemaking. Nevertheless, there are some general factors to consider, such as: purchase cost, cost of repairs and replacement parts, durability, ease of repair, ease of cleaning, and versatility.

Below is a very short description of the most common winery pumps, followed by a brief discussion of each type:

1. Flexible impeller
Rotating blades alternately spring forward to create suction, then are bent back to expel the liquid.

- Typical uses: liquid transfer, must pumping, filtering, bottling.
- Virtues: versatile, inexpensive, self priming, easy to repair. This is a decent (but not a great) pump for many purposes.
- Defects: being a generalist, it does not excel in any of these tasks. The pumping action is moderately rough on wine. It is extremely rough on crushed grapes, pulverizing them and creating large amounts of undesirable solids, thus lowering the yield of finished red wine by approximately 10%. Also, its maximum output pressure is limited. When pressure rises (example: a filter becoming partially clogged), pumping almost ceases, and the liquid is

subjected to violent agitation inside the pump body. If run dry, the impeller quickly heats up from friction, damaging itself and the wine.

2. Low speed centrifugal
A propeller freely rotating inside a chamber creates areas of low and high pressure, sucking in liquid toward the center, and expelling it from the perimeter.

- Typical uses: liquid transfer, agitation, some bottling.
- Virtues: relatively inexpensive. With good design, can be relatively gentle on the wine. Can be run dry, and the outlet can be closed off for a short period to stop flow. Maintenance and repair are relatively easy.
- Defects: low speed generally means limited pressure. That's a problem when, for example, filling tall tanks, or filtering when the filter is partially clogged. Centrifugals are not self priming. They are not suitable for crushed grapes or must. When pumping against pressure, the wine moves only slowly through the pump, risking degradation.

2a. High speed modified centrifugal
This pump type avoids some of the limitations of low speed centrifugals, but at a cost. The higher pressures enable filtration. Indeed, they are often used in pressure leaf filters. But the pumps are more expensive, and the more complex design is subject to damage from foreign objects.

3. Piston pump
A piston sucks liquid into a cylinder, then expels it. One-way valves (typically rubber balls) keep the liquid flowing in the proper direction.

- Typical uses: crushed grapes, fermented must, wine transfer, dosing of filter-aid slurries.
- Virtues: piston pumps are good at moving wine and juice, but they really excel at moving crushed grapes and other slurries. Where other pumps clog, the surging motion of pistons tends to keep the solids moving. They are self priming, and can build up to high pressures.
- Defects: piston pumps are bulky and heavy. They are more expensive to purchase and to repair than centrifugals or flexible impeller designs. Complex internal architecture requires careful attention to sanitation, especially when pumping crushed grapes. Their surging action makes them unsuitable for filtration or bottling.

4. Rotary lobe (Waukesha type)
Heavy rotors create suction when they separate, and expel the wine when they come together.

- Typical uses: transfer, filtration, bottling, crushed grapes and must.

- Virtues: versatile, performing well in a variety of tasks, they self-prime, easily build pressure, and can be run dry. They handle wine more gently than the preceding types.
- Defects: rotary lobe pumps are precision-machined, and are therefore relatively expensive to buy and repair. Because of the tight clearance, even small foreign objects (stones and staples, for example) can cause significant damage.

5. Progressive cavity

An internal rotor moving inside a rubber casing forms spaces which move from one end of the pump to the other.

- Typical uses: most often used for moving crushed grapes and must, this pump also works for transfer, filtration and bottling.
- Virtues: progressive cavity pumps are very gentle, versatile, self-priming, and can build high pressures.
- Defects: these pumps are expensive and heavy. If run dry, damage results, and is expensive to repair. After prolonged storage, they may be difficult to start. Many winery models are hopper-fed, and can only be used for moving crushed grapes.

6. Diaphragm

A membrane flexes back and forth. One-way valves cause the liquid to move forward, preventing backflow.

- Typical uses: moving crushed grapes and heavy lees, dosing of filter-aid slurries.
- Virtues: self-priming, able to move very heavy slurries. Relatively gentle handling of wine. Only the diaphragm contacts the liquid, so abrasive mixtures can be pumped without damaging the pump.
- Disadvantages: surging action is unsuitable for filtration or bottling. There are anti-surge designs, but these increase size, expense and complexity. Most are operated by air, and so require an air compressor.

7. Peristaltic

Wide rollers squeeze a flexible tube, forcing the liquid inside to move.

- Typical uses: wine transfer, moving crushed grapes and must, filtration, bottling.
- Virtues: very gentle handling of wine and must, self-priming, versatile.
- Defects: bulky and expensive.

8.6 Why isn't my pump priming and why isn't it pumping fast enough?

H. Bursen

One of the most essential winemaking tasks is moving the wine from place to place. This is normally accomplished by pumping. There are many different types of pump, including: centrifugal, flexible impeller, rotary lobe, piston, progressive cavity, peristaltic, diaphragm and more! In this troubleshooting section there is no space to describe them all. Each has its advantages and disadvantages for a particular task. Sometimes problems result when an unsuitable pump is employed.

Common problems
1. Failure to prime
For a pump to prime by itself, it must generate suction by pumping air. Some pump types (examples: piston pump, flexible impeller) self-prime easily, and others do not (example: true centrifugal pumps). But many pumps which self-prime when new, lose this ability with age. New pumps are tight, having little clearance between the moving impeller and stationary pump body. Once significant gaps develop, air can slip back instead of being pumped through. No suction is generated, and the pump will not prime itself.

Note that priming is much easier when the wine in the source tank is higher than the pump, and the destination tank is empty. It's much more difficult when pumping into a tank that is partially full. Let's put ourselves in just that situation: tank A has 500 gallons of Chardonnay, and tank B, with a capacity of 1000 gallons, is half full. We hook up a pump to tank A, to fill tank B. Let's assume further that it is a pump which, in good times, will self prime. But the pump won't prime. Why?

Cause: The pump is 'air bound'. Because tank B is half full, the pump would have to pump air against the pressure of the wine in tank B, otherwise it won't prime.
Cure #1: disconnect the outlet hose from tank B. The pump will no longer have to fight the back-pressure from tank B. Once hoses and pump are full, reattach to tank B; the pump will function normally.
Tip: This is good practice anyway, to prevent unwanted aeration of the wine. Best done with a valve on the end of the outlet hose.

Cure #2: (to ease priming in many situations) Squirt water into suction hose, and tip hose to move water to the pump. Start pump at high speed, shut off as soon as wine reaches past the pump. Dump water from far end.

Some other causes of priming failure
Cause: impeller is worn or damaged
Cure: inspect and replace

Cause: rear seal is worn or damaged
Cure: inspect and replace

Cause: suction-side tank valve not fully open
Cure: open valve fully

Cause: discharge-side tank valve not fully open
Cure: open valve fully

Cause: suction hose fittings are loose or improperly connected
Cure: inspect, tighten and reconnect

Cause: pump bypass (if fitted) is open
Cure: close bypass

Cause: suction hose diameter is too small
Cure: replace with wider-diameter hose

2. Pump is slowing down, or losing pressure
Cause: worn pump allows backflow when there is too much backpressure. Examples:
- waukesha-type lobe pump with worn rotors
- flexible impeller pump with torn impeller, missing vanes
- piston pump with damaged one-way valve (often a rubber ball)

Cure: replace damaged parts

Cause: stems may prevent ball from seating (piston pump)
Cause: solids buildup may prevent ball from moving (piston pump)
Cure: clean pump chamber

Cause: loose suction-side connections allow in air, slowing output; worse, unwanted oxygen damages wine
Cure: check and retighten connections

Cause: tank bottom valve may clog with seeds and yeast during pumpover
Cure: reverse pump to clear blockage.

8.7 How do I care for the winery pump?

M. Sipowicz and B. Trela

The three main types of pumps found in wineries are positive displacement pumps, centrifugal pumps and air pumps.

Positive displacement pumps

Positive displacement pumps or PD pumps, in this case most commonly the rotary lobe pump, are capable of producing very high pressure. It is imperative to never stop the product flow while the pump is on. Excessive backpressure caused by a closed valve, crimped winery hose or unnecessarily long hose runs may result in a ruptured hose or the cause of a hose end to blow off. This style pump is usually configured to run effectively forward or in reverse. Rotary lobe pumps can move must, lees, juice and wine.

Maintenance matters

This pump usually contains both body and shaft 'O' rings. Additionally oil levels within the pump head gearbox and motor reduction gearbox must be maintained. The pump head and occasionally pump wheels have grease (zerk) fittings. Establish correct service intervals for your specific piece of equipment. See owner's manual or contact your equipment dealer for more information.

Centrifugal pumps

Centrifugal pumps rely upon fluid acceleration within the pump head to create pressure differential and thus product flow. These pumps are best suited for transferring fairly clear liquids, i.e. wine and juice (not must or heavy lees). They are not able to run in reverse nor should they EVER run dry (this will destroy the head seal). Common uses are for pump-overs, general wine or juice moves, and tank cleaning. Centrifugal pumps are not considered positive displacement pumps and thus may not be the best choice for applications requiring steady backpressure delivered from the pump as required during, for example filtration operations. Cavitation is a common occurrence with centrifugal pumps. This occurs when the pump outlet liquid is moving faster than the liquid entering the pump at the inlet. Cavitation within the pump head causes a very loud rattling, crackling noise, and damages the pump impeller. Generally one of the following operations will alleviate this problem:

- Ensure that there are no restrictions on the suction side (a closed or partially closed valve, kinked hose, object within product flow, etc.).
- Slow the pump speed.
- Apply backpressure on the outlet side (e.g., partially closing a valve on the outlet line).

Maintenance matters

This pump style usually has an 'O' ring between the pump head (impeller casing) and the pump body. There is also a head seal located behind the pump impeller (between the impeller and the pump body). These seals and 'O' rings may vary from one pump manufacturer to the next. Centrifugal pumps may have grease fittings on the ends of the drive motor. Occasionally these fittings are hidden on the underside of the unit. Zerk fittings, if present on wheels, should also be greased. Establish correct service intervals for your specific piece of equipment. See owner's manual or contact your equipment dealer for more information.

Air pumps

Air pumps convert applied air pressure into fluid pressure. This is a direct correlation therefore as the air pressure that is applied increases, the fluid pressure increases. This pump utilizes internal diaphragms connected by a center shaft. There are also rubber balls which act as one-way valves. The diaphragms have an average service life of about one year. If, however, wine is seen coming out of the exhaust muffler, the diaphragm(s) has failed and must be replaced as per manufacturer's guidelines (owner's manual). If product flow has decreased during normal run speeds, there is likely an obstruction at the point where the internal rubber balls seat (grape stems, vine matter, gaskets). Obstruction can usually be extracted by removing first the inlet and outlet ports and inspecting ball seat. If the pump operates slowly or sporadically, the center shaft is likely operating under excessive friction load. Disassemble unit as per manufacturer's guidelines (owner's manual) to determine if the center shaft requires replacement or simply needs to be cleaned and lubricated. Air pumps are commonly used for pumping must, juice and wines. They are ideal for filtration operations and barrel work and are able to move lees quite well. These pumps require a source of clean, cool compressed air. Minimum supply pressures are manufacturer specific.

Maintenance matters

Air pumps may or may not require oil for operation. Some units have a combination water trap/regulator/oiler which introduces a small amount of oil to internal air actuated parts during use. This oil should be H1 food grade lubricant oil. Regardless if oil is required or not, the combination water trap/regulator (oiler) must be removed and cleaned periodically with soap and water to prevent clogging of the filter. As condensation water collects in this trap it must periodically be emptied. Replace pump diaphragm as needed or at recommended service interval (whichever comes first). Clean and lubricate center shaft and as well clean balls and ball seats as needed or at recommended service interval (whichever comes first). Do not forget to grease zerk fittings if present on wheels. Establish correct service intervals for your specific piece of equipment. See owner's manual or contact your equipment dealer for more information.

Lubrication

A properly maintained winery pump will provide many years of trouble-free service, making worth the substantial and necessary investment that a set of good pumps entails. A poorly maintained winery pump, however, can be the source of numerous wine ailments including oxidation, microbial spoilage, contamination, and excessive heating of the wine. Given the numerous types of pumps available, the focus here will be on general preventative maintenance.

Wet conditions within a winery make timely maintenance of grease ('Zerk') fittings imperative. Most winery pumps have numerous lubrication points. Keeping these lubrication points at the proper level (see owners manual) will displace moisture from the region and minimize corrosion and friction within moving parts. All the grease nipples of a pump (or any piece of equipment) are often hard to locate, and owner's manuals are sometimes missing. Make a location sketch and list of all fittings for each of your pumps so that new cellar workers or interns can easily do the lubrication job. Placing caps over each fitting will prevent rust/corrosion build up between grease applications, ensuring proper function. The amount of grease applied and frequency of application are pump-specific. Refer to the owner's manual for each pump type used. If a pump has a gearbox or closed drive mechanism, maintenance of proper fluid level is additionally required. Make sure you use only food-grade lubricants if any part of the pump may come in contact with your juice, must or wine. Grease fittings should be wiped clean before grease is injected. This will prevent dirt and other contaminants from entering along with the grease application. Between applications, foreign material/dirt is prevented from entering the system by a spring loaded steel (usually) ball on the internal side of the grease fitting inlet. During each grease application, inspect that this ball fully seats within the opening. If the ball is not fully seated or function is otherwise impaired, replace with a new grease fitting immediately. These are commonly found at better hardware stores.

Seals and O-rings

Proper seal and 'O' ring function will prevent wine contamination and leaks, as well as minimize pump wear and wine oxidation from air intrusion. Frequently inspect pumps for signs of leakage. Most leaks are 'two-way', meaning that if wine can leak into the environment or gearbox, for example, then air or lubricant can also be pulled into the pumped wine during operation. Whether seal and 'O' ring inspection is part of routine maintenance or due to an observed malfunction, all seals and 'O' rings within a component should be inspected and if necessary replaced, best on a routine schedule every year. Rubber or elastic 'O' rings that have become brittle or cracked should be replaced immediately. Keep all owner's manuals in one accessible place in the winery such as the lab, just like your material safety data sheets (MSDS). Schedule ahead of time the proper maintenance inspection and replacement intervals, and have plenty of spare 'O' rings and lubricants available before harvest starts. Keep in mind that higher operating pressures as well as higher temperatures (e.g., sterile bottling operations, cleaning operations) will accelerate 'O' ring and gasket degradation. Likewise be aware of ozone incompatibility issues with certain 'O' rings, gasket material and one-way valve seating balls (air pumps). Ozone will rapidly degrade components made from rubber-based elastomers or Nitrile/Buna-N polyamides. Some pump manufacturers recommend that 'O' rings and gaskets be replaced numerous times per year, whereas others suggest this action on an annual or service-driven basis. Consult with your pump manufacturer and/or owner's manual for proper service intervals.

The characteristics of the fluid pumped (e.g., must versus juice), pump speed, back pressure, etc., will affect the wear rate of the pump impeller, lobes, ball (diaphragm pumps), and peristaltic tube. Additionally, pump cavitation will significantly reduce service life. Timely inspection of these parts will maximize pump efficiency as well as minimize down time. Do not use pumps that are not specifically made or specified for handling wine. Corrosion and contamination are most likely to occur!

Electrical considerations
Preventing electrocution and minimizing component damage depend upon a well-maintained, water-tight electrical system. Pump power cords should be in excellent condition with damage-free insulation (sheathing). Do not wait to replace damaged insulation or wiring. All wire connections must be properly enclosed in a plug boot and component housing, and must include a ground wire. The pump power switch/variable speed controller must be housed in a water-tight enclosure for protection. The door gasket should be checked periodically to ensure proper water-tight seal.

Sanitation issues
Pumps that are left unclean between uses may exhibit reduced service life due to material buildup and subsequently, increased working friction. Additionally, pumps left in this condition become a source of wine contamination. Pumps should be stored clean and dry between uses. Special steps should be taken to eliminate 'standing water' within the pump head. To deal with this issue, one option is to install a small ball valve at the lowest point of the pump head in order to drain residual water. Remember to cool the pump down after sanitizing with hot water or steam, before turning it on. Unless stated in the owner's manual, winery pumps, especially positive displacement or lobe ones, may not be able to handle the expansion and subsequently reduce clearance of different metal parts due to high rinse temperature. While often laborious, taking time to disassemble a pump and its moving parts and cleaning them by hand rather than just by pumping sanitizer through is generally a good maintenance practice.

Further reading

Maintenance Manual, Yamada Air-Operated Diaphragm Pumps, Document Number NDP027M-31, Yamada America, Inc., Arlington Heights, Illinois.

Operation and Maintenance Manual, Universal I Series, Positive Displacement Pumps, Waukesha Cherry-Burrell (2006), Delavan, Wisconsin

User's Guide for Carlsen and Associates Pumps, Revision 1.0, July 2002, Carlsen & Associates, Healdsburg, California.

8.8 What is bubble point membrane filter integrity testing?

B. Trela

Definition and principle

The bubble point method is the most widely used non-destructive technique for determining sterile membrane filter integrity and surface filter pore sizes. The technique can indicate the proper nominal pore size (for example, 0.45 vs. 0.2 μm), a kinked gasket, an improperly placed membrane; minor filter defects and pores too large. It is based on the properties (surface tension, solubility, and capillary action resp.) that, for a given fluid (water or wine), gas (typically nitrogen) and pore size with constant wetting (membrane housing filled with fluid), the pressure required to force the liquid and eventually a gas bubble out of the pore is inversely proportional to the size of the largest pore. Note that each filter membrane has a range of pore sizes, i.e. not all pores of a 0.45 μm filter are indeed exactly 0.45 μm in diameter. The so-called nominal size of the filter (for example '0.45') indicates that the average size of its pores is 0.45 μm. This means that there is a distribution of pores some of which are smaller or larger than the nominal size. However, the bubble point pressure depends on the largest pore (or hole) in the entire filter. Practically, we use 0.6, 0.45 or 0.2 μm nominal pore sizes because the pertinent microorganisms in wine are still larger than even the pores with the largest absolute size in those cartridges (yeast is about 10 μm in diameter, bacteria about 1.0 μm). In recent years, some wine bacteria have been observed to shrivel under certain stress conditions which could allow them to slip through pore sizes that a normal vegetative cell could not. Typically the bubble point test is performed before use, but should also be used after filtration to detect if the filtration process compromised the integrity of the filter and thus the bottling run.

In practice, the largest pore size of the filter cartridge can be established by flooding the cartridge inside the housing ('wetting') with a fluid (always measure the bubble point pressure with the same liquid and gas, e.g. water and nitrogen) and measuring the minimum gas pressure (applied to the underside of the housing) at which a first rapid stream of bubbles is emitted from the upper outlet of the cartridge. These bubbles can be easily observed through a sight glass integrated into the pipe right before or after the valve at the filter cartridge outlet. At prolonged gas pressures below the bubble point, gas molecules diffuse through the water-filled pores of a wetted membrane following Fick's Law of Diffusion. The gas permeation rate for a filter is proportional to the differential pressure and the total surface area of the filter. When using large surface area filters ($\geq 0.19\,\text{m}^2$), a diffusion test such as a forward flow, pressure hold or pressure decay test may be more effective than the bubble point test to determine a filter's integrity since the volume of gas diffusion through the membrane during testing could lead to a false bubble point determination. Each sterile membrane that is purchased will have the approximate bubble point pressure indicated in the instructions. Since it varies based not only on the fluid and gas

Winemaking equipment maintenance and troubleshooting   219

used, but also on the length (and therefore surface area) of the cartridge, the pressure at which the first bubbles are observed should be determined and recorded for each individual setup at the winery.

It is important to realize that the bubble point test only determines the integrity of your filter setup. It does not at all guarantee that the setup is sterile. Sanitation of the tested filter setup from filter inlet all the way to the filler spouts can be performed by hot (at least 82°C at the outlet) water or ozonated water (if all gaskets are ozone-safe and no air pockets exist that cannot be contacted directly).

Further reading

ASTM. 2005. *Standard Test Methods for Pore Size Characteristics of Membrane Filters by Bubble Point and Mean Flow Pore Test. Method F316-03*. American Society for Testing and Materials Standard.

BROCK, T. D. 1983. *Membrane Filtration: User's Guide and Reference Manual*. Madison, WI, Science Tech.

DICKENSON, C. T. 1997. *Filters and Filtration Handbook*. Elsevier.

TB039. 1999. *Filter Integrity Test Methods, TB039*. Millipore Technical Publications, Millipore Corporation.

8.9 How do I perform a bubble point membrane filter integrity test?

B. Trela

Procedure

The clean filter cartridge is placed in its filter housing. The filter is wetted by flushing it with water – thereby filling the pores. The bottom of the filter holder is pressurized with a regulated source of air, or an inert gas such as nitrogen from a gas bottle cylinder. The gas pressure is gradually increased within a few seconds to about 80% of the pressure mentioned on the membrane packaging, then slowly until a steady stream of bubbles appears on the liquid side above the filter. The bubble point is observed when the gas pressure is high enough to displace liquid from the largest pores, bulk flow of gas begins and rapid continuous bubbling will be seen until most water has been pushed out of the cartridge (ASTM 2005). In a winery the bubble point test is usually performed in-process, with water, after steam or hot water sterilization procedures, and after the system is cooled to operation or product stream temperatures (usually by pumping cool water through the same system which is now sterile). Nitrogen gas is often administered through a valve at the bottom of the filter housing and the bubble point on water is observed through a sight glass near the (open) top valve. The actual bubble point pressure is measured by a pressure gauge at the inlet side of the filter or can be read at the secondary regulator on the gas bottle. To determine if a filter has a filter pore size at the manufacturer's rating, the bubble point must occur at the minimum pressure for which the filter is rated. The initial bubble test pressure determines the size of the largest hole in the filter element. As an example, a bubble point pressure of about 2.0 bar (30 psi) of nitrogen can be expected for a 100 cm (40 inch) filter cartridge of 0.45 μm nominal pore size wetted with water. For any given sterile filtration setup, cartridge and pore size, the observed bubble point pressure should be recorded as a reference for the next bottling run. The observance of a significantly higher pressure does indicate any one of the problems described below.

With filter fouling occurring during the bottling run or over time as the membrane is reused, the bubble point pressure may increase. If the integrity of the membrane is tested again as recommended at the end of the bottling run, a slightly lower bubble point pressure may be observed if the test is performed with wine wetting the membrane because of alcohol's lower surface tension. The bubble point for standard dry table wines with 12% v/v ethanol at 20°C would be approximately 49.5/72.8 times that of water based on the relative surface tension values of 12% v/v aqueous ethanol solution and water respectively. Given the variable surface tension values of product streams, most manufacturers recommend bubble point determinations with water.

To avoid filter fouling and growth of microorganisms inside the cartridge while it is not in use, the membrane should be stored fully wetted in a 200 to 300 mg/L SO_2 in citric acid solution inside a separate housing which can be easily built out of PVC piping or custom-made from stainless steel. At this point, it is also helpful to mark the manufacturer's pore size on the membrane and the housing to avoid getting a false bubble point pressure reading at the next test.

In summary, since the bubble point pressure is a function of the surface tension of the liquid and the filter pore size, a bubble point value lower than the filter manufacture's specification is an indication of one or more of the following:

- cracked filter element-to-filter housing seals
- incorrect filter or gasket installation
- system leaks
- comparing a reading taken with water to a reading taken with wine as the wetting fluid
- incorrect pore size filter
- high water temperature
- incompletely wetted membrane
- physically damaged membrane.

A bubble point value higher than the filter manufacture's specification is an indication of one or more of the following:

- comparing a reading taken with wine to a reading taken with water as the wetting fluid

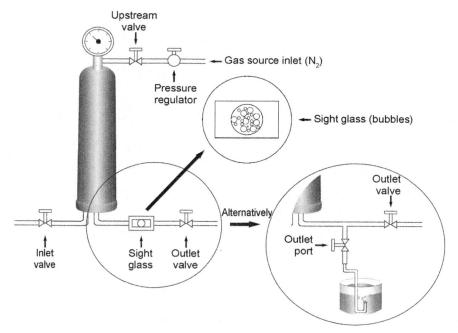

Fig. 1 In-process bubble point test illustration (Illustration, © Ratree Trela, 2010).

- membrane fouling due to wine deposits during the bottling run or improper storage of the membrane when not in use.

Advantages of bubble point testing
- Can be performed on filters under actual use conditions and with any membrane filter.
- It is a non-destructive test, thus it does not contaminate the filter and so can be used to determine the integrity of a filter at any time.
- Performing the test should not take more than five minutes of the winemaker's time but will help avoid sleepless nights worrying about refermentions in the bottle – provided that the rest of the bottling line is sterile!

References and further reading
ASTM. 2005. *Standard Test Methods for Pore Size Characteristics of Membrane Filters by Bubble Point and Mean Flow Pore Test. Method F316-03*. American Society for Testing and Materials Standard.

BROCK, T. D. 1983. *Membrane Filtration: User's Guide and Reference Manual*. Madison, WI, Science Tech.

DICKENSON, C. T. 1997. *Filters and Filtration Handbook*. Elsevier.

TB039. 1999. *Filter Integrity Test Methods, TB039*. Millipore Technical Publications, Millipore Corporation.

8.10 What's the best procedure to clean or store a used barrel?

C. Butzke

Here is a simple barrel cleaning procedure:

- Cold water rinse with a well-designed barrel washer spray head for 3 min to remove lees.
- Moderately hot (62°F) water rinse for 3 min. This is not a sanitation and as previously noted, the solubility of potassium tartrate is ten times higher in hot than in cold water. Don't cook the barrel and the residual wine soaked in its staves!
- Soda ash (Na_2CO_3) addition optional at 2 g/L.
- Cold water rinse for 3 min.
- Sodium peroxycarbonate (Na_2CO_4) addition optional at 2 g/L.
- Citric acid (5 g/L) rinse to neutralize.

A barrel that has been infected with *Brettanomyes* yeast cannot be sterilized and should be put out of commission. Remember that handing down a *Bretty* barrel to homewinemakers is not a fair practice. Note that wood conducts heat very poorly, 30 times less than stainless steel. This insulation power means that the heat even from prolonged hot water exposure doesn't penetrate the staves deeply enough to kill all microorganisms that live in the cracks and blisters of a toasted oak barrel. Neither will the use of ozonated water provide absolute sterility. In the worst case, the Brett bloom will occur later in the barrel aging process, i.e. closer to bottling.

Ideally, the winemaking schedule should be cyclic so that re-usable barrels get emptied, cleaned and re-filled with new wine within a day of two. If this is not possible, e.g., when the wine is becoming too oaky, barrel can be stored without wine either wet or dry. To prepare for dry storage, the cleaned barrel needs to drain out completely and be allowed to dry in a clean, TCA-free, low relative humidity environment without fruit flies getting access through the open bung hole. The use of air-purifying systems in the cellar that employ UV light or photocatalytic oxidation (AiroCide®, etc.) may be recommended. Wet storage avoids the barrels from drying out too much and becoming leaky, or getting unpredictably depleted of SO_2 gas after burning of an elemental sulfur wick or direct gassing with bottled SO_2. The disadvantages of wet barrel storage are additional depletion of precious oak aroma components and increased waste water BOD load when emptied.

Following are basic barrel storage procedures for both options:

Dry:
- 1.7 to 3.4 g elemental sulfur burnt/bbl
- Every 4 weeks!

Wet:
- 1125 g tartaric acid/bbl (5 g/L)

- 45 g potassium metabisulfite/bbl (200 mg/L)
- In cold, chlorine-free water!

Both:
- Silicon bungs should be wrapped in smaller freezer bags to prevent hardening and discoloration. Alternatively, they can be replaced with water-filled freezer bags covering the bung hole.
- To bleach a barrel that contained red wine, a dose of 450 g potassium metabisulfite/bbl for three weeks may be necessary.

8.11 How do I manage my barrels?

T. J. Payette

Barrel care

Barrels can be one of the best places to store certain wines and some of the worst. This is mostly subject to what style of wine one wants to produce, how one cares for the full barrels and how one stores barrels that are empty. The best rule of thumb for many is: a full barrel is a 'happy' or 'safe' barrel. The author does want to disclose he represents Demptos Cooperage on the East Coast of the United States.

Water will be mentioned many times in this article. It is assumed the water is always chlorine free water. **Do not use chlorinated water on barrels.**

When?

Every winemaker has his own way of taking care of his barrels. In essence, there is no right way to take care of a barrel, just many wrong ways. If one ignores the timely applications of some simple processes, the result may be a certain spoilage bacteria becoming established in the porous grain of the wood that will be a cumbersome battle for the future. If we are timely and diligent in our barrel care, we save ourselves much time, work and dollars keeping many oak barrels for up to a decade or more if desired. Always visually inspect barrels when they are received from the cooperage and smell each barrel to know what each barrel smells like. This is true before each fill also.

New barrels

Most winemakers have several different ways to prepare their barrels for their first fill. This can range from no treatment (not recommended), quick rinse, head soaking to full fill. Do visually inspect and smell every barrel before filling just to know what your hard-earned juice/wine is about to go into and extract. Always remember it is better to find a leaky barrel with water, not wine or juice!

No treatment
Some winemakers will simply receive the barrel and fill the barrel up with juice or wine. In some cases the barrel will leak, so most winemakers at commercial wineries do generally broadly not recommend this. (*Not recommended*)

Quick rinse
This can be acceptable in some cases if the barrel is allowed to actually soak up the water in order to swell beyond the leaking point if that may be an issue. The author still finds risk with this process unless potentially using a form of pressure test to know that the barrel will not leak. (*Not recommended*)

Head swell
Some winemakers fill their barrels with about 10–15 gallons of water and allow them to sit on one head for 12–24 hours and then roll them over to allow the

opposite head to have contact with the water. This, in my opinion, is still better than the above solutions because the time needed for the wood to swell is taken before placing juice or wine in them.

Full fill
This involves either using cold or hot (120°F) water to completely fill a barrel about 24–48 hours prior to needing the barrel. This time can be shortened if all looks well and leaks are not discovered. Cold water can be used also. Do keep in mind not to extend the time since the water can turn smelly and certain slims/films may develop. Freshen the water or make a solution of citric acid water (pH to near 3.5 or lower) and SO_2 (near 60–70 ppm). This solution will allow longer contact time with the wood if that is desired.

Depending on the cooperage used, the author prefers to fill his barrels with fresh water at 120°F, bung solid and allow to cool overnight before emptying the water, allowing to drain completely for one hour (bung hole facing upside down) then righting, another visual and smell check, then filling.

Freshly emptied barrels
Barrels that have been freshly emptied of wine are often overlooked too long. A barrel that is empty will start to dry reasonably rapidly so we must take the bull by the horns, even though we may be tired or distracted, to man handle these empty barrels as soon as possible. As with any clean up, if one can get to the dirt before it dries the clean up goes much easier. Try to at least give the barrels a good solid rinse as soon after emptying as possible. A good solid rinse may be a reasonably high pressure rinse to loosen and rinse the sediment from the barrel. Then start any other cleaning regimens from then on or perhaps if re-filling one may be fully prepared for this action at this time. If further cleaning is needed this can also be a time to start a more serious high pressure washing regimen, ozone or any other sought after method to address a specific need. Be sure to take the time to clean the bung area of the barrel both inside and out. The bung area can be a place for solids to dry and for large numbers of bacteria to proliferate, if uncared for.

Storing of empty barrels
Most readers are perhaps scanning this article for answers just to this issue. Wide-ranging experiences have winemakers storing empty barrels in many different fashions. The author prefers to rinse a freshly emptied barrel as described above, allow the barrel to drain overnight bung downward and then re-right the barrel – with bung facing skyward.

Liquid sulfur dioxide
Fresh pure liquid sulfur dioxide can be the best choice, in my opinion, yet care must be taken when using this liquid gas. Please follow all MSDS and handling instructions. Another common sense helpful task is to only do this outdoors, upwind from the barrels and away from others. The author typically will meter

10 grams of pure liquid sulfur dioxide per 60 gallon (225 liter) barrel for the initial dose and then re-dose at 5 grams every month thereafter. If the author has ever been in doubt what may be too much pure sulfur dioxide he always leans toward more in the empty barrel because he has not seen any literature suggesting the negatives of slightly higher use. This is not to say to go higher than recommended or to become excessive. Please be careful here.

Wicks and disks
This is another form of introducing sulfur dioxide into the barrel. Most sulfur wicks contain roughly 6 grams of actual sulfur dioxide in them and yield this upon burning. This is another great way to take care of your barrels when empty. Please remember to re-burn a wick/disc in the barrel every month.

Do your absolute best to store barrels, full or empty, in the best conditions possible. One should strive to store barrels full or empty at or near 50°F, out of direct sunlight, avoiding excessive temperature swings and in a mold-free environment.

Dixie cup, Styrofoam or bung?
Some winemakers allow barrels to be stored unbunged after treatment with sulfur dioxide. These winemakers are few. Others use different methods of sealing the barrels to keep the risk of insects and small animals out of the barrels. The most popular methods are as follows.

Dixie^{TM} cups
These fit nicely in most 50 mm bung whole openings and do a fine job at sealing the barrel just enough to keep the gas inside the barrel. They do tend to pop out as some actually like to store the empty barrel bung down while empty. (*Not recommended*)

Styrofoam^{TM} cups
These are similar to the Dixie^{TM} cup above with essentially the same function. (*Not recommended*)

Bungs
The author prefers this seal to the barrel to best keep the gas in. It is also recommended to wrap these bungs in Seran^{TM} wrap to protect the bung silicone from the harsh sulfur dioxide environment. This is functional and the clear wrapping is a great way to identify empty wine barrels in your cellar. (*Recommended*)

The author has little experience with glass bungs or other forms of barrel closures for this purpose. These may need further review in your own cellar.

Types of rinse
Water
All types of rinse generally use water. Please keep in mind this should be Chlorine free and have enough pressure to rinse the back and top of the vessel in question. This can be difficult with larger barrels such as puncheons.

High pressure
Nice to have but not essential with overall good cellar practices. This will aid in tartrate removal and it has much more logical mechanical pressure removal than just typical water pressure in a winery.

Ozone
More and more popular but this process must be used with caution and in well-ventilated areas. With best cellar practices these machines should not be needed but they are a great tool to fight back when bad spoilage bacteria become troublesome.

Hot water vs. cold water
The author feels it best to rinse barrels with hot water after emptying. Water at a temperature of 170–180°F is preferred for this step and the rinse generally takes about 1–2 minutes to complete the task of mostly clear water coming from the barrel. Please note hot water would not be used with ozone.

Monthly management
Whether full or empty, each barrel needs monthly attention. Try to store full and empty barrels at or near 50°F when possible.

When full
One will need to taste and check the free sulfur dioxide and pHs of these wines monthly at a minimum. After this check, additions can be made to the wines in the barrel and then they can be topped. Topping is one of the keys to keeping a barrel safe from bacteria spoilage. The wines must be of sound chemistry to make this a successful statement and the topping wines need to be 'clean'. The author will often use wine from a tank that has been filtered and nearing bottling to know the bacteria load is greatly reduced from that vessel.

When empty
Once again we will need to visit each barrel monthly. In all cases the barrel will have been rinsed and sulfured prior. We should return to these same barrels and, retreat with sulfur dioxide by gas or wick (see above) to ensure the barrels' integrity will continue to be sound.

Where
Most work with empty barrels will happen outside on a crush pad or strung out in caves and warehouses. When possible, try to have as much space and

ventilated air as possible moving through the workspace if using liquid sulfur dioxide or wicks/discs. When possible, the author prefers to unstack the barrels, remove the bungs, look inside the barrel with a flashlight and then smell the barrels at each visit.

If the barrels must stay in place, one can easily work with them also but some of the more critical reviews of sight and smell become more cumbersome.

Rain

At certain times at certain winery locations the author likes to use Mother Nature. It is not uncommon to plan a day's barrel work around the weather. If the weather forecast is for rain the author finds advantages, with uncovered crush pad areas, to do a day's barrel work, rinse the interior and then allow the barrels to remain on racks, bung down, in the rain to get a nice soaking cleaning on the exterior as well. This can also apply in conjunction to the swelling procedure above but with barrels full of water and bung upward. Please try this experimentally first on a small batch of barrels since some wood discoloration may take place and not be visually to ones liking.

Always clean the bung opening area and when needed one can cauterize/burn that area again. This process may be needed about every 5 years or so at the maximum and a special tool is needed for this process.

Tartrate removal

Tartrate removal can be a nuisance for those that focus on it. In general it should not be a huge issue. *Do note some winemakers care to cold stabilize their wines before placing in barrel for this reason.* Most do not, however. Also note that when one looks inside the barrel one will see more tartrates because they typically 'fall out' and go to the bottom. In the case of sur lie wines in the barrel the yeast layer does a great job of protecting the bottom of the barrel from tartrate adhesion to the wood.

A high-pressure rinse may remove these tartrates effectively. A hot water rinse my help them 'flake off' and dissolve more readily.

Some winemakers use a high pH (warm water helps here) soak followed by a light citric acid soak. This can be very effective in tartrate removal. Makes sure the soda ash, the high pH solute, and citric acid, the low pH solute, dissolve completely before adding any one of them to the barrel.

Some more European-trained winemakers will insert a stainless steel chain and have the less stipend 'summer help' roll the barrels with the chain inside to knock the tartrates free – then rinse and sulfur. Be sure to devise a way to retrieve the chain from the barrel.

In most cases, however, the tartrate removal is not a huge focus for the majority of the winemakers due to practical applications.

Tools needed

Many tools for barrel care may be purchased at winery supply stores, cooperage houses and other specialty suppliers specific to these types of products. Research your needs and then contact these companies to see what they offer. In general, only a good barrel rinser, good to great water pressure (chlorine free) is needed and a way to introduce the SO_2 – wick or gas.

Glass head barrels
A great tool to be able to see inside the barrel when performing certain tasks from burning sulfur wicks, filling, rinsing, lees stirring, etc. Watching fermentation and malolactic with these glass head barrels can be fascinating beyond the other features. Every cellar should have just one of these glass head barrels to better know what is happening inside their barrels with certain specific functions.

Humidity
Humidity is undoubtedly a factor when dealing with barrels. The author prefers a less humid cellar to make sure the vacuum needed inside the barrel is fully established on barrels with wine in them. If barrels are kept full and production practices to store a few barrels empty for any length of time are employed, this can be the best way to use barrels to their fullest and best capacity. There is some 'angel's breath' evaporation loss but that is a part of the process. If a barrel is stored for less than three months empty most will have few to no issues with reswelling.

Spicing it up!
Some winemakers prefer to cold stabilize their wines before placing them in barrels to prevent tartrate build-up in the barrels. (reference above) This can be effective but most winemakers do not do this in large practice.

Burning a sulfur wick in a barrel (5–6 grams) does two things. It puts sulfur dioxide in the barrel as well as displaces oxygen with carbon dioxide. This practice may lend toward mimicking by the winemaker using a carbon dioxide flush on their barrels then using pure liquid sulfur dioxide, following. Many large wineries use liquid sulfur dioxide and this may apply to their needs best. Many large wineries also could make dry ice, on site, and this could be used as the carbon dioxide source. This may be the way of the future to help combat spoilage bacteria growth in barrels. This may well be the future established standard for proper barrel care.

Wrapping it up
Tackling the barrel care issue is a trying one but actually an easy one. There is only one way to do it right. The way it works for your cellar. There are many off branches and combinations of what has been described above. Please take from this article anything you think may help your current process and refine, for the better, what will work best for your winery and wines. As can be seen these are some must do processes but most recommendations have some variation.

In all be diligent and respectful of the barrels in your cellar and they will provide many years of service to you, your cellar and your wines.

Other helpful tips
- Alcohol is less dense than water or juice. It is not all too uncommon to fill a barrel with water to validate the barrel will not leak, only to find a barrel may develop a leak later on. These leaks can typically be fixed on sight of the winery with little effort.
- Do not fill a barrel with wine you know to be bacterially unsound. This will only start the further spread of the unwanted bacteria.
- Smelling the barrel is one of the best ways to acknowledge its condition and readiness to potentially help or harm your wines in the cellar.
- Whether full or empty, each barrel normally requires some form of monthly maintenance.
- Topping on time is critical and resulfuring, on time, is critical.
- Lower storage temperatures (50°F) can be a very useful tool and one more winemakers should try to strive to use.
- Resist the temptation to store barrels outside. There are many wood boring insects that may take fancy to this easy target leaving the winemaker with leaky barrels. Small periods of time outside may be acceptable.
- American oaks tend to need reswelling more than European woods. The author has noticed American oak may develop more ethyl acetate type aromas when stored empty due to a reaction of the wood, moisture and sulfur dioxide. This is generally not a bacterially generated ethyl acetate aroma if sound procedures are followed and not a concern – just an observation winemakers may notice in their cellars and to be aware.
- Uprights and ovals beyond the 600-liter capacity are beyond the scope of this article and care should be taken to establish contact with appropriate sources to secure proper methods of working with these wood vessels.

Acknowledgements
Based on a verbal discussion with Mr Jacques Boissenot and Mr Jacques Recht. Many thanks to Mark Heinemann and all the Demptos Cooperage team for their help.

8.12 Why are my bottles underfilled?

H. Bursen

Many wineries use gravity-fill or vacuum-assisted bottling equipment. The most common automatic style is a rotary filler which puts the bottles onto a rotating carousel. We'll concentrate on that type of machine. These machines are used for high quality production. The bottle enters the carousel and is sealed to the filler spout. By the end of the rotation, the bottle should be full.

If you are having trouble filling your bottles, the first step in successful troubleshooting requires close observation of individual spouts over several filling cycles. Are all the bottles consistently underfilled? Or are only some bottles underfilled?

Causes of low fill level

1. *Symptom*: Bottles are full at end of carousel, but are *consistently* low when spout is removed.

Cause: Filler spout enters the bottle too deeply, displacing too much wine.

Cure: Adjust the spouts so that they don't penetrate so far into the bottle.

2. *Symptom*: Bottles are *consistently* not filled all the way by end of filling cycle. This problem may have several causes:

Cause: Machine speed adjusted too high, beyond rated capacity (in bottles per minute).

Cure: Time the number of bottles filled per minute. Adjust to within specification.

Cause: Filler spouts are not open wide enough.

Cure: Adjust opening aperture to within spec.*

* *Note*: Higher viscosity wines require larger spout aperture. A given setting may work for one wine, but not another. That's why the specifications are in the form of a range. Sweet wines and botrytised wines often have high viscosity.

Foaming and *turbulence* often cause *inconsistent* filling. A particular spout may underfill one bottle, yet fill the next.

Symptom: Inconsistent filling.

Cause: Turbulence. Filler spouts open too wide, causing turbulent flow.
Cure: Adjust aperture.

Cause: Turbulence. Defective spout tip seal causing turbulent flow.
Cure: Replace worn or damaged spout tip seal.

Cause: Foaming. Air exits the bottle through a narrow opening. Foam can't escape as rapidly, slowing the filling process. Foaming has multiple causes, therefore multiple cures:

Cause: Excessive residual CO_2 in wine.
Cure: Warm wine to around 60°F and splash through inert gas before bottling.

Cure: Apply filler bowl vacuum (only some models). This *may* allow foaming to occur before wine enters the bottle.

Cause: Glucans or other foamy compounds (often in *botrytis*-infected grapes).
Cure: Lower machine speed until foaming subsides.
Cure: Treat wine with glucanase enzymes well before bottling (where allowed). Always bench-test a small volume before treating the whole batch.

Cause: Turbulence. Turbulent flow can lead to foaming in a susceptible wine.
Cure: Replace worn spout tips and/or narrow spout aperture to reduce turbulence (see 'turbulence', above).

Cause: Partially clogged filter causing high pressure differential; can cause CO_2 release, even in a wine with normal levels of residual CO_2.
Cure: Replace clogged filter.

Occasional empty bottles
Cause: Bottle is misaligned and doesn't seal to spout.
Cure: Check and adjust guides along the bottle's path as follows:

- Rails may need adjusting; 2 mm space between bottle and rails is good.
- Guide pieces may be loose, or may be misshapen.
- Star wheel must deliver bottle to place it under the spout. Check and adjust star wheel timing.
- Spout may be loosening.
- Pedestal may be loosening.

Now a word about *manual fillers* usually used by wineries whose production is limited to a few thousand cases per year. These relatively simple machines usually have a shelf which holds the bottles under the spouts. The height of the shelf is adjustable, and that's how bottle fill-height is regulated; the lower the shelf, the higher the wine in the bottle.

Dedication
This article is dedicated to my two troubleshooting heroes: Richard Turner and Rod Salisbury.

8.13 What should I consider for winery preventative maintenance

M. Sipowicz and B. Trela

Make a plan

Properly functioning winery equipment requires timely preventative maintenance. The effective preventative maintenance (PM) program will integrate a winery's equipment, machine and system manufacturers recommended maintenance procedures into a logical, useful tool that will help prevent mechanical/system failure through timely, planned maintenance actions. Such a program is only effective if it is thorough. The final PM program can comprise several smaller 'sub-plans'. Examples of this are *pre-harvest PM* and *pre-bottling PM*. To design a winery PM plan, an inventory of all winery processes performed during the year is compiled along with associated equipment, machinery and systems. These processes are then grouped into seasonal events i.e. harvest, bottling, cellar operations, etc. The next step is to identify all associated 'mission critical' equipment, machinery and systems within each seasonal event or sub-plan. Based on the manufacturer's maintenance procedures and service intervals for each system or piece of equipment, preventative maintenance is scheduled within the annual PM program. The time at which each sub-plan is implemented during the year is a very important aspect. Inspection and maintenance must be scheduled with ample lead time prior to each processing event, allowing enough time for parts ordering and equipment repair if necessary. Special lead time consideration must be given for foreign and difficult to obtain repair parts. Very successful programs will have the aforementioned sub-plans arranged chronologically within a master PM schedule. Each event will have a 'trigger date' or 'perform no later than' date. The required PM tasks should then be integrated into a winery's monthly work schedule. It is also advisable for each serviceable piece of equipment to have an affixed maintenance tag. This strategically placed service sticker or tag has lines for the service operator to sign and date stating that the piece of gear was indeed serviced and when. This finally leads to perhaps the most important part of any PM plan, or any plan in general: accountability. It is imperative that someone is ultimately responsible for the implementation of the plan. All too often it is an amorphous management style not poor lubricity that is to blame for the loss of a bearing or seal.

Considerations: often overlooked

Winery refrigeration system

System leaks are common in the winery glycol system. The addition of a food grade dye to the glycol reservoir will make leaks very apparent as soon as they develop anywhere within the system. Additionally, it is advised to simulate a heavy system work load prior to heavy use events (e.g., harvest, cold stabilization). This can be carried out by turning on refrigeration to numerous empty tanks within the cellar, simulating a full cellar of fermenting fruit. Problems with the system, plumbing and solenoids can be preemptively addressed. Missing, wet or moldy glycol line insulation often compromises the capacity of the cooling system especially at warmer cellar temperatures during harvest. Emptying strainer screens, adjustment of fluid levels/concentrations, as well as other pertinent maintenance tasks should also be performed. Investing in an electrical back-up generator that can run the cooling system's compressor and pump is crucial to any wineries operation during and after harvest when cooling capacity is essential.

Calibrate tank temperature controllers
Remove the temperature probe from the thermo-well and place it into a container of ice water, calibrating the controller to 0°C (32°F) as per unit's user guide. For better heat transfer, repack thermo-wells with food grade grease when probe is reinstalled. Thermostats often fail, so keep two extra units around to replace failing ones immediately.

Winery hoses and pumps
Poorly maintained hoses and pumps can lead to wine oxidation and microbial contamination. Cracked hoses, loosely banded and 'mushroomed' hose ends should be replaced. Leaking pump seals and 'O' rings should be replaced immediately. Follow pump manufacturer's service intervals and procedures. Develop a rack system in the winery that allows the hoses to completely drain and air dry after use. Sanitizing/rinsing them is not enough to keep mold or vinegar bacteria from growing inside the hoses when not in use. Hoses should be hung or placed straight on a slightly sloped rack to drain and dry. Avoid using hoses of excessive length (> 8 m; 25 ft), instead combine shorter ones that you can store properly without reeling them. In general, the need to use excessively long hoses to move wine or juice is an indication of poor winery layout and design.

Tank gaskets
Rubber or silicon gaskets around manholes or other tank openings are major sources of microbial contamination of musts and wines. Remove them completely even if cumbersome on a routine schedule and sanitize them with suitable clearing agents (no ozone water on rubber; no soaking silicon in SO_2 solutions!), preferably just with hot water to avoid the risk of sanitizer residues in hidden pockets of the gasket.

Lubrication
Timely lubrication of machinery/equipment accounts for a majority of winery PM operations. Most moving parts require some form of lubrication. A thorough inventory of winery equipment/systems manuals will indicate what needs to be lubricated and how frequently. Currently the properties of food grade lubricants rival those of their non-food grade counterparts. Many wineries have chosen to

use food grade lubricants even in situations where non-food grade lubricants will suffice. Where functionality is not sacrificed, such a move can significantly simplify lubrication schedules and inventories. Growth of microorganisms such as bacteria, yeast and fungi is possible within food-grade lubricants. An ever increasing number food grade lubricant manufacturers are adding antimicrobial agents to their lubrication products. Such additions are intended to prevent spoilage and fouling of the lubricant, and in some cases to prevent microbial contamination via incidental food contact.

Special considerations
Certain atmospheric conditions can accelerate winery equipment wear or cause untimely surface degradation. Cellars with high humidity can cause rapid surface decay of susceptible materials and as well jeopardize the long term integrity of unenclosed electrical devices/circuitry. This scenario is exacerbated in facilities which utilize liquid/gaseous sulfur dioxide. Elevated ozone levels in cellar atmosphere will speed degradation of certain susceptible materials as well. Ozone reacts with and eventually destroys anything containing natural rubber. O-rings, gaskets, winery hoses, or anything else that contain rubber-based elastomers are subject to ozone degradation, either directly (e.g., hose feeding ozonated water to sanitize a tank) or indirectly (e.g., off-gassed atmospheric ozone in contact with rubber wheels on the pump pushing the solution of ozonated water). Fiberglass/fiberglass resins are subject to ozone degradation as well. Other materials with poor resistance to ozone include Nitrile/Buna-N, polyamides nylon & PA, magnesium, mild as well as high-strength low-alloy (*HSLA*) steel, and zinc.

In environments such as these, it is advisable to weather guard/enclose electrical components where applicable. Moisture/acid resistant surface coatings can prevent atmospheric degradation. Choose coatings specifically designed to resist acid forming gases such as sulfur dioxide. Materials susceptible to ozone degradation should be replaced with ozone compatible materials wherever ozone is used. The supplier of your ozone generating system will have the most up to date list of materials. Finally, whenever possible, store unused equipment in less harsh conditions.

Further reading
Material Compatibility with Ozone, Ozone Solutions Inc., www.ozonesolutions.com, Hull, Iowa.

8.14 What is corrosion?

B. Trela

Corrosion is the chemical or electrochemical reaction between a material, usually a metal, and its environment that produces a deterioration of the material and its properties. Corrosion also includes the dissolution of ceramic materials and can refer to discoloration and weakening of polymers by the sun's ultraviolet light.

Corrosion usually results in a change in the oxidation state of the material, such as the loss of electrons in metals reacting with water and oxygen, acids or other strong oxidants. The rusting of iron to form iron(III) oxides is due to oxidation of the iron atoms and is a process of electrochemical corrosion. This electrolytic corrosion consists of two partial processes: an anodic (oxidation) and cathodic (reduction) reaction. Electrochemical corrosion of metallic materials typically produces oxide(s) and/or salt(s) of the original metal.

The anodic reaction of the corrosion process results in the loss of metal dissolving as ions and the generation of electrons can be represented by Equation (1), Fig. 1.

Equation 1. Corrosion of metal

$$M \rightarrow M^{n+} + ne^-$$

Where the oxidation of a metal (M) from the elemental (zero valence) state to an oxidation state of M^{n+} generates n moles of electrons (e^-). The anodic reaction may occur uniformly over a metal surface or may be localized to a specific area.

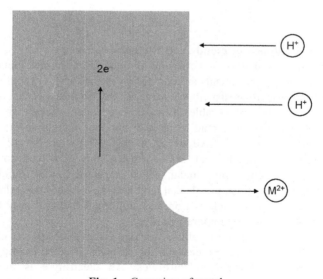

Fig. 1 Corrosion of metal.

In the absence of applied voltage, the electrons generated by the anodic reaction (Equation 1) must be balanced and are consumed by the secondary, cathodic reaction. In most situations, the cathodic reaction is either a hydrogen-evolution reaction (Equation 2) or an oxygen-reduction reaction (Equations 3 and 4).

Equation 2. Hydrogen-evolution reaction

$$2H^+ + 2e^- \rightarrow H_2$$

Equation 3. Oxygen-reduction reaction at low pH

$$O_2 + 4H^+ + 4e^- \rightarrow 2H_2O$$

Equation 4. Oxygen-reduction reaction at high pH

$$O_2 + 2H_2O + 4e^- \rightarrow 4OH^-$$

The hydrogen-evolution reaction results in the formation of hydrogen (H_2) gas when protons (H^+) combine with electrons. This reaction is often the dominant cathodic reaction in systems at low pH (higher hydrogen ion concentration). The hydrogen-evolution reaction can itself cause corrosion-related problems. Atomic hydrogen (H) may diffuse into the metal, attenuating its mechanical properties in a process called embrittlement that can result in catastrophic failure.

The oxygen-reduction reaction favored in acidic solution is represented by Equation 3, whereas alkaline solutions favor the reaction represented by Equation 4. The oxygen-reduction reaction is usually rate limited by the amount of oxygen present at the reaction site. In other words, for the reaction to occur, oxygen must be present at the metal surface and replenished for the reaction to continue.

Corrosion resistance in metals and the galvanic series

The resistance of metals to corrosion is dependent on the metal's thermodynamic properties. No metals are completely corrosion proof, but some metals are more intrinsically resistant to corrosion than others because corrosion is thermodynamically unfavorable. Corrosion resistant metals are sometimes called noble metals, these tend to be also precious metals due to their rarity. Gold is the most non-reactive of all metals. Corrosion of gold is thermodynamically un-favorable and any corrosion products tend to decompose spontaneously back into pure metal, which is why it can be found in metallic form on Earth. In contrast, metals that tend to oxidize or corrode relatively easily are called base metals, such as iron and aluminum that can only be protected from corrosion by more temporary means.

Most metals and metallic alloys corrode from exposure to moisture and oxygen in the air, but the process can be strongly affected by the presence of certain substances such as electrolytes. An electrolyte is any substance containing free ions that behaves as an electrically conductive medium.

Electrolytes commonly are materials such as acids, bases or salts that dissolve in water to give a solution that conducts an electric current. Corrosion can be concentrated locally to form a pit or crack, or it can extend across a wide area to produce general deterioration. Corrosion treatments such as passivation will increase a material's corrosion resistance.

Galvanic corrosion
The galvanic series (or electropotential series) is a list of metals and alloys arranged according to their relative corrosion potentials in a given environment. Table 1 shows various metals in order from most active to most noble.

The greater the separation between two metals listed, the greater the corrosion potential. When two metals are submerged in an electrolyte, while electrically connected, the less noble (base) metal will experience galvanic corrosion. The galvanic corrosion potential of dissimilar metals can be measured as the difference in their voltage potential in a given environment. Galvanic reaction is the principle on which batteries are based.

Because corrosion is an electrochemical process involving the flow of electric current, corrosion can be generated galvanically through contact of

Table 1 Galvanic series of metals and alloys in low oxygen content seawater

Anodic – least noble (more likely to be attacked)
  Magnesium
  Zinc
  Galvanized steel
  Aluminum
  Mild steel
  Low alloy steel
  Iron
  Cast iron
  Lead

Cathodic – most noble (less likely to be attacked)
  Tin
  Brass
  Copper
  Bronze
  Copper-nickel alloys
  Monel
  Stainless steel (410)
  Stainless steel (430)
  Stainless steel (304)
  Stainless steel (316)
  Silver
  Titanium
  Gold
  Platinum
  Graphite

dissimilar metals in an electrolyte. There are three mutually dependent conditions required for galvanic corrosion to proceed; the two metals must be widely separated on the galvanic series (Table 1), they must be in electrical contact and their surfaces must be bridged by an electrolyte. Removal of any of these three conditions will prevent galvanic corrosion. The obvious means of prevention is therefore to avoid mixed metal fabrications. Frequently this is not practical, but galvanic compatibility can be managed by finishes and plating, e.g. galvanizing (zinc plating) steel, or by removing the electrical contact, e.g. through the use of non conductive insulators, or by preventing contact with electrolytes through drainage or protection from the elements.

Galvanic compatibility is also dependent upon the relative areas of the dissimilar metals. If the area of the less noble (anodic) material is large compared to that of the more noble (cathodic), then the corrosive effect is greatly reduced, e.g. galvanized steel where the zinc acts as a barrier coating and in the event the coating is scratched and the steel is exposed, then the zinc acts as a sacrificial anode. Conversely, a large area of noble metal in contact with a small area of less noble metal will produce a high electrical current and greatly accelerate the galvanic corrosion rate of the anode. This is significant when choosing the metal of bolts or screws if dissimilar to the metal to be bolted.

The rate of corrosion is determined by the electrolyte, the difference in nobility and the relative surface areas of the respective metals. For example, if aluminum, an active material, and stainless steel, a nobler metal, were in direct contact in the presence of an electrolyte (water), the aluminum would galvanically corrode.

References and further reading

JONES, D.A. 1996. *Principles and Prevention of Corrosion*, 2nd edition. Prentice Hall, Upper Saddle River, NJ, pp. 50–52.

ROBERGE, P.R. 2006. *Corrosion Basics: An Introduction*, 2nd edition. Nace Press Book.

8.15 What is stainless steel?

B. Trela

Modern winemaking and stainless steel are practically inseparable. Today stainless steel is used almost exclusively for most harvest, storage and processing materials that come in contact with product in the winery, and in many cases is also used for esthetic effect. In order to better understand stainless steel, it might help to first understand more about iron and steel.

Iron and steel

Iron is an element that, like most metals, is not usually found in nature in its elemental state. Iron is typically found in nature as an extractable ore that is a mineral alloy with oxygen (oxides) or sulfur (sulfides). Common mineral ores include hematite (Fe_2O_3), magnetite (Fe_3O_4), and pyrite (FeS_2), the last of which is also known as fool's gold. Iron is extracted from ore through a process called smelting. Smelting removes oxygen through chemical reduction at high temperature such as in a blast furnace. Historically, smelting used coal or coke as a carbon source, the burning of which produced carbon monoxide (CO) that acted as the reducing agent to remove oxygen as CO_2. Molten iron (melting point 1535°C) readily dissolves carbon, and the process of direct smelting with coal often results in an iron alloy (pig iron or cast iron) containing too much carbon to be called steel. Subsequent processing such as with the Bessemer and Basic oxygen (Linz-Donawitz) methods introduces air or oxygen to the pig iron to oxidize the impurities, e.g. silicon, manganese, and carbon as oxides. These oxides either escape as gas or form a solid slag with the clay refractory lining or through the addition of limestone.

Traditionally, steel is an alloy consisting mostly of iron, with carbon content between 0.2% and 2.14% by weight, depending on grade. Other steel classes may replace carbon with other alloying materials if carbon is undesirable. Alloys usually have properties different from those of the component elements. Alloying one metal with other metal(s) or non metal(s) often enhances its properties. Cast iron is a generic term for alloys of iron that has higher carbon content (2–4% by weight) than steel and also contains silicon (1–3% by weight) (Walker 2002). Cast iron differs from steel in that it has a lower melting point, castability, and excellent machinability, among other attributes, but tends to be brittle. Wrought iron differs from steel in that it is mostly pure iron, containing only a very small amount of other elements, but has 1–3% by weight of slag 'fiber' inclusions that give the iron a characteristic grain. Slag, composed mainly of limestone or dolomite, is often added as a flux to absorb oxide impurities such as iron silicate (Fe_2SiO_4), iron oxide and iron phosphate ($FePO_4$) in the smelting process resulting in purer refined metal and a partially vitreous by-product sometimes referred to as cinder. Wrought iron is more rust-resistant than steel and welds more easily.

All metals are crystalline materials that have specific crystal structures that are dependent on temperature. These structures are referred to as phases. There are three main phases of iron-carbon alloys. At room temperature, the most thermodynamically stable form of iron is the body-centered cubic (BCC) structure ferrite or α-iron, a fairly soft metallic material that can dissolve only a small amount of carbon (no more than 0.021 wt% at 910°C). Above 910°C ferrite undergoes a phase transition from BCC to a face-centered cubic (FCC) structure, called austenite or γ-iron, which is a soft metallic solid solution that is non-magnetic and can dissolve considerably more carbon, up to 2.14% by weight at 1149°C. Higher concentrations of carbon in iron will produce cementite or iron carbide (Fe_3C), a very hard material that can be classified as a type of ceramic.

Stainless steel
Stainless steel is a family of alloy steels containing low carbon steel with a minimum chromium content of 10% or more by weight. The name originates from the fact that stainless steel does not stain, corrode or rust as easily as ordinary steel; however, 'stain-less' is not 'stain-proof' in all conditions. It is important to select the correct type and grade of stainless steel for a particular application. In many cases, manufacturing rooms, processing lines, equipment and machines will be subject to requirements from authorities, manufacturers or customers.

The addition of chromium gives the steel its unique stainless, corrosion-resistant properties. The chromium, when in contact with oxygen, forms a natural barrier of adherent chromium(III) oxide (Cr_2O_3), commonly called 'ceramic,' which is a 'passive film' resistant to further oxidation or rusting. This event is called passivation and is seen in other metals, such as aluminum and silver, but unlike in these metals this passive film is transparent on stainless steel. This invisible, self repairing and relatively inert film is only a few microns thick so the metal stays shiny. If damaged mechanically or chemically, the film is self-healing, meaning the layer quickly reforms, providing that oxygen is present, even if in very small amounts. This protective oxide or ceramic coating is common to most corrosion resistant materials. Similarly, anodizing is an electrolytic passivation process used to increase the thickness of the natural oxide layer on the surface of metals such as aluminum, titanium, and zinc among others. Passivation is not a useful treatment for iron or carbon steel because these metals exfoliate when oxidized, i.e. the iron(III) oxide (rust) flakes off, constantly exposing the underlying metal to corrosion.

The corrosion resistance and other useful properties of stainless steel can be enhanced by increasing the chromium content and the addition of other alloying elements such as molybdenum, nickel and nitrogen. There are more than 60 grades of stainless steel, however, the entire group can be divided into five classes (cast stainless steels, in general, are similar to the equivalent wrought alloys). Each is identified by the alloying elements which affect their microstructure and for which each is named.

Types of stainless steel
There are different types of stainless steels and they can be magnetic or non-magnetic. When nickel is added, the austenite structure of iron is stabilized and this crystal structure makes such steels non-magnetic and less brittle at low temperatures. For more hardness and strength, carbon is added. When subjected to adequate heat treatment these steels are used as razor blades, cutlery, tools, etc. Manganese preserves an austenitic structure in the steel similar to nickel, but at a lower cost, as a result, significant quantities of manganese have been used in many stainless steel compositions. The type most commonly encountered in the food and beverage industry is the Austenitic 300 series stainless steel, in particular type 304. Specialty grades of stainless steel are more likely to be used in specific instances as components of pumps (shafts) or machined parts for example based on suitability to the application and cost.

The international designation for common steel materials is 1.XXXX (e.g. 1.4301). The Society of Automotive Engineers (SAE) designates SAE steel grades. The American Iron and Steel Institute (AISI) and SAE numbers refer to the same alloy, but the AISI system used a letter prefix to denote the steelmaking process. In the SAE system, numbers are used to designate different chemical compositions: the class to which the metal belongs, the predominant alloying agent, and the average carbon content percentage are given. A four-digit number series designates carbon and alloying steels according to the types and classes. This system has been expanded, and in some cases five digits are used to designate certain alloy steels. Stainless steels always have high chromium content and are identified by a three-digit number beginning with 2, 3, 4, 5, or 6.

Stainless steels are classified by their crystalline structure:

1. Ferritic (magnetic)
- This group of ferrous alloys generally contain only chromium at between 10.5% and 27% and very little nickel, if any. Most compositions include molybdenum, and some contain aluminum or titanium. Ferritic stainless steels are highly corrosion resistant, but far less durable than austenitic grades and cannot be hardened by heat treatment.
- Typical applications are cutlery, kitchen sinks and drums for washing machines.
- Examples: EN 1.4016/AISI 430

2. Austenitic (non-magnetic)
- Austenitic stainless steels comprise over 70% of total stainless steel production. Of this, EN 1.4301/AISI 304 and EN 1.4401/AISI 316 constitute the greatest part. They contain a maximum of 0.15% carbon, a minimum of 16% chromium and sufficient nickel and/or manganese to retain an austenitic structure at all temperatures from the cryogenic region to the melting point of the alloy. Type 304 is a stainless steel that is 18% chromium and 8% nickel (18/8), one of the most versatile and widely used stainless steels, including in

winery, brewery and dairy industries. Alloys with high molybdenum content (>6%) exhibit greater resistance to chloride pitting and crevice corrosion. Nitrogen additions and higher nickel content provides better resistance to stress-corrosion cracking. These alloys are not hardenable by heat treatment but can be work hardened from cold deformation. The popularity of austenitic stainless steel is due to its corrosion resistance, weldability and shaping properties. Likewise, their high-temperature and low-temperature properties tend to be good. The austenitic series of stainless steels are more acid neutral than ferritic stainless steels which makes them the stainless steel of choice in the food services industry.
- Examples: EN 1.4301/AISI 304, EN 1.4401/AISI 316, EN 1.4547/254SMO

3. Martensitic (magnetic)
- The members of this family of stainless steels may be hardened and tempered just like alloy steels. Martensitic stainless steels are not as corrosion resistant as the other two classes, but are extremely strong and tough as well as highly machineable, and can be hardened by heat treatment. Martensitic stainless steel contains chromium (12–14%), molybdenum (0.2–1%), no nickel, and about 0.1–1% carbon. It is also known as 'series-00' steel. These alloys are magnetic.
- Typical applications are knives, motors and pump shafts.
- Examples: EN 1.4057/AISI 431

4. Austenitic-ferritic (duplex) (magnetic)
- Duplex stainless steels have two distinct microstructure phases – ferrite and austenite. The Duplex alloys have higher strength than the other austenitic or ferritic grades and greater resistance to localized corrosion, particularly pitting, crevice corrosion and stress corrosion cracking due to chlorides. They are characterized by high chromium (19–28%) and molybdenum (up to 5%) and lower nickel contents than austenitic stainless steels.
- Examples: EN 1.4462/SAF 2205, EN 1.4410/SAF 2507

Stainless steel grades
- 200 Series – austenitic chromium-nickel-manganese alloys
- 300 Series – austenitic chromium-nickel alloys
   o Type 301 – highly ductile, for formed products. Also hardens rapidly during mechanical working.
   o Type 303 – free machining version of 304 via addition of sulfur. More easily attacked by chlorinated cleaners than type 304 or 316.
   o Type 304 (EN 1.4301) – the most common; the classic 18/8 stainless steel. The most common in the winery in destemmer/crushers, presses, tanks, and bottling lines.
   o Type 316 (EN 1.4404) – the next most common; for food and surgical stainless steel uses. The addition of molybdenum to the alloy prevents specific forms of corrosion. Also known as 'marine grade' stainless steel

Winemaking equipment maintenance and troubleshooting 245

due to its increased ability to resist saltwater corrosion compared to type 304. Low carbon (<0.03%) versions of stainless steel are denoted with the letter 'L' such as type 316L, which avoid corrosion problems caused by welding. In the winery this is commonly used in the upper sections and roof of wine tanks since it is more corrosion resistant to pitting caused by sulfides than type 304. Type 316 contains more nickel (10–14%) than type 304 (8–10%) and costs about 25% more than type 304 depending on alloy constituent prices. Cost is primary reasons for the prevalence of type 304 vs. 316 in the food and wine industries.

- 400 Series – ferritic and martensitic chromium alloys
    o Type 408 – heat-resistant; poor corrosion resistance; 11% chromium, 8% nickel.
    o Type 409 – used for automobile exhausts; ferritic (iron/chromium only).
    o Type 410 – martensitic (high-strength iron/chromium).
    o Type 416 – has the highest machinability of any stainless steel.
    o Type 420 – 'Cutlery Grade' martensitic.
    o Type 430 – decorative, e.g. for automotive trim; ferritic.
    o Type 440 – a higher grade of cutlery steel, with more carbon in it, which allows for much better edge retention when the steel is heat treated properly.
- 500 Series – heat resisting chromium alloys
- 600 Series – martensitic precipitation hardening alloys
    o Type 630 – the most common precipitation hardened stainless steel, better known as 17-4; 17% chromium, 4% nickel.

References and further reading

ASHBY, M. F. and D. R. H. JONES. 1992. *Engineering Materials*, 2nd (with corrections) edn. Oxford: Pergamon Press.

BRINGAS, J. E. 2004. *Handbook of Comparative World Steel Standards*, 3rd edn, ASTM International, p. 14.

LLEWELLYN, D. T. and R. C. HUDD. 1998. *Steels: Metallurgy and Applications*, 3rd edn. Butterworth-Heinemann, Woburn, MA.

OBERG, E., F. JONES, H. RYFFEL, C. MCCAULEY and R. HEALD. 1996. *Machinery's Handbook*, 25th edn, Industrial Press Inc., New York, pp. 406, 411–412.

WALKER VII, R. 2002. The production, microstructure, and properties of wrought iron. *Journal of Chemical Education* 79 (4): 443–447.

8.16 What is corrosion in passivated materials?

B. Trela

Passivation reduces the reactivity of a chemically active metal surface by electrochemical polarization that resists further oxidation or rusting. The ability to form this passive protective surface coating gives stainless steel its corrosion resistance. In oxidizing atmospheres such as air or liquids containing oxygen the formation of the passive film is instantaneous. However, even a high-quality alloy such as stainless steel will corrode if there is an overall breakdown of the passive film on the metal or its ability to form a passivating film is hindered. When stainless steel is scratched or damaged and is in an anoxic atmosphere it lacks the ability to re-form a passivating film. Additionally, the halogens: fluorine, chlorine, bromine, iodine, and astatine (listed in order of reactivity from high to low) and their salts, such as chlorides can prevent repassivation through ion competition, or can directly penetrate the passive film and corrosively attack the metal. Halogen salts are also good electrolytes that can promote galvanic corrosion.

Halogenated cleaners such as chlorine or iodine are best kept out of the winery for additional reasons. Halogens and their salts can also attack and degrade Teflon materials such as Teflon coated or encapsulated O-rings, seals and/or similar coated materials including those in pumps. Chlorine in particular is problematic in its ability to react with phenols in wood, especially barrels, cork, cardboard and cellar structural materials to form chlorinated phenols, the primary precursor to odiferous low sensory threshold (parts per trillion) wine tainting chlorinated anisols (such as 2,4,6-trichloroanisole or TCA) that are difficult and costly to rectify.

Corrosion in metals is also affected by numerous additional conditions such as electrolyte concentration, temperature (the corrosion rate will generally double with every 10°C rise in temperature), fluid velocity, and mechanical stress in the metal subject to attack. Some additional types of corrosion commonly encountered in stainless steel are described below.

Fretting corrosion
Fretting corrosion is corrosion that is induced under load and surface motion at the contact surfaces between mating materials. It is caused by the combination of corrosion and the abrasive effects of corrosion product debris in equipment with moving or vibrating parts, particularly in pumps. The passive film on the metal surfaces is removed and prevented from permanently reforming by the repetitive abrasive rubbing action that exposes fresh, active metal to the corrosive action of the atmosphere. Heat generated through the friction will increase the corrosion. The 'passive' chrome oxide debris in this scenario now become abrasive ceramics that can also imbed in the soft elastomer of pump seals to become grinding surfaces that perpetuates the process. In the winery, fretting corrosion is typically seen on the pump shaft or sleeve under the seals

leading to a visible groove cut into the shaft or sleeve that can cause seal leakage. Seals commonly affected include the following:

- bearing grease seals or lip seals
- fluid packing seals
- rotating metal bellows seals (vibration damper)
- rubber boot seals.

Along with general deterioration in the efficiency and performance of products affected by this corrosion, it also leads to increasingly difficult to control sanitary conditions through corrosion pockets that can harbor and promote microbial contamination and further corrosion.

Fretting corrosion can lead to other problems such as surface pitting, seizing and galling of the mating surfaces. Galling is typically seen when stainless steel, titanium or aluminum parts such as nuts and bolts are forced together scraping off the oxide layers and the interface metal high points shear or lock together through increased adhesion. When disassembled, the cold welded material may be torn and pitted.

Concentrated cell or crevice corrosion

Crevice corrosion is a localized form of corrosion that occurs when liquid flow stagnates. It is typical in stagnant microenvironments such as those found in crevices formed between nut and bolt surfaces, washers, threads, clamps, under gaskets and O-rings, insulation material, and surface deposits. The mechanism of crevice corrosion can generally be described as a process initiated by changes in local chemistry within the crevice, particularly in salt water environments or where halogenated cleaners are used that cause:

1. Chloride compromise and pitting of the passivated stainless steel surface.
2. Oxygen depletion in the crevice. Deoxygenation of the crevice solution due to initial corrosion in the crevice and the diffusion rate of oxygen into the crevice is not sufficiently rapid to replace its rate of depletion due to the local cathodic process. As a result oxygen is not available to passivate the stainless steel.
3. Concentration of chloride or other aggressive electrolyte ion species in the crevice. The electropotential differential changes between the microenvironment of the crevice to the external surface. The cathodic oxygen reduction reaction cannot be sustained in the crevice, creating an anode imbalance in the concentration cell. This causes chloride ions to migrate to, and build up inside the crevice, lowering the pH in the crevice solution and enhancing galvanic corrosion.
4. Acidification in the crevice. The build-up of chlorides and decreasing pH create a crevice solution that causes a breakdown of the passive film on the alloy.
5. Depletion of passivation in the crevice. Once the passive film is compromised, corrosion continues to spread.

Pitting and crevice corrosion remain among the most common and damaging forms of corrosion in stainless alloys. It can lead to increasingly difficult to control sanitary conditions through corrosion pockets that can harbor and promote microbial contamination and further corrosion. Prevention is possible by ensuring oxygen exposure and eliminating crevices and exposure to chlorides or halogenated cleaners.

Microbial corrosion

Microbial corrosion is corrosion caused or promoted by microorganisms, usually bacteria, particularly chemoautotrophs. It can occur in aerobic or anaerobic environments. Most microbial corrosion occurs underneath microbial colonies or biofilms, and mineral and biodeposits from these colonies. These biofilms and deposits can create protective environments that become concentration cells inducing and accelerating galvanic corrosion and crevice corrosion. Inorganic and organic acids can both be produced that will initiate corrosion at the colony to metal interface. In anaerobic conditions sulfate-reducing bacteria can produce corrosive hydrogen sulfide that reacts with iron to form iron sulfide. In aerobic environments, such as headspace environments, some bacteria directly oxidize iron to iron oxides and hydroxides, while other bacteria cause biogenic sulfide corrosion through the oxidation of sulfur (as hydrogen sulfide) to produce sulfuric acid.

Further reading

JONES, D. A. 1996. *Principles and Prevention of Corrosion*, 2nd edn. Prentice Hall, Upper Saddle River, NJ, pp. 50–52.
ROBERGE, P. R. 2006. *Corrosion Basics: An Introduction*, 2nd edn. Nace Press Book.

8.17 How do I clean and protect stainless steel?

B. Trela

Passivation and pickling

Passivation is the process of making a chemically active metal surface 'passive,' i.e. reducing its reactivity. Passivation in some metals including stainless steel involves the spontaneous formation of a hard non-reactive surface film (typically an oxide) that inhibits further corrosion. This layer is usually an oxide or nitride that is a few atoms thick. In contrast to passivation, the process of pickling or chemical descaling is typically performed as a superior cleaning operation to remove tightly adherent oxide films resulting from hot-forming, heat treating, welding and other high temperature operations. Pickling is one of several pretreatment steps available for preparing an article for further processing such as passivation or electropolishing.

Among stainless steels, passivation is defined, according to ASTM A380, as 'the removal of exogenous iron or iron compounds from the surface of stainless steel by means of a chemical dissolution, most typically by a treatment with an acid solution that will remove the surface contamination, but will not significantly affect the stainless steel itself.' In addition, it also describes passivation as 'the chemical treatment of stainless steel with a mild oxidant, such as a nitric acid solution, for the purpose of enhancing the spontaneous formation of the protective passive film.'

In other words, passivation removes contaminants: free iron, iron particles, rust, metal chips, oxide scale or other nonvolatile deposits, left behind on the surface of the stainless steel by machining and fabricating. These residues might adversely affect the metallurgical, sanitary condition or stability of the surface, the mechanical operation of a part, component or system, or contaminate the process fluid. If not removed, these contaminants are potential sources of oxidation in the wine and are potential corrosion sites that could result in deterioration of the wine tank or equipment itself. Additionally, passivation facilitates the formation of the oxide film that protects the stainless steel from corrosion.

Under normal conditions of pH and oxygen concentration, passivation is spontaneous in such materials as aluminum, iron, zinc, magnesium, copper, stainless steel, titanium, and silicon. Ordinary steel can form a passivating layer in alkali environments, as rebar does in concrete. The conditions necessary for passivation are recorded in Pourbaix diagrams.

Methods of pickling and cleaning

The process of passivation typically begins with a thorough cleaning cycle. It is intended to remove oils, greases, forming compounds, lubricants, coolants, cutting fluids, fingerprints, organic and other undesirable organic residues left behind as a result of fabrication and machining processes.

General degreasing and cleaning can be accomplished by a variety of commonly accepted methods, including alkaline cleaning common in wineries.

If solvents are used, they should be non-chlorinated in order to avoid leaving residues of chloride ions in crevices and other locations where they can initiate crevice attack, pitting, and/or stress corrosion. Chlorinated cleaning compounds should also be avoided in the winery because of their ability to form chlorophenols such as 2,4,6-trichloroanisole (TCA) that can lead to wine taint.

After fabrications and degreasing, metallic surface contaminants such as iron embedded in fabrication forming and handling, weld splatter, heat tint, inclusions and other metallic particles must be removed in order to restore the inherent corrosion resistance of the stainless steel surface. This is accomplished through a variety of methods:

- Nitric and hydrofluoric acid pickling
  o 10% by vol. HNO_3, 2% by vol. HF at 49°C to 60°C for a period of time ranging from minutes to hours.
  o This is the most widely used and effective method for removing metallic surface contamination.
  o Conducted by immersion, bath, or locally using a pickling paste.
  o Removes a surface metal layer several atoms deep.
- Electropolishing
  o Metal is immersed in a temperature controlled bath of electrolyte such as concentrated oxalic, sulfuric or phosphoric acids and connected to a DC power supply with the electrolyte installed to the anode (−) side of the power source and the metal part to be electropolished attached to the cathode (+) side.
  o Achieves results similar to nitric-hydrofluoric acid pickling, but in the case of rough metal surface, protrusions dissolve faster than the recesses in a process called anodic leveling that results in a bright polished surface.
  o May be done locally to remove heat tint alongside of welds or over the whole surface.
  o Removes a surface metal layer several atoms deep.
  o Often referred to as 'reverse plating.'
- Frictional methods
  o Glass bead or walnut shell blasting.
    – Effective and non-damaging.
    – Use clean blasting material.
  o Sand blasting.
    – Effective for heavily contaminated surfaces.
    – Must be clean and not too coarse to roughen or pit the surface.
  o Stainless steel wire brushing.
  o Aluminum oxide abrasives.
    – Grinding wheels or belts.
    – Exercise care not to overheat the metal surface.
  o Steel shot – DO NOT USE.
    – Iron deposits can contaminate the surface.

Passivation considerations

Control of the passivation process involves three major variables that include time (20 min–2 hr), temperature (20–70°C), and concentration of the passivating solution (generally 20% to 50% by volume of nitric acid). Additional control of water purity, metal and other impurities in the passivating solution are critical for reproducible success. To aid in the formation of a chromic oxide film, many specifications include the use of sodium dichromate in the passivation solution, or as a post-passivation rinse.

After general cleaning to remove manufacturing residues, passivation is usually performed using variations of nitric acid based passivating solutions. New or repaired stainless steel (type 300 series) wine tanks can also be cleaned and passivated through a process involving cleaning with NaOH and citric acid followed by nitric acid passivation (typically 20% by volume HNO_3 at 49°C for 30 minutes) and a complete water rinse. This process will remove dirt, metal particles, welding generated compounds (e.g. oxides) and restore the passive film.

Pickling and passivation of new materials are often performed by the manufacturer, but any time repairs are made, equipment is damaged or contaminated, these processes could be used to restore the metal.

The use of hot concentrated acids is not to be undertaken casually. Given the potential safety hazards and environmental protection requirements associated with passivation processes, it should only be performed by trained and experienced technicians familiar with the hazards and possessing proper permits, safety equipment and environmental treatment capabilities.

Passivation verification

A common passivation verification test for austenitic stainless steels such as types 304 and 316 is the copper sulfate test. Passivated parts are swabbed (or immersed) with a copper sulfate and sulfuric acid solution, wetness is maintained for six min, and then the part is rinsed and visually examined for copper plating. Any copper (pink) color indicates the presence of free iron. This test should not be used on food processing parts due to the potential for copper leaching by the foodstuffs. An alternative passivation verification evaluation is the 5 percent salt spray test at 35°C; however, interpreting salt spray test results is somewhat subjective.

References and further reading

ASTM A 380, *Recommended Practice for Cleaning and Descaling Stainless Steel Parts, Equipment and Systems*, ASTM, 1916 Race Street, Philadelphia, PA 19103.

ASTM A967, *Standard Specification for Chemical Passivation Treatments for Stainless Steel Parts*, ASTM, 1916 Race Street, Philadelphia, PA 19103.

ASTM B912-00, *Passivation of Stainless Steels Using Electropolishing*, ASTM, 1916 Race Street, Philadelphia, PA 19103.

JONES, D.A. 1996. *Principles and Prevention of Corrosion*, 2nd edn. Prentice Hall, Upper Saddle River, NJ, pp. 50–52.
LYMAN, T. 1964. *Metals Handbook, Volume 2: Heat Treating, Cleaning, and Finishing*, 10th edn. American Society for Metals, Materials Park, OH.
TUTHILL, A. H., *Fabrication and Post Fabrication Cleanup of Stainless Steel*, NiDI literature, Item 10 004.

8.18 How often do I need to lubricate my destemmer, press, corker jaws, etc.?

M. Sipowicz and B. Trela

Lubricant is lubricant, right?
When choosing a lubricant, it is advisable to consult the operator's manual for the piece of equipment to be serviced. Types of lubricants vary tremendously, and depend on the operating environment in which it will be used. All lubricants, however, share several properties in common. The main function of all lubricants is to minimize friction and wear by separating contact surfaces. Lubricants may possess additional beneficial traits:

- Ability to remove foreign/wear particles.
- Possess sufficient stability in order to perform predictably for its designed lifetime.
- Ability to mix with chemical additives to improve performance, corrosive resistance, detergency, etc.
- Ability to cool contact surfaces and carry heat away.
- Through hydraulic action, may be drawn between moving parts.
- Forms seal against external environment protecting equipment from ingress of wine, juice, dirt, water, etc.

Pay close attention to equipment manufacturer's lubricant viscosity recommendations. Lubricant viscosity must be high enough to provide a continuous film layer between contacting surfaces, but not so high as to cause excess fluid friction (viscous shear).

What are food grade lubricants?
Food grade lubricants must perform at least as well as non-food grade lubricants in similar environments. Additional traits that may be required of food grade lubricants:

- physiologically inert
- tasteless
- odorless
- ability to resist degradation by specific food products or chemicals
- ability to resist degradation by water, steam or high temperatures
- comply with international food/health and safety regulations.

Common lubricants such as WD-40 are not food grade, but petrolatum (Vaseline) and glycerine-based personal lubricants such as K-Y Jelly are food grade. None of these is a suitable choice for winery equipment lubrication. Petroleum products such as petroleum jelly can soften rubber O-rings or gaskets and may result in premature failure. Glycerine is water-based and water-soluble, properties that cause it to either dry out or be dissolved in the product or during cleaning; it also has poor equipment lubricity, limiting its industrial application.

Categories of food grade lubricants

The US Department of Agriculture (USDA) Food Safety Inspection Service (FSIS) established three categories of food grade lubricants; H1, H2, and H3 lubricants. In 1999, The National Sanitation Foundation International (NSF) established the NSF Nonfood Compounds Registration Program, thus assuming the duties of the FSIS authorization program for nonfood compounds, which includes food grade lubricants. NSF International is the accepted international authority for food grade lubricant standards.

- **H1 lubricants** are lubricants used where there is some possibility of incidental food contact (up to 10 mg/L). The lubricants in this class must use only United States Food and Drug Administration (FDA) approved base-stocks, additives and thickeners as listed in the US Code of Federal Regulations, Title 21. H1 basestock may be either mineral or synthetic. An ever increasing number food grade lubricant manufacturers are adding antimicrobial agents to their lubrication products. Such additions are intended to prevent spoilage and fouling of the lubricant, and in some cases to prevent microbial contamination via incidental food contact.
- **H2 lubricants** are used in situations where there is no possibility of food contact. These lubricants do not have a defined list of ingredients, but must not contain heavy metals or substances which are carcinogens, mutagens, teratogens or mineral acids.
- **H3 lubricants** are referred to as edible or soluble oils. H3 lubricants are the only food grade lubricant classification approved for direct food contact. This class of lubricants is primarily used to clean and prevent rust on food contact surfaces.

What lubricants should I use in my winery?

Currently the properties of food grade lubricants rival those of their non-food grade counterparts. Many wineries have chosen to use food grade lubricants even in situations where non-food grade lubricants will suffice. Where functionality is not sacrificed, such a move can significantly simplify lubrication schedules and inventories. Due to their ability to resist breakdown along with excellent load carrying and anti-wear performance, synthetic based food grade lubricants are rapidly becoming the lubricant of choice. A lubricant based on a synthetic fluid such as poly alpha olefins (PAO) will save a winery money because it resists degradation and will therefore stay in place longer, in the long run requiring less product and less labor to apply it.

What should I know about lubricating my winery equipment?

- **Stemmer/crusher:** grease fittings located at the distal end of the pin shaft should be lubricated as per manufacturer's recommendation, and more frequently with heavy use. Grease fittings on drive gears for the pin shaft, destemming cylinder and crusher rollers are often overlooked. Many models

have sealed bearings. H1 food grade lubricant should be used on pin shaft and destemming cylinder grease nipples if present.
- **Sorting/vibrating tables, inclines, belted fruit elevators:** improper belt tracking can cause accelerated wear on the belt as well as on belt rollers and greased bearings/bushings. Most belt rollers have two grease nipples per roller. H1 food grade lubricant should be used on these at manufacturer's recommended intervals.
- **Wine press:** horizontal bladder style grape presses have a front and rear bearings which require greasing as often as every 2–3 press cycles. Additionally the central filling (axial) flange should be greased as per manufacturer's recommendations. An H1 food grade lubricant is used for these applications. Depending on press model, the filling door may require occasional lubrication to prevent sticking. Use an H3 food grade lubricant here. Lastly, drive chains, drive motors, compressor, vacuum pump, etc., require use-specific lubrication (check with manufacturer for specific recommendations). An H2 lubricant will suffice here.
- **Pumps:** waukesha-style positive displacement pumps require service 2–3 times per year. At this time it is advisable to check head gearbox and motor reduction gearbox oil levels. It is also recommended at this time to replace body as well as shaft O-rings. Head gearbox oil need not be food grade but is normally synthetic SAE 140. The motor reduction gearbox commonly uses 80w–90 gear oil. Oil in both gearboxes should be replaced approximately every 2 years. Centrifugal pumps most often will have a grease fitting(s) on the ends of the drive motor. Occasionally these fittings are hidden on the underside of the unit. See manufacturer's recommendations for unit-specific frequency of lubrication. Air pumps may or may not require oil for operation. Some units have a combination water trap/regulator/oiler which introduces a small amount of oil to internal air actuated parts during use. This oil should be H1 food grade lubricant oil.
- **Bottling equipment:** for filler, check for grease fittings on the filler bowl axis (above and below the deck). All gearbox lubrication as well as shaft bearings below deck need not be food grade type. Below deck lubrication should comply with manufacturer's recommended timelines. If the filler is heat sterilized, lubrication interval will be shorter. Once cleaned and sanitized and while the filler bowl is rotating, lubricate each pedestal with mineral oil or other comparable food grade oil. Lubrication for corker gearbox and shaft bearings below deck need not be food grade. Corker guide bars as well as corker jaws should be lubricated with light food grade oil after each cleaning cycle before each use. Internal cams and tacks should maintain a thin coating of lubricating oil at all times.

On all winery equipment, drive motors may or may not have user serviceable grease fittings or lubrication access. This does not imply that they do not require lubrication. It is advisable to contact equipment representatives for specific maintenance recommendations for all winery equipment at your facility. This information should be compiled into a winery service manual specific to your equipment. Integrate this data into your winery's annual operating procedures.

Further reading

BANNISTER, K. E. 2006. *Lubrication for Industry*. Industrial Press, Inc., New York.
BOOSER, E. R. 1997. *Tribology Data Handbook*. CRC Press, Boca Raton, FL.
HAMROCK B. J., JACOBSON, B. O. and SCHMID, S. R. 2004. *Fundamentals of Fluid Film Lubrication*. McGraw-Hill, New York.
ROBERTSON, W. S. 1984. *Lubrication in Practice*, 2nd edn. Macmillan, New York.
THE NATIONAL SANITATION FOUNDATION INTERNATIONAL. NSF Registration Guidelines. Section 5.9.1. Ann Arbor, Michigan.
US CODE OF FEDERAL REGULATIONS. Title 21 CFR 178.3750. Office of the Federal Register, Washington DC.

9

Winery microbiology and sanitation

9.1 What are the essential elements for an operational sanitation program?

K. C. Fugelsang

Microorganisms arriving on fruit, and/or already present in the winery can not only survive, but, as conditions permit, proliferate during the winemaking process. Once established, undetected/uncontrolled population 'blooms' can pose a significant threat to wine quality. Thus, proactive control becomes the cornerstone in minimizing the risk of spoilage and the single most important control point is sanitation.

As with any processing operation, the winery should develop site-specific Standard Operating Procedures (SOP) for sanitation. SOPs should include specific protocols as well as schedules that document implementation. These records become part of the overall Quality Points Program, as described by Fugelsang and Edwards (2007).

An effective sanitation program consists of at least three and, potentially, four levels of effort and vigilance. The first step is preparation of the surface to be sanitized/sterilized. Here, the goal is removal of as much of the visible (or first-level) debris as possible. Although this may be accomplished manually, the more efficient (and preferred) approach is by use of high pressure (i.e., 600 to 1200 lb/in^2 /4000 to 8000 kPa) cleaning systems. Such systems achieve the goal in a shorter time while using less water than traditional approaches. Further, they are cost-effective in terms of employee time and reducing the potential liability associated with working in confined spaces. It is recommended that workers use warm (38°C/100°F to 43°C/109°F) water for first-stage cleaning. Application of heat to organic materials adhering to a surface may only serve to cook-on (rather than wash-off) debris, thereby requiring greater effort and cost to effect removal.

Where stains/bloom remain, it may be necessary for workers to enter the tank. Fermenters and/or storage tanks are regarded as 'confined spaces,' and as such, special health and safety regulations apply. These include preliminary forced-air ventilation of the tank to reduce carbon dioxide levels followed by verification that safe levels have been reached.

Carbon dioxide and oxygen are easily measured by use of readily available and moderately priced meters and probes. Current National Institute for Occupational Safety and Health (NIOSH) definitions of 'safe' for oxygen is >19.5% (v/v) whereas for carbon dioxide the permissible exposure level (PEL) is <5000 ppm (on-line: 2008). Once cleared for entry, employee(s) must be equipped with harnesses and tethered to the outside through a side/bottom manhole where at least one worker remains on-station during the entire operation.

To prevent compromising the integrity of the protective oxide coating on stainless steel, only soft-bristle brushes should be used in cases where manual scrubbing is required. Fiber or metal 'scratch' pads or brushes should never be used for removal of tenacious deposits.

Once visible debris and film have been removed, detergents/cleaners are incorporated into the process to solubilize any remaining deposits. Strong alkalis or 'caustics,' such as NaOH (caustic soda or lye) or KOH (caustic potash) are most commonly used at this stage. Although concentrations and temperatures vary, a typical protocol calls for 1–2% (w/v) caustic dissolved in warm (approx 38°C/100°F) water followed by heating the solution to 75°C/167°F to 80°C/176°F and application to contact surface for 15–20 minutes. Increasing the concentration of caustics beyond recommended levels generally provides little additional benefit and may damage metal surfaces. Since high-pressure hot caustic is being used, it is strongly recommended that side and bottom as well as top manholes and vents be tarped to minimize exposure of workers to spray.

Where the relative amount of organic material is not heavy, mild alkalis such as sodium carbonate (soda ash) or trisodium phosphate (TSP) find application. However, regular use of sodium carbonate may contribute to precipitate ('scale') formation when prepared in hard water. Additionally, sodium *ortho*- and *meta*-silicates may be used. Although less caustic than NaOH, silicates possess better detergent properties, and are less corrosive towards equipment.

In many wineries, the caustic wash cycle described above, followed by a thorough citric acid-water rinse, fulfills the sanitation requirements for fermenters and wine storage tanks. Depending upon biological stability of the wine, it may be necessary to follow-up using of one or more sanitizing/sterilizing agents including halogens (iodine and chorine-based compounds), quaternary ammonium salts, hot water/steam, peroxides and, increasingly, ozone. Depending upon application (e.g., winery hose), acidulated sulfur dioxide is also widely used.

Iodine-based formulations ('iodophors')
Consisting of iodine, acid (commonly phosphoric acid) and nonionic surfactants, iodophores are broad-spectrum antimicrobials. However, since I_2 volatilizes at >49°C/120°F, these formulations are not compatible with hot water applications. Further, they may foam excessively and stain polyvinyl-chloride and other surfaces.

Quaternary ammonium compounds
Better known as 'QUATS', the group exists as cationic surface-active agents (surfactants), of the general structure NR_4^+. Among members of the group, nitrogen is covalently bonded to four alkyl or aromatic groups. Unlike other ionic species, QUATS are permanently charged and, thus, independent of the pH of solution. The antimicrobial properties of QUATS reside in their strong affinity for negatively charged surfaces, such as bacterial cell surfaces, where they function by disrupting cell membrane operation.

As noted, QUATS have extended activity over a broad pH range and are heat stable and noncorrosive. In addition, activity is not compromised by hard water or poorly prepared surfaces. QUATS are often employed for the control of mold on the outside surfaces of tanks as well as winery walls, floors and in drains. Here, the formulation is sprayed onto the surface and left without rinsing.

Built detergents
Specially formulated ('built') preparations containing varying proportions of caustics and sanitizers as well as surfactants, chelating agents and acids have become increasingly popular in winery applications. Such 'built' formulations fill the combined roles of detergent and sanitizing agents in classic sanitation protocol. Where wineries choose to prepare their own blends of detergents, workers should consult specific suppliers for information related to proper use of chemicals.

Peroxides
Also known as 'proxy' compounds, the group, including its best known example, hydrogen peroxide, is characterized by having at least one pair of highly reactive covalently bonded oxygen atoms (—O—O—) that breaks down to generate toxic singlet or superoxide (O_2^-) oxygen.

Sodium percarbonate is a stabilized powder containing hydrogen peroxide. The product is widely used as the active component in laundry detergent and all-fabric bleach. In the wine industry, sodium percarbonate is sold under the trade name Proxycarb™ and is widely used to treat barrels believed to be contaminated with spoilage microorganisms and/or to neutralize offensive odors that may be present.

Peroxyacetic acid (PAA or 'peracetic acid'), is also a highly reactive peroxide-based sanitizer. Like hydrogen peroxide, concentrated PAA (40% w/v) is a highly toxic oxidant which, upon dilution, is often used in barrel and bottling line applications. Compared with peroxides, PAA has better stability at application concentrations (100 to 200 mg/L), improved compatibility with hard water, and reduced foaming and corrosive properties.

Ozone

Ozone (O_3) is one of the most potent sanitizers available and, thus, is finding increased use as a replacement for other sanitizers in the food and wine industry. As an oxidant, ozone's properties reside in its inherent instability and rapid release of oxygen in transition from O_3 to O_2 upon contact with soluble or particulate oxidizable substrates. In that decomposition is accelerated by heat, the gas is best used in conjunction with cold water applications.

Whereas early uses of ozone centered on bottling line sterilization, many wineries are now using it for domestic and wastewater treatment as well as in barrel washing.

Hot water and steam

Despite a long history in the wine and food industries, both hot water (>82°C/180°F) and steam continue to be regarded as near-ideal sterilants (Wilker and Dharmadhikari 1977). In addition to their universally lethal impact on microbes, both are noncorrosive and, properly used, leave no residue.

Major concerns with the use of either/both include costs associated with generation and delivery of sufficient volumes to meet the time and temperature requirements for sterilization. Further, regular use of either/both may degrade gaskets more rapidly than other agents. Where 'hard' water is used, salt precipitates (scale) may be a concern during rinse and cool-down cycles.

The most frequent application for hot water/steam is bottling line sterilization. Hot water sterilization requires temperatures greater than 82°C/180°F for no less than 20 minutes as monitored at the point most distal to the steam source (i.e., the end of the line, fill spouts, etc.). The sterilization cycle begins when the target temperature is reached.

Acidulated sulfur dioxide

Acidulated SO_2 is an effective sanitizing agent especially for hoses and other enclosed systems. Because the antimicrobial activity of SO_2 is pH dependent, the solution is usually made up as 100 mg/L SO_2 (or 200 mg/L potassium metabisulfite) in cold water acidulated with citric acid at 3 g/L. Due to its volatility and corrosive properties as well as employee health concerns, acidulated SO_2 solutions should never be prepared in hot water. Once prepared, these should be stored and used only in well ventilated areas away from metal surfaces. Employees should also be cautioned to avoid direct contact or inhalation of SO_2 and, when not in use, SO_2 solutions should be stored in clearly identified, sealed containers to minimize volatilization.

Ultraviolet light (UVL)
Although UVL is directly effective against microbes, it has very low penetrative capabilities and even a thin film of water will serve as an effective barrier between radiation and microbes. Thus, its use is generally restricted to laboratory applications for surface sterilization. Skin and eyes must be shielded from continued exposure to UVL.

Finishing operations
Once the cleaning/sanitation cycle is completed, contact surfaces should be thoroughly rinsed, using either hot or cold water, to remove residual chemicals. Citric acid is often incorporated into the rinse to neutralize alkaline detergent residues. Kits and/or test strips are available to screen for residual detergent/sanitizer upon completion (Zoecklein *et al.* 1995).

Monitoring
Effectiveness of the sanitation program is, necessarily, an ongoing concern. Efforts to detect poorly cleaned and sanitized surfaces may range from simple subjective sensory evaluation (i.e., presence of odor and stains) to the use of surface swabs and enzymatic assays.

Sampling sites should be selected to include all points that may escape treatment and, thus, harbor microorganisms. Important among these are contact surfaces of processing equipment as well as the interior of pipelines, conveyors and tanks. Other areas subject to indirect contamination should also be reviewed. These include condensate from ceilings or equipment, aerosols, and lubricants.

References
FUGELSANG, K. C. and EDWARDS, C. G. 2007. *Wine Microbiology: Practical Applications and Procedures*, 2nd edn. Springer, New York.
ON-LINE 2008. The MSDS Hyper-Glossary. http://www.ilpi.com/msds/ref/stel.html. 6 June, 2008.
WILKER, K. L. and Dharmadhikari, M. R. 1977. Treatment of barrel wood infected with acetic acid bacteria. *Am. J. Enol. Vitic* 48: 516–520.
ZOECKLEIN, B., FUGELSANG, K. C. and GUMP, B. H. 1995. *Wine Analysis and Production*. Springer, New York.

9.2 What are biofilms and are they important in winery sanitation?

K. C. Fugelsang

Properly carried out, the sanitation process will significantly lower or eliminate microbial populations. However, survivors may remain and, over time, proliferate on contact surfaces. In some instances, the developing colony may secrete a protective exopolysaccharide sheath which confers increased resistance to chemical cleaning agents and sanitizers (Kumar and Anand 1998). Further, the physical properties of biofilms make them difficult to remove and, as noted, often refractile to routine sanitation efforts. In time, these may become large enough to trap debris as well as a reservoir for spoilage organisms entering the winemaking process.

Wine microorganisms reported to produce biofilms include *Brettanomyces bruxellensis* (Joseph et al. 2007) and *Oenococcus oeni* (Nel et al. 2002). Although most studies have focused on development of a single microbe, biofilms typically are more complex, including two or more species. In a recent study, Kawarai et al. (2007) identified *Lactobacillus casei* and *Saccharomyces cerevisiae* present as a mixed-species biofilm in rice wine brewing cultures.

Sanitation cycles using caustics (75°C/167°F for 30 minutes), followed by a citric acid rinse have proven successful in removing biofilms (Parkar et al. 2004). However, Jones (1994) observed rapid reformation after treatment with disinfectants. Other studies suggest that the use of bacteriocins may be effective in control of biofilm formation (Bauer et al. 2003).

References

BAUER, R., NEL, H. A. and DICKS, L. M. T. 2003. Pediocin PD-1 as a method to control growth of *Oenococcus oeni* in wine. *Am. J. Enology and Viticulture* 54: 86–91.

KAWARAI, T., FURUKAWA, S., OGIHARA, H. and YAMASAKI, M. 2007. Mixed-species biofilm formation by lactic acid bacteria and rice wine yeasts. *Appl. Environ. Microbiol.* 73(14): 4673–4676.

KUMAR, G. G. and ANAND, S. K. 1998. Significance of microbial biofilms in food industry: a review. *Int. J. Food Microbiol.* 42: 9–27.

JONES, M. 1994. Biofilms and the food industry. In: *Bacterial Biofilms and their Control in Medicine and Industry*, ed. Wimpenny, J., Nichols, W., Stickler, D., and Lappin-Scott, H., pp. 113–116. Cardiff: Bioline Press.

JOSEPH, C. M. L., KUMAR, G., SU, E. and BISSON, L. F. 2007. Adhesion and biofilm production by wine isolates of *Brettanomyces bruxellensis*. *Am. J. Enol. Vitic.* 58(3): 373–378.

NEL, H. A., BAUER, R., WOLFAARDT, G. M. and DICKS, L. M. T. 2002. Effect of bacteriocins PediocinPD-1, Plantaricin 423 and nisin on biofilms of *Oenococcus oeni* on stainless steel surfaces. *Am. J. Enol. Vitic* 53(3): 191–196.

PARKAR, S. G., FLINT, S. H. and BROOKS, J. D. 2004. Evaluation of the effect of cleaning regimes on biofilms of thermophilic bacilli on stainless steel. *J. Appl. Microbiol.* 96: 110–116.

9.3 What is cross-contamination?

T. J. Payette

Cross-contamination and the winery cellar

Cross-contamination is something we have all been made aware because of the food industry. We have learned that using the same plate to take food to the grill as well as to serve food from the grill, prior to proper cleaning, may result in a salmonella outbreak causing discomfort to many. Using a cutting board to prepare a meat or fish and then to cut a vegetable for a salad may result in similar reactions due to a bacterial contamination from an uncooked meat source to a product eaten raw. Once one focuses on these same principles and perhaps has training in microbiology one keenly 'tunes in' to the principles of cross-contamination.

An everyday less seen example outside the food industry is easy to illustrate. After using the rest room we wash our hands diligently only to turn around and place our hand on the door handle to exit the rest room. At that instant we have contaminated our hand with microorganisms from other individuals that have placed their hands on the same door handle prior to us. Perhaps we have all seen the individual that continues to use the paper towel to maneuver the door handle and then throwing away the paper towel. This is a microorganism conscientious individual that understands the above principle.

By definition from the Foods Standards Agency: 'Cross-contamination occurs when bacteria spread between food, surfaces or equipment.' One may go to http://cleanup.food.gov.uk/data/cross-contamination.htm# to play the video clip 'Bacteria Bites Business'.

Looking at our own cellars, we may find many areas that need work to prevent microbiological cross-contamination. This article will explore some areas that are culprits in the spread of microorganisms. Every winemaker needs to have great hygiene and sanitation in the cellar to have the control a winemaker needs to make sound wines. After reading this article, the cellar will become a different place as other sources of contamination become evident to the cellar team.

Sampling

Most wineries, with sound wines, may taste from vessel to vessel while returning the left-over portion back to the vessel sampled with no worries. This is one of the major areas that may need tightening up if the winery is experiencing problems. Winemakers sample from one vessel to another perhaps expressing discontent in one form or another. Often the discontent is directly linked to a spoilage bacteria or yeast that is growing unchecked. The winemaker will rapidly move through the cellar's containers in hopes of quickly reaching a vessel that has not progressed negatively. What some winemakers don't catch on to is that *they* are indeed the culprits to the spread of the very element with which they are not happy. When sampling a container, look inside the vessel for a potential surface film. This may indicate a spoilage position for that wine. Know the sulfur dioxide and pH of the wine. When experiencing spoilage yeast or bacteria, be sure to sanitize/sterilize the sampling instrument and wine glass. Do not return the leftover portion to the vessel and be careful to discard the leftover in an area to be cleaned. Do not dump it in the drain or on the floor for reasons to be explained later. (In clean cellars where sound wines are made it is not usually a problem to sample and pour back wines – only in unsound conditions should one avoid this habit.)

Transfers

If working with wines that are known to have some risk of infection – always move them last in the day of the transfers. Clean the hoses, pumps and other areas of wine contact between movements. An example: if 40 barrels need to be racked and one barrel is suspected to have some spoilage, rack the 39 barrels first then rack the last barrel separate to another tank – do not mix it into the blend. If the wine is to be returned to barrel give serious consideration to returning the suspect wine to the same barrel from which it was removed to 'contain' the spoilage and create a quarantine type situation. Once the movements of any suspect wines have been made, thoroughly clean the pumps and hoses before resuming to the next transfer. Be sure to clean the racking wand or any other devices that have had contact with the suspect wine. Mark the exterior of these suspect vessels so others will be aware of the problem and cross-contamination will be minimized during sampling.

Topping

Another area of great concern for cross contamination is topping. Make sure to top wines with only clean wines of the same type or variety. Often the topping wine of choice may be a recently sterile filtered dry wine that the winemaker has prepared for bottling. This wine should have a greatly reduced yeast and bacterial load. Always use clean wines for topping because the risk of spreading organisms is great here.

Blending

If potential spoilage wines have been caught early, quarantined, and arrested they may still be used in a final blend in small quantities. If the wines have been cared for and kept 'in check' they may add to the complexity of the wine. This should always be determined by a blending trial first. The trick with blending is to wait to the last possible moment to make the blend to achieve protein, color and tartrate stability of the wine prior to bottling. This should be done in stainless steel because it is easier to clean and sanitize after removing the wine from the vessel. After blending, the wine should be filtered as soon as possible to eliminate the bacterial load.

Hands and clothing

As with many food processing and preparation operations, always wash your hands frequently especially after handling wines that are suspect. Be certain not to wipe your hands on your clothing, prior to washing them, after handling suspect wines. This is the main reason that early in this article it is recommended to move suspect wines last in the day. Always wear clean clothing from day to day. Think in terms of what to do when. If starting a yeast culture for sparkling wine production and bottling a sweet wine all in the same day, use common sense to work with the bottling first and then to work with the yeast starter culture. Otherwise a major cross-contamination could occur resulting in a re-fermentation of the bottled wines.

Insects and creatures

Insects and other mobile creatures are a large source of contamination that is more difficult to regulate. For this reason a strong sanitation program is always recommended. Fruit flies and other flying insects are always a difficult battle during crush and throughout the year. Incorporate the elimination of these creatures, as best as possible, as a major part of the sanitation program. These insects fly from the drains to open vessels and handling tools such as: hoses, fittings, buckets, racking wands, pumps, filler spouts and many other areas. Every surface they land on will have a cross-contamination residue left on it from their previous landings! This was the reason under 'sampling' it is recommended best not to pour samples known to inhabit spoilage yeast or bacteria on the floor. These areas may become a food source for the insect or simply may be an area of contact for an insect or other creature.

Chemicals and dry goods

Chemicals and other dry goods are often an overlooked source of potential problems. Using scoops for one material and then using them for another before cleansing will result in a cross contamination. Soiled scoops will always transfer one material to the other as they are used. Open containers of chemicals such as acids, bentonite and sugar bags must be avoided. Cross-contamination is not always microbiological! Seal the bags after using them because contamination from insects and other potential rodents may result in problems. If not already a standard procedure, reseal all open cork bags and other dry goods materials immediately after opening and partial use. A classic example of this is one who uses a soiled scoop from citric acid and then places that same scoop into a container of metabi-sulfite. This will result in a huge sulfur dioxide aroma cloud near the incident.

Airlocks and bungs

Airlocks and bungs need to be thoroughly cleaned after each use. Airlocks are exposed to moisture and liquids. This moisture will support bacterial and yeast growth, which must be eliminated before placing them on another vessel. Since containers may pull a vacuum during a cool down in the cellar after fermentation and draw some of the water into the container, clean them thoroughly before storage and before use. When storing airlocks be sure to blow out any water and allow them to air dry. Bungs are similar. When working with barrels remove the bungs and clean them with a cleaning solution. Rinse them in a low pH water solution to rinse and neutralize the cleaning solution and then replace them on the vessel. If possible it is best to have a large number of extra clean bungs available to use with the current day's barrel work. If so – one can collect the bungs off the barrels for that day's work and soak them in the cleaning solution. Clean and rinse them at your leisure after the day's work. Allow them to dry and they will be ready for the next day of barrel work.

Pomace
Remove all pomace from the winery as soon as possible. It is a food source for yeast and spoilage bacteria. Try to take the pomace as far from the winery as possible and consider treating it with copious amounts of hydrated lime to elevate the pH and to keep odors in check. This elevation in pH will prevent lower pH bacteria from growing and result in safer pomace as far as cross-contamination is concerned. Birds, insects and animals may visit this pomace pile before traveling to other areas, perhaps near or in your winery, carrying spoilage microbes with them.

Filter pads and DE
Removing filter pads from a filter and placing them in an indoor trash receptacle that is emptied only once a week has never made microbiological sense. Instead remove them as rapidly as possible from the cellar and get them to a trash receptacle outside off the property to avoid spoilage yeast from growing and being transferred to other areas in or near your winery. Not only are they growing unwanted microbes – but also left long enough they will become very pungent! Diatomaceous earth should be treated the same way or disposed of properly for bacteria growth reasons.

Tanks
Clean the wine tanks just after emptying. Once emptied the vessel will be open for insects to fly and move about freely inside the vessel so it should be cleaned. If residuals of wine are left in the tank they will spoil and become cross-contamination sources.

Summary
The above examples are just some areas to consider. Each winery cellar is different and each cellar has unique areas that need attention with regards to the above practices. Take some time to walk around the cellar and out on the crush

pad to explore possible areas to tighten up the sanitation regime to minimize and eliminate cross-contamination sources from the cellar.

It should be the desire of every winemaker to have and keep a spoilage bacteria-free cellar. Wines are easy to make and to keep in a healthy condition. If the wines are kept free of spoilage conditions the workload is less. Once spoilage conditions exist, the winemaker's efforts are complicated and more time and effort are needed to focus on extreme sanitation measures. Every winemaker should employ good winemaking practices to avoid such situations, which are easily avoided with proper cellar management.

Cross contamination is the number one reason for wine spoilage, as the microbe *has* to come into your winery from one source or another to begin to grow.

You will find your winery a different place after you review your cellar and identify sources of cross-contamination. The wines will improve as a result of your diligence to remove cross-contamination sources, once identified.

9.4 What are viable but non-culturable organisms and should I be concerned with them during winemaking?

J. P. Osborne

The term 'viable but non-culturable' (VBNC) refers to a state some microorganisms can exist in where they fail to grow on microbiological media but yet display low levels of metabolic activity. There are a number of microorganisms that are able to exist in this state and there is recent evidence that some microorganisms found in wine are able to enter the VBNC state. In particular, *Acetobacter aceti*, *Brettanomyces bruxellensis*, *Candida stellata*, *Lactobacillus plantarum*, and *Saccharomyces cerevisiae* are believed to be able to enter a VBNC state. This is important from a wine production stand point because traditionally wine microbial analysis is performed by colony counting using suitable media and conditions for incubation. However, microorganisms in a VBNC state will not grow which may lead to the false conclusion that there are no particular microorganisms in the wine. For example, you may plate a wine sample on selective '*Brettanomyces*' media and observe no growth indicating that *Brettanomyces* is not present. However, because *Brettanomyces* is capable of existing in a VBNC state there may in fact be viable *Brettanomyces* cells in your wine.

The VBNC state is typically induced due to stressful conditions such as extremes in pH, temperature, O_2 concentration, or the presence of antimicrobials such as SO_2. When conditions become more favorable, cells may begin to grow again. In a hostile environment such as wine the combination of ethanol, low pH, limited O_2, and SO_2 may cause microorganisms to enter the VBNC state. Recent research has pointed to the lack of oxygen in wine causing *Acetobacter aceti* to enter a VBNC state. The bacteria may later recover when oxygen is introduced into the wine. From a practical point of view, this may mean that although during aging in barrel the wine will contain very few viable *Acetobacter* cells (due to the anaerobic environment) it may contain many *Acetobacter* cells in a VBNC state. When oxygen is introduced through racking these cells may be able to grow again and cause spoilage of the wine. Other evidence also points to the role of SO_2 in causing microorganisms to enter the VBNC state.

So how do you account for microorganisms that are capable of existing in the VBNC state? Being aware of the fact that certain spoilage microbes can be undetectable by standard culturing methods is a good start. This will prevent you from coming to false conclusions regarding the microbial populations in your wine. By continuing to follow sound practices to control microbial growth in your wine (SO_2 management, pH management, rigorous sanitation practices) you will continue to control spoilage microbes be they actively growing or VBNC. There are also some new methodologies that testing labs are now adopting that allow the analysis of wines for VBNC cells. These methodologies include the use of molecular biology tools (PCR based techniques), as well as epifluorescence microscopy. Utilizing analyses that can tell you the populations of all the cells in you wine, both growing and VBNC, will aid you in decisions regarding the control of microbial growth.

In summary, viable but non-culturable cells exist in a state where they are unable to grow on microbiological media but still have some metabolic activity. These cells are able to be resuscitated under favorable conditions and examples of microorganisms in wine that can enter this state include *Acetobacter aceti*, *Brettanomyces bruxellensis*, and *Lactobacillus plantarum*. Failure to detect these organisms by traditional culturing methods may lead to false conclusions regarding their presence in the wine. New methodologies are becoming more readily available that allow the detection of microorganisms in the VBNC state and include PCR based methods and epifluorescence microscopy.

Further reading

DU TOIT, W. J., PRETORIUS, I. S. and LONVAUD-FUNEL, A. 2005. 'The effect of sulphur dioxide and oxygen on a strain of *Acetobacter pasteurianus* and a strain of *Brettanomyces bruxellensis* isolated from wine', *J. Appl. Microbiol.* 98, 862–871.

FUGELSANG, K. C. and EDWARDS, C. G. 2007. *Wine Microbiology: Practical Applications and Procedures*, 2nd edn. New York, Springer Science and Business Media.

MILLET, V. and LONVAUD-FUNEL, A. 2000. 'The viable but non-culturable state of wine micro-organisms during storage', *Lett. Appl. Microbiol.* 30, 136–141.

9.5 I am buying a microscope for the winery. What features should I look for and how can I utilize it best in the winery?

J. P. Osborne

A microscope can be a very valuable addition to a winery laboratory. It can allow a winemaker to quickly monitor the progress of the alcoholic and malolactic fermentation as well as tentatively determine the source of microbiological problems. However, without proper training in the use and maintenance of a microscope this piece of equipment will be of little use. In addition, the type of microscope that is being used may also limit its effectiveness in the winery setting.

There are a number of different types of microscopes with the basic compound microscope being the most widely used in wineries. The most basic compound microscope, such as a brightfield microscope, requires staining of the sample in order to see microbial cells. Light passes through the slide and if the sample is not stained there is usually insufficient contrast to allow visualization to occur. Typically a Gram stain is used to visualize bacteria and methylene blue is used to stain yeast cells. However, staining can distort cell structure and/or cause death of the cells and may impact the ability to determine cell shape which can be a critical factor in identifying different microorganisms.

An alternative method to obtain contrast is a technique called phase contrast. This technique makes use of the fact that light passing through the cell is diffracted differently compared to light passing through the surrounding medium. Phase-contrast equipped microscopes enhance these differences and allow visualization of the cell. The major advantage in the use of a phase-contrast microscope is that it allows you to visualize live cells and determine cell shape and size. In addition, there is no need for the tedious and time-consuming step of staining that is required for visualization using a brightfield microscope. This is why it is recommended that a winery purchase a phase-contrast microscope rather than just a basic light microscope. The phase-contrast microscope will be more expensive but will also be much more useful and effective given the role it will have in the winery.

An additional feature that is recommended is high quality objective lenses. This would include a $10\times$, $40\times$, and $100\times$ (oil immersion) lens. The eyepiece lens of most microscopes is typically $10\times$ meaning you can view images at $10 \times 10 = 100\times$, $10 \times 40 = 400\times$, or $10 \times 100 = 1000\times$ magnification. While use of the $40\times$ objective lens is usually sufficient to visualize yeast cells, the much smaller bacterial cells need to be viewed using the $100\times$ oil immersion lens. This is the only lens where oil immersion is used. Special immersion oil is placed between the lens and the cover slip which yields higher resolution images during viewing using the $100\times$ lens. The need for high quality lenses, especially the $100\times$ oil immersion lens, means that you should be careful buying used microscopes or poor quality microscopes produced by obscure brands. This may initially save you money but will cause you a lot of frustration when you are unable to clearly view bacterial cells under $1000\times$ magnification.

The microscope has a multitude of uses during the winemaking process beginning with when grapes are received at the winery. Must samples can be inspected for populations of undesirable microorganisms such as *Lactobacillus* and non-*Saccharomyces* yeast allowing appropriate actions to be taken such as SO_2 or lysozyme addition. Both alcoholic fermentation and malolactic fermentation can be monitored and the microscope can also be used to check starter culture yeast viability. If MLF is problematic you can look for the presence of spoilage microorganisms such as *Brettanomyces*, *Lactobacillus*, or *Pediococcus*. If these microbes have begun to grow in your wine you can take steps to prevent their growth at an early stage. Finally, you perform some tentative identification of these microorganisms you find in your wine. In this role the microscope can be particularly useful in conjunction with a book containing microscope photos of wine microorganisms (see further reading section) or, if you have a camera mounted on your microscope, you can build your own photo record of microbes for comparison.

In summary, a phase-contrast microscope can be a powerful tool for monitoring microbial populations during winemaking. While more expensive than a brightfield light microscope, the phase-contrast microscope will allow you to quickly visualize live cells and give you real time data regarding what is present in your wine. To best utilize your microscope in the winery, it is advised that you purchase one with high quality optics including a 100× oil immersion lens and that you are trained in its use.

Further reading
EDWARDS, C. G. 2005. *Illustrated Guide to Microbes and Sediments in Wine, Beer and Juice*. Lancaster, PA, Cadmus/Science Press.
LUTHI, H. and VETSCH, U. 1981. *Practical Microscope Evaluation of Wines and Fruit Juices*. Schwäbisch Hall, Germany, Heller Chemie- und Verwaltungsgesellschaft.

9.6 Should I use or continue to use bleach or chlorinated cleaners in my sanitation program?

K. C. Fugelsang

Chlorine, in its active form (hypochlorous acid), is a powerful oxidant and antimicrobial agent. Historically, chlorine and chlorine-based sanitizers were widely used for sanitation in the wine and food industries. Although relatively inexpensive, their long-term use may damage stainless steel and other metal surfaces as well as pose a potential health risk to workers. Given these concerns and their potential involvement in environmental chloroanisole formation, use of chlorine and/or chlorinebased sanitizers in the winery is not recommended.

9.7 We use chlorine bleach to clean the floors and walls of the crush area and cellar once a season, and rinse real well afterwards. Is this advisable?

C. Butzke

No, it's not. Cleaning products that contain hypochlorite (OCl⁻) should not be used anywhere near the winery, especially the entire production area and the hospitality areas, specifically the tasting room.

Presence of chlorine is one of the two major contributors to the production of 2,4,6-trichloroanisole (TCA), the impact components for the moldy, musty *cork taint*. Its sensory threshold is one of the lowest in nature at around 1 to 5 ng/L. The second requirement for TCA formation is the presence of molds. They are a common sight even in water-tight caves and cellars due to frequent rinsing of tanks and floors, and the desirably high relative humidity (80+%) in barrel rooms which minimizes evaporative losses of wine. Chlorinated and mold-methylated phenolics from materials such as wood or cork bark, are known as *chloroanisoles*, their equally potent bromine analogues, *bromoanisoles*.

Dirty floor drains in particular can become a potential source for TCA formation in the winery as they combine chlorine residues from rinses with the rich microbial activity needed for its formation.

If TCA is subsequently present in the cellar air, it can get introduced into the wine when barrels or tanks are emptied and re-filled. The tiny amounts of TCA that it takes to spoil a wine lot correspond to equally small residues of chlorine from sanitizing operations. TCA is also easily absorbed by corks stored in the bottling line hopper, open bags of bentonite or filter pads, suggesting that proper and separated storage of all processing aids is pertinent.

Unfortunately, it is not always easy to immediately recognize that a product contains hypochlorite. Look closely at the ingredient list in detergents for (tasting glass) dishwashers, kitchen and bathroom cleaners, disinfecting wipes, anti-allergen and sanitizing sprays, and watch out for novel fabrics and textiles that were treated with proprietary coating techniques that bind hypochlorite and prolong the presence of chlorine bleach. Because it is easily inactivated on contact with organic matter, chlorine often just bleaches the dirt without removing it, while leaving a 'clean' (only by association) smell behind.

In addition to eliminating hypochlorite-based cleaning products, wineries should not use chlorinated municipal water for processing grapes or wine, e.g. when re-hydrating yeast or ML bacteria, rinsing destemmer-crushers, tanks, or hoses, etc. If not, other options exist, the water needs to be pre-treated with high capacity in-line carbon filters that are maintained on a very regular base and exchanged frequently.

In recent years, the use of chlorine dioxide (ClO_2) has been introduced to sterilize containers in the food industry. So far, there is not enough evidence that the purity or use of ClO_2 could not contribute traces of hypochlorite that are minute enough to produce relevant amounts of TCA in the winery.

9.8 What should be used instead of chlorine bleach to clean and sanitize the winery?

C. Butzke

> *Winemaking is 10% inspiration and 90% sanitation.*
> (T. A. Edison, modified)

The winemaker has some options, though:

- *Mechanical cleaning* – hands-on scrubbing – is a pre-requisite to any sanitation regimen and the best way to avoid the use of significant quantities of harsh chemicals in the winery. Do not confuse sanitation with cleaning, as they are two separate beasts, and one will not succeed without the other. Beware of proprietary mixtures of different cleaning and sanitizing agents as their exact composition may be difficult to discern.
- *Hot* water (> 82°C = 180°F) is the most effective, cheapest and greenest cleaner and sanitizer in the winery. The solubility of potassium bitartrate is ten times higher in hot water than in cold. Note that residual protein from yeast lees, etc., must be rinsed away first with cold water, as heat-denatured proteins will stick to any surface.
- *Per(oxy)carbonate*-based (B-Brite™, Oxiclean™ Free, etc.) products at 4 g/L (0.4%) dissolved in hot water quickly remove tartrate residues because of their basic pH. Due to the release of hydrogen peroxide they have good bleaching power, but their antimicrobial activity is uncertain.
- *Iodine* (ZEP-I-Dine™, etc.; 0.7 mL of 1.75%-I iodophor/L = 12.5 mg/L iodine) kills microorganisms but is odorous as it is chemically similar to chlorine and bromine but apparently without forming more potent iodoanisoles. Crush equipment and tanks sprayed with iodophors (iodine plus non-ionic surfactants) resist mold growth but must be carefully rinsed with clean water to avoid any residual getting into the wine or must.
- *Per(oxy)acetic acid*, another oxidizer, (200 mg/L = 0.02%) sanitizes well but may not be every winemaker's choice because of its strong acetic acid (vinegar) smell. It shall be noted that the correct use of peracetic acid or even distilled vinegar as a sanitizing agent in the winery is perfectly fine does *not* contribute to a volatile acidity (V.A.) problem in wine even though its smell may be suggesting it.
- *Potassium metabisulfite* solutions (865 to 1730 mg/L = 500 to 1000 mg/L SO_2) must be acidified (with 5 g/L citric acid) to release any molecular SO_2 which creates a odor and breathing nuisance for the winemaker and potentially asthmatic visitors touring the winery. Their use for equipment rinses is discouraged, but they may be used to store emptied barrels. Strong SO_2 solutions also irreversibly harden silicon, rubber and plastic parts such as bungs and press membranes/bladders, and corrodes 304 (EN 1.4301) stainless steel. Note that it is illegal to use the much less expensive *sodium* metabisulfite as a straight winemaking ingredient.

- *Citric acid* solutions (5 to 12 g/L) work well to prep filter pads and rinse hoses and barrels. A 7% solution can also be used to passivate a new stainless steel tanks, or remove rust stains from 'stainless' steel. However, while biodegradable, is considered the main contributor to an increased biological or chemical oxygen demand (BOD5/COD) of winery waste water which may be costly if drained into municipal sewer systems.
- *Quaternary ammonium compounds* (Quats) at 200 mg/L (= 0.02%) are frequently used as antimicrobial agents in the food and beverage industries but their residues are hard to assess as they are colorless and do not smell. Like chlorine, they quickly lose their efficacy in the presence of large amounts of organic matter, so manual pre-cleaning of soiled equipment and biofilms in tanks is essential. Given the hygienic nature of wine, there should be little need in a winery to apply them. It is good to know that no human pathogen can grow in wine!
- *Trisodium phosphate* (TSP) at 4 g/L helps emulsifying dirt on equipment or floors but an increase in the phosphate load of the winery effluent may cause additional algae growth in your irrigation pond or wherever your waste water is going.
- *Ozonated water* readily kills molds, yeast and bacteria but has little cleaning power and is very hazardous to the cellar staff's health even at concentration that cannot be detected by nose. In bottling lines, it may not reach the dead spots that convective heat – via hot water or steam – can sanitize. Ozone (O_3) is destructive against rubber and will damage gaskets and seals (and your winery truck's tires) unless they are specifically ozone-proof. The liberal use of ozone and ozonated water in the winery remains troubling and often substitutes for proper cleaning procedures.

It would be wise for a winery for both economic and environmental reasons to invest in a cleaning-in-place (CIP) system that has been the sanitation standard in related industries such as dairy and brewing for decades. CIP systems – commercially readily available to wineries – use concentrated cleaning solutions that are kept separate and are recycled for multiple uses throughout the season.

9.9 What is environmental TCA?

K. C. Fugelsang

In recent years, chlorine and chlorine-based products, have been increasingly implicated in microbial formation of chloroanisoles. Causative species range from mold (Kaminski et al. 1974) to actinomycetes (Silva Pereira et al. 2000), cyanobacteria (Darriet et al. 2001) and algae. Chloroanisoles arise from microbial methylation of chlorophenols which may be produced by reaction of the halogen and the lignin component of wood (Pena-Neira et al. 2000) or from biocides used in treating wood. Whereas chlorophenols are fungicidal, the corresponding anisole is relatively benign but, unfortunately, sensorially offensive at vanishingly low concentrations. For example, the detection threshold for 2,4,6-trichloroanisole in white wines may be 1–2 ng/L (parts-per-trillion), while in reds and dessert wines it is higher (4–5+ ng/L). By comparison, the detection threshold for 2,4,6-trichlorophenol may be an order of magnitude higher.

Other polychlorophenol biocides such as tetra- and pentachlorophenol, have a long history of use as wood preservatives worldwide. Thus, it is not surprising that they are commonly found in foods and packaging as well as domestic water. These may also serve as substrate for microbial formation of choloroanisoles.

Chloroanisoles are readily volatilized and carried via air circulation to distant locales within the winery where they may enter the winemaking process either directly or via absorption into fining agents and packaging materials. In this regard, the winery drainage system may play a significant role. Drains constructed without vapor traps ('T-Drains') are unexpected, but important vehicles for spreading contamination.

Increasingly, 2,4,6-tribromoanisole (TBA), a musty-smelling metabolite with sensory thresholds similar to that of TCA is also being reported in wine. Tribromophenol (the precursor of TBA) is a flame retardant and fungicide also widely used to treat wood and wood products.

References

DARRIET, P., LAMY, S., LA GUERCHE, S., PONS, M., DUBOURDIEU, D., BLANCARD, D., STELIOPOULOS, P. and MOSANDI, A. 2001. Stereodifferentiation of geosmin in wine. *Eur. Food Res. Technol.* 213: 122–135.

KAMINSKI, E., STAWICKI, S. and WASOWICZ, E. 1974. Volatile flavor compounds produced by molds of *Aspergillus, Penicillium* and *Fungi imperfecti. Appl. Microbiol.* 27: 1001–1004.

PENA-NEIRA, A., FERNANDEZ, DE SIMON, B., GARCIA-VALLEJO, M. C., HERNANDEZ, T., CADAHIA, E. and SUAREZ, J. A. 2000. Presence of cork-taint responsible compounds in wines and their cork stoppers. *Eur. Food Res. Tech.* 211: 257–261.

SILVA PEREIRA, C., FIGUEIREDO MARQUES, J. J. and SAN ROMAO, M. V. 2000. Cork taint in wine: scientific knowledge and public perception – a critical review. *CRC Crit. Rev. Microbiology* 26: 147–162.

9.10 What is ozone and how is it used in the winery?

K. C. Fugelsang

Ozone (O_3) is the unstable allotrope of oxygen which undergoes rapid decomposition to O_2. Because of its short half-life (approximately 30 min), ozone cannot be stored and must be generated on demand. This is accomplished by use of equipment that exposes a stream of dry air to either ultraviolet light (185 nm) or electrical discharge.

The most frequent application for ozone is clean-in-place (CIP) operations such as the bottling line or for treating in-house water for off-odors and/or discoloration. Given its limited solubility and potential for out-gassing, O_3 should not be used to clean tanks or in other applications where volatility is encouraged.

Since O_3 rapidly degrades in warm (>35°C/95°F) water, its use is limited to cold water systems. There is some evidence that ozone may accelerate deterioration of soft plastics and rubber requiring more rigorous preventative maintenance.

Despite its efficacy and the need for special equipment to generate and deliver the gas, ozone has advantages, compared with other sanitizing agents, that may justify the additional cost. As an oxidant, ozone is effective against all viable cells. Further, since the primary site of attack is the cell coat and membrane, development of resistant strains is, compared with chemical preservatives such as SO_2, unlikely. Secondly, within a short time or, upon reaction with susceptible substrates, ozone breaks down to molecular oxygen leaving no residue. It is relatively non-corrosive towards stainless steel when compared to chemical sanitizers and is less expensive to generate than hot water/steam.

From a health and safety point of view, ozone is a recognized hazardous substance. Winery staff should be trained in its chemical and physical properties and safe use. Long- and short-term exposure limits, as set by the Occupational Safety and Health Administration (OSHA), should be known and provided to cellar workers (online: www.OSHA.gov). Currently, the maximum continuous exposure over an 8 hour period is 0.1 mg/L. Short-term exposure is 0.2 mg/L for 10 min (Khadre *et al.* 2001). Where O_3 is used, appropriate monitors should be on-site and operational. These may include hand-held and wall-mounted units as well as personal ozone-specific badges worn by personnel.

Reference

KHADRE, M. A., YOUSEF, A. E. and KIM, J.-G. 2001. Microbiological aspects of ozone applications in food – a review. *J. Food Sci.* 66: 1242–1252.

9.11 How do I clean my wine tanks?

T. J. Payette

Cleaning in the winery is one of the most important tasks the winemaker has the most control over in the cellar. It is often said, and very nearly true, winemaking is 95% cleaning. Data are shy when it comes to how to clean certain parts of the operation; yet, here is a step-by-step process of how to clean a stainless steel wine tank in the cellar. Please keep in mind every cellar visited may have some conditions that may need to have this plan altered. If developing a HACCP (Hazard Analysis and Critical Control Points plan) like plan – use the data below to build on the correct procedure for your specific winery's needs.

Chemistry

There is some chemistry to cleaning a wine tank that will be addressed briefly to have an understanding of what one would like to achieve. Simply put, one must have physical cleanliness first. This is the removal of all solid particles from the tank's interior surface(s). Examples of these items may be seeds, skins, spent yeast, bentonite and so on. This may not include tartrate removal because this can be assisted chemically if desired. Once the solids are removed, the tank cleaning person will use a high pH cleaning material to remove the tartrates and to clean the surface of the stainless steel. This high pH will not only remove tartrates but also kill and eliminate a broad range of wine spoilage micro-organisms. Once this high pH operation is completed, the operator will always come back with a light citric acid rinse to neutralize the high pH cleaner and to have some killing power due to this solution's low pH value.

Items needed

All safety material to include but not be limited to:

- Safety goggles
- Rubber gloves
- Rubber boots
- Hat and/or rain gear
- Procedure
- Eyewash station or portable eyewash
- A light citric and water solution (2 tbsp per gallon).

Other items needed will include:

- Pump that will handle hot water and the chemicals desired to be used
- Hoses that are food grade and will stand up to heat and all chemicals used
- pH meter (optional but the winery really should have one anyway)
- High pH cleaner – such as soda ash
- Low pH cleaner – such as citric acid
- Material safety data sheets (MSDS) – on all chemicals used

- Flashlight(s)
- Cover for the tank such as a shower curtain, towel, bed sheet
- Distribution system such as a spray ball.

Preparation

If the winery's tanks are equipped with automatic solenoids on the tanks, be sure to override the chilling system or to generally isolate the tank to be cleaned, so the chilling system will not engage to cool the hot water that will be added to the tank. Overriding the system may be performed by moving the temperature dial setting all the way up so the solenoid will not engage, thus preventing cooling from circulating through the tank jackets. Shutting the system down may be achieved by simply shutting a valve on the supply side of the cooling system, once again disrupting cooling from entering the jackets.

Allow any ice to fall off the exterior/interior of the wine tank jacket so that it will not fall on the operator, other staff or persons or any equipment used to clean the tanks.

Rinse the tank and physically remove all of the solids possible.

Procedure

Once the tank is free of all solids, the chilling is turned off and the tank has de-iced, if applicable, one may start the cleaning process.

1. Apply all safety gear necessary to be safe while doing the tasks described. This is an internal winery decision that the winery will need to address.
2. Disassemble the tank of all valves, fittings and gaskets that may be easy to remove. Rinse these parts with fresh water first then soak them in a high pH solution to help with the cleaning process. Remove the doors unless these are needed to prevent splashing of the chemical cleaning solutions outside of the tank. In this case, close the doors loosely to allow the cleaning solution to cover all parts of the door.
3. Take a brush and clean all of these orifices thoroughly. Inspect them to make sure they are free of solid debris. Be certain if any threads exist, a microbial hazard in itself, that these threads are cleaned using a brush or a toothbrush. Be careful since many tanks have sharp threads that will lacerate one's skin easily. Use a brush to remove any other hardened dirty areas on the exterior of the tank.
4. Inspect the tank visually to see if any solids remain in the tank and rinse them from the tank.
5. Set up the pump with hoses in a strategic area that will not interfere with any part of the tank cleaning process. This area could include away from the front of the tank should a ladder need to be placed in that workspace or away from an opening in the tank where water and chemicals could splash/slosh out onto this piece of electrical equipment.

6. Fill the tank with enough water that one may be able to circulate the water from the bottom port/valve of the tank to the top of the tank with ample extra so that the water will not deplete itself. This amount could be near 120 gallons for a 3000-gallon tank or less depending on the tank's configuration. If the tank has a conical bottom one may need to avoid a vortex. To combat a vortex, one may place a 5-gallon food grade plastic or stainless bucket or two into the tank. These will break the development of the vortex. Be careful they do not clog the outlet of the tank supplying the pump for the circulation efforts. Hot water is recommended but not an absolute.
7. Attach the suction side of the circulation setup to the bottom valve with a hose.
8. Attach another set of hose to go to the top of the tank. This piece should be placed where it will strategically spray water back toward the top of the tank to give maximum distribution of the water and/or cleaning solution.
9. Cover the top opening with a sheet or towel to prevent splashing of the cleaning water outside of the tank's opening.
10. Open the bottom valve and allow the water to circulate to prove to the operator that this action will work as desired. Look for splashing hazards one may want to avoid if this solution were to contain a cleaning chemical. Always play with water first!
11. Once comfortable with the mechanical portion of this process and the operator feels comfortable, one may turn the pump off.
12. Open the side door and make a cleaning solution to clean the tank. This is very dependent upon the size of the tank and the amount of potential tartrates that may be present. To make the cleaning solution, always dissolve the powdered cleaning solution in a bucket before placing into the tank. *(This is performed to make sure no caking of the solution may happen. The solution should be fully and carefully made into a liquid.)*
13. As soon as the cleaning solution is made in the bucket, be sure to add the solution to the wine tank, shut the side door and start the circulation. Step back from the tank just in case the cleaning solution should want to splash; yet, be able to operate the pump to shut the operation down if needed. Observe the operation from a distance and listen to make sure the process is working as designed. Look for open valves to show signs of the cleaning solution and any other areas.
14. Continue to monitor the process from a distance and always keep your ears on the operation. The sound of a tank cleaning can be just as important as visually watching the operation.
15. One can let this process go on for 20–30 minutes or more depending on other operations in the cellar. The author likes to start the tank cleaning process while working on other projects as long as each process can be monitored properly. The time is largely affected by the size of the tank and only experience will help the cellar crew in this estimation.
16. Once the process has been allowed to work, one may turn off the pump and wait 4–5 minutes for the extra dripping of the cleaning solution to cease.

17. One may carefully open the door. With safety goggles on and a flashlight in hand – one may inspect the tank to see if the process was effective. Look for areas or patches of tartrates that may not have been dissolved or other areas visually not looking clean. Take appropriate actions to correct any of these.
18. Feel the cleaning solution or take a pH reading. Is the pH still high and does it still feel slippery?
19. Once one deems the tank to be clean, one can dispose of the spent cleaning solution in the proper manner.
20. Rinse the tank and empty all hoses of the cleaning solution. If a bucket was placed in the tank to prevent the vortex – remember to empty its contents.
21. Add fresh water back to the tank to circulate one more time.
22. To this water add enough citric acid to get the water at a low pH – perhaps 10 cups into 120 gallons. (Dissolve in water first, as always.)
23. Circulate this solution to contact all parts of the tank the high pH cleaner contacted. This will neutralize any places back to a reasonable pH level. This circulation may only take about 5 minutes versus the previous step.
24. Once finished, open the tank door and feel the water. It should not be slippery. Run a pH. The pH should be below 5.5. If not – add more citric.
25. Allow this spent water to drain from the tank and dispose of properly.
26. Break down the circulation system or move it to another tank.
27. Rinse the tank one more time with fresh water.
28. Inspect the tank one more time after the cleaning and make sure to remove the bucket or other tools used in the vortex preventions.
29. Take the fittings out of the soaking tub and give them a light citric rinse or do this when appropriate and on your timeline.
30. Always inspect the tank again before filling with wine or juice.
31. Always look at and smell all the fittings before reinstalling on the tank. Fittings that smell bad more than likely have bad microbes in them.
32. Remember to reengage the chilling to that tank so it is ready.
33. Label the tank cleaned, the date and by whom so others will know what the last process with that tank was.

Summary

Tank cleaning is extremely important. It can be performed easily just after the tank has been emptied. The author reports better progress and success with tartrate removal especially if the tank is cleaned within two hours of emptying. Set your tank cleaning system up to be as easy as possible and make sure the cellar staff is keenly aware of your expectations. Tanks that are not cleaned properly should not be used and instructions to clean them again would be prudent. Remember, wine is a food product that others will drink. Use tanks that are cleaned with the same amount of dignity that you want your foods and beverages prepared in.

Helpful hints
- It is not recommended to enter the tank to do any of these processes. If tank entry is needed, that could require a completely different set up for safety reasons.
- Crack doors and valves to allow the cleaning solutions to coat all areas. Try these areas first with water and then perform this action with the cleaning solution added. Remove all gaskets, where appropriate, to allow cleaning them.
- Always check on the interior of the tanks temperature probes and inside manway doors to make sure all is clean, both above and below them.
- Try to have two people around at all times just in case.
- If a certain tank orifice has trouble getting clean try and place a brush or rag in the orifice to absorb the cleaning material so it will 'wick' to the upper areas of the orifice. Then clean the area again physically, rinse the brush or rag and replace for the low pH rinse portion of the cleaning.
- Have a bucket of a light citric solution close by to have access to neutralize any high pH cleaners.

Acknowledgments
Based on verbal conversations with Jacques Boissenot and Jacques Recht.

9.12 How do I clean my winery transfer hoses?

T. J. Payette

In other sections of this book we addressed how to clean a wine tank. In reality a clean wine tank is of little benefit if the means of getting the juice or wine to that tank is a contamination source in itself. Just as much diligence needs to be applied to the wine transfer hoses to insure a wine arrives at its destination in as microbial-free state as it left the previous storage container. Build on the correct procedure for your specific winery's needs and perhaps incorporate them into your winery HACCP plan (Hazard Analysis and Critical Control Points).

Chemistry

The chemistry of cleaning the winery hoses is very similar to cleaning the wine tanks or most anything else in the winery for that matter. One must have physical cleanliness first. This means all of the solid particles are removed from a surface prior to or in conjunction with a high pH cleaner. Once dirt is removed from a surface the chemical may react on that surface to clean and kill certain microbes that will not survive in the harsh environment of a higher pH. After physical cleanliness is achieved and the high pH cleaner has cleaned the surface, a low pH cleaner such as citric acid may be used to neutralize the high pH cleaner and to kill certain microbes that will not live in those lower pH environments. Make sure all cleaners used are suitable for the food processing industry and are safe for the winery.

Items needed

All safety material to include but not be limited to:

- Safety goggles
- Rubber gloves
- Rubber boots
- Hat and/or chemical resistant rain gear
- High pH cleaner (such as soda ash)
- Low pH rinser (such as citric acid)
- Material safety data sheets (MSDS) for all chemicals used.
- Eyewash station or portable eyewash
- A light citric and water solution (2 tbsp per gallon).

Other items needed will include:

- Pump that will handle warm water and the chemicals desired
- Wine transfer hoses that will resist warm water and all chemicals used
- pH meter (optional but the winery really should have one anyway)
- Flashlight(s)
- Sponge balls at diameter(s) needed for hose inside diameter

- Tub for water circulation
- Water source.

Preparation

Apply all safety gear and prepare a light citric and water solution in a bucket to set aside. This is a light 'lemonade strength' water that may come in useful should some of the high pH cleaner come in contact with your skin. (Roughly two tbsp. of citric in two gallons of water depending on the tap water pH.) Select a good positive displacement pump from the cellar that will help power a sponge ball through the wine hoses to be cleaned. Collect all of the wine hoses you want to clean. The author prefers to do this on the crush pad just after harvest, in the spring and just prior to harvest at a minimum.

Procedure

1. Apply all safety gear necessary to be safe while doing the tasks described. This is an internal winery decision that the winery will need to address.
2. Move all equipment outside that needs to be used to clean the hoses.
3. Have the 'lemonade strength' bucket of water mixture mentioned above placed close by and in a spot that can be easily located.
4. On the suction side of the pump assemble a short section of hose. This hose should be long enough to span from the tub of cleaning water to the pump.
5. Assemble the other remaining sections of hose on the pressure side of the pump from the largest diameter size to the smaller diameter size. An example may be having all the 2″ sections connected, then a reducer to the 1.5 inch sections down to 1 inch and three quarters and so on.
6. Once all the connections are made, start to fill the tub with clean fresh water. Warm, not hot, may be the best water for this process. A good target temperature should be in the 90°F temperature range.
7. While the tub is filling, one may start and stop the pump to fill the lines with water. Memorize the direction of the flow since we will always run the pump in that direction for this exercise. (This is important so we do not suck the sponge balls we will be using back into the head of the pump.)
8. Once the lines are full be sure to pump about 10 gallons of water out on the floor to eliminate any obvious solids that may have collected in the hoses during storage. (This is especially true if wine hose is stored curled up on the floor – not a recommended way to store winery hose.)
9. Once the winery lines are completely filled with water one may stop the pump.
10. Gently disconnect the discharge side of the pump from the pump head fittings and insert the proper size sponge ball to clean the smallest size diameter of hose assembled in this set up.
11. Reconnect the discharge side of the hose back to the pump.

12. Turn the pump on in the direction to push the sponge ball through to the lines to be cleaned. Leave the discharge end of the hose in the tub for the time being to conserve water.
13. Follow the sponge ball visually, if possible, through the maze of hose making sure the suction line has a continuous source of water supplied.
14. Once the sponge ball reaches the specified diameter of hose it is designed/sized to clean, keep an eye on the hoses since one may see a slight pressurization and accordion type movement in the hoses at this time. *Be aware fittings could be blown off under pressure.*
15. As the sponge ball makes its way through the lines and the ball has about 7 feet more to go, remove the discharge line from tub of water and allow the water to exit onto the floor or crush pad. You will notice a 'tea like' to 'coffee like' colored water will start to exit the discharge line just before the sponge ball exits. This is true for even any well kept hoses that have not been cleaned in this fashion for over one year. It is inevitable beyond anyone's sanitation programs.
16. Recapture the sponge ball and run the ball through again. It will still clean a bit more on the second and third pass.
17. Once one feels this section of hose has 'mechanical cleanliness' one may disconnect that size diameter line from the assembly.
18. Select the proper size sponge ball to clean the next diameter size section of hose near the end of the assembly and repeat the procedure gaining mechanical cleanliness on each diameter size hose working your way up to the largest size line.
19. Once all of the lines are cleaned be sure to swap out the suction side supply line with a cleaned section and run the proper sized sponge ball through that section.
20. Now that mechanical cleanliness is achieved, one may reassemble all of the hoses and start the pump for a circulation.
21. Once the circulation is started in the clean tub of water, one may add a high pH cleaner. Always dissolve any solid cleaners in water first before adding to a tub of water. (This will take some trial and error on the operator's part to establish just how much may be needed.) (Use a pH meter to determine this strength needed.)
22. Allow this high pH solution cleaner to circulate for an adequate time. This may be near 15 minutes depending on the length of hose line, sizes, speed of pump and the amount of water in the circulation tub.
23. Once the operator feels the hoses' interiors are well exposed to this higher pH water, the operator may then flush the hoses out with copious amounts of fresh water.
24. After a fresh water rinse one should continue to circulate water and add a low pH cleaner, such as citric acid, to the mixture to ensure the high pH water has been neutralized.
25. After this neutralizing step, it is best, once again, to do a fresh water rinse.

286   Winemaking problems solved

26. The hoses should now be clean, but not considered sterile, to the satisfaction of most wineries' sanitation programs.
27. One may disconnect all the winery hoses and store them properly to drain dry. Resist rolling hoses up on the floor and laying them flat because water, moisture and insects/rodents may have a better opportunity to become an issue for them.

When selecting hoses for use at any given time, it is best to make the assembly of the hoses and to flush the hoses or clean them in some fashion just prior to pumping juice or wine. This will clean out any items from the hoses or pump that should not have been in them.

Just prior to harvest consider performing this operation on the hoses but perhaps take the step a bit further. Once the hose lines are cleaned, remove the fittings from the ends of the hoses and either clean them vigorously or cut off the portion of the hose that was in contact with the fitting. Clean the stainless fittings until they are sparkling and then re-install the fittings and tighten the clamps properly. (Note: when putting hose clamps on have them pull and installed in opposing directions to get a better tightening grip.) Also, apply the clamps as close as possible to the end of the stainless fitting that is inside the hose line. If this is not carried out, wine may seep between the fitting and the hose line, especially when ballooning under pressure, and force wine between them. Over time, spoilage will occur which will result in a cross-contamination source for every transfer or operation performed with that set of hoses in the future.

Summary

Set up your hose cleaning operation to be as easy as possible and make sure the cellar staff is keenly aware of your expectations. Hoses that are not cleaned properly should not be used and instructions to clean them again would be prudent. Remember, wine is a food product that you and others will drink. Use hoses that are cleaned with the same amount of dignity that you would want your foods and other beverages prepared in.

Helpful hints

- Mono type pumps have been known to pass the mentioned sponge balls easily provided the pumps are not smaller than the actual ball diameter used.
- Be sure to keep a watchful eye on the diameter of the sponge ball and the diameter of the wine line you are trying to clean.
- Do not run the pump while dry or damage may occur.
- Use a pH meter to determine the pH of your cleaning solutions.
- Smell your hose before you use them for a wine transfer.
- Look inside your hoses before using. What do you see?
- As with tanks and other items, don't let dirt, juice or wine dry on them. Clean immediately after use (inside and out).

- Always store the hoses so they will drain and dry completely. Hoses should not be curled up on the floor with potential standing moisture inside them. Being on the floor also makes them easily available to any winery critters or insects.
- If one has cleaned the winery hoses and removed the fittings, the author recommends a way to test their strength. Assemble all the hoses together with a valve at the very end of the discharge side. Circulate fresh water with no chemicals for cleaning. After about 5 minutes of circulation take the discharge side of the hose from the circulation bucket and start to slowly move the valve toward the closed position for a brief moment. Be aware at this moment pressure will be building inside the transfer lines and to be clear of any hoses that may pop off their fittings. Be very careful with this procedure and use common knowledge not to shut the valve all the way creating extreme pressure.
- Have two people around at all times for safety.

Acknowledgments

Based on verbal conversations with Jacques Boissenot and Jacques Recht.

9.13 How can I determine if my sanitation program is successful?

K. C. Fugelsang

As discussed by Fugelsang and Edwards (2007), a variety of tests are used for site evaluation. These generally include sampling a defined area with a sterile cotton swab or adhesive strip, before and after sanitation, followed by isolation and culture of the microbes. Given the time-lag associated with classic laboratory methods of culture (e.g., plate counts), 'real-time' methods are finding increased application in wine and juice industries. Among these, the use of ATP assay by bioluminescence is the most widely used.

Reference

FUGELSANG, K. C. and EDWARDS, C. G. 2007. *Wine Microbiology: Practical Applications and Procedures*, 2nd edn. Springer, New York.

9.14 What are the safety issues associated with sanitation operations?

K. C. Fugelsang

Owing to the use of strong oxidants (ozone, peroxides), caustics (NaOH, KOH), and acidic chemicals (phosphoric acid) as well as hot water and/or steam and high pressure delivery systems, there are significant safety issues associated with winery sanitation. Because of these, personal protective equipment (PPE) including gloves, goggles, appropriate footwear, and waterproof aprons are essential. Where workers use respirators, it is strongly recommended that supervisors review regulations regarding, qualification, training and medical certifications associated with these devices.

Federal and state regulations require employees receive ongoing and documented training in the proper use of winery chemicals as well as methods for their application and required PPE. Further, workers must have full access to Material Safety Data Sheets (MSDS) that contain the necessary health and safety information associated with each on-site chemical. Copies must be on file and accessible at each work site where chemicals are stored and used. Additionally, a master copy that includes all winery chemicals must be maintained in the production office. Although MSDS typically accompany shipment, they may be mailed separately. They are also readily available from suppliers' websites.

10
Brettanomyces infection in wine

10.1 What is 'Brett'?

L. Bisson

'Brett' is a colloquialism referring to the yeast *Brettanomyces*. *Brettanomyces* is a budding yeast also called *Dekkera* and is perhaps the most controversial organism of wine production. *Brettanomyces* is frequently isolated from wines. If these isolates display sexual spore formation they are classified as *Dekkera*. If no spores are formed they are classified as *Brettanomyces*. DNA sequence comparisons indicate that *Dekkera* and *Brettanomyces* are identical, but the use of the two terms still persists in taxonomic classification.

Brettanomyces infection of wine is widespread and this yeast has been isolated from all wine-producing regions on six continents. This yeast is able to form biofilms, coatings of tanks, hoses and other winery surfaces that are difficult to eliminate and organisms in biofilms can resist sanitation agents and survive. Because wood is porous and provides wood sugars barrels are difficult to clean and sanitize and often host populations of *Brettanomyces*. Another reason *Brettanomyces* is such a problem is that it can exist in what is called a viable non-culturable (VNC) state. Most sanitation practices are monitored for effectiveness by determining if viable cells remain on the surface that was treated. Organisms in a VNC state do not grow under these conditions and give the false impression that the sanitation regimen was effective, but are in fact still viable and capable of growing later on.

Brettanomyces is known for growing under harsh environmental conditions and has an interesting metabolism. This organism makes both ethanol and acetate from sugar, and which form is produced depends upon the strain and the conditions of growth. More importantly, *Brettanomyces* is able to produce vinyl phenols from wine phenolic compounds. The characters made by *Brettanomyces* are described as: sweaty horse, animal, earthy, wet wool, burnt plastic, Band-Aid, barnyard, smoky, soy sauce, spicy, mushroom, putrid, leather, mushroom, tobacco, lilac, wet dog, pharmaceutical, soapy, and baby diaper just to name a few. Not all strains produce all characters, and there is a great variation in the levels and spectrum of characters produced. It is difficult to predict how a given strain will perform in a specific wine, as the factors leading to the appearance of these compounds are not well known. This is the heart of the controversy surrounding this yeast – some winemakers feel that the conversion of grape varietal characters to specific *Brettanomyces* end products is central to the development of varietal character and the expression of terroir. Other winemakers consider the development of these characters to detract from true varietal character giving the wine an off-taste and defective aroma.

10.2 What is the history of *Brettanomyces* and where does it come from?

L. Van de Water

Taxonomy

With sincere apologies to Shakespeare, 'What's in a name? Would 'Brett' by any other name smell as nasty?' Some taxonomists use the name *Brettanomyces* (anamorph, nonsporulating) for these yeasts, others use *Dekkera* (teleomorph, sporulating) for all of them, and some use a combination. For simplicity here we will use the name *Brettanomyces* to mean *Dekkera/Brettanomyces bruxellenensis*, excluding *Dekkera anomala*.

Of the five *Brettanomyces/Dekkera* species currently recognized (*bruxellensis, custersianus, naardenensis,* and *nanus*, plus *D. anomala*), only *bruxellensis* is consistently isolated from wine. Former species that were consolidated into *bruxellensis* after genetic and other tests include *B. custersii, B. intermedius, B. lambicus, D. abstinens,* and *D. intermedia* (Smith *et al.*, 1990; Smith, 1998; Barnett *et al.*, 2000; Egli and Henick-Kling, 2001). *Dekkera anomala* has been isolated in Europe from cider (Delanoe and Suberville, 2005) soft drinks (Smith and van Grinsven, 1984), beer and wine (Oelofse *et al.*, 2009). In the USA it has been isolated from cider and fruit juice (personal observation).

Because the same yeast's name changes when its taxonomy changes, and because different researchers continue to call the same yeast by different names, someone who is not a yeast taxonomist is likely to be confused. If you check the references cited here, you will find a plethora of different names, many of which have since been consolidated. However, a 'Brett-type' yeast in wine is almost certain to be *Brettanomyces/Dekkera bruxellensis* (it could be *D. anomala* but that is less likely), so you are dealing with the same yeast, no matter what its current taxonomic name(s) is/are.

Many different strains of *Brettanomyces/Dekkera bruxellensis* have been isolated from wine and studied to one extent or another. Genetic and physiological differences among strains are profound, and will be discussed in **10.3**.

History and distribution

In *Études sur le vin*, the original treatise on wine microbiology, Pasteur does not describe any malady that seems to refer to *Brettanomyces*, nor do his meticulous drawings include any yeasts resembling them microscopically (Pasteur, 1866). *Brettanomyces* was identified in 1904 in British beer (Claussen, 1904), and subsequently in beer and wine from other countries. 'Lambic beer' and some ciders are traditionally produced with *Brettanomyces* yeast. By now, *Brettanomyces* infection of wine has been confirmed in wines on six continents, and in nearly every region where wine is made (Licker *et al.*, 1998; Arvik and Henick-Kling, 2002). Our labs have cultured *Brettanomyces* from wines from Argentina, Australia, Canada, Chile, France, Italy, Moldova, New Zealand,

Thailand, Uruguay, and USA, which is to say, all countries whose wines were cultured.

Interestingly, the genetics of the strains studied by Conterno *et al.* (2006) indicated a loose geographical association among isolates, but there were many exceptions, with some New World strains in groups of Old World strains, and vice versa. For example, their largest grouping on the basis of 26S rDNA included strains from France and Belgium but also strains from California and Chile, plus a morphologically distinct strain from Thailand. Complicating the picture, physiological characteristics did not match the genetic groupings. Dias *et al.* (2003) and Curtin *et al.* (2005, 2007) also described strains belonging to one genotype found in wines from widely separated countries.

Recent history of *Brettanomyces* identification
In California, a 1972 Napa Valley Gamay spoiled in the bottle. The change was appalling; a fruity wine with lots of mid-palate flavors lost its fruitiness, developed strong horsey odors, became simple-tasting, and had a metallic aftertaste (personal observation, 1973). *Brettanomyces* was isolated and identified in 1974 from that Gamay by Dr Ralph Kunkee, microbiology professor at U.C. Davis (Kunkee, personal communication; Kunkee, 1996). Shortly afterwards, the yeast was cultured and identified in a number of other California wines (personal experience; personal communication, names withheld on request). Before then, *Brettanomyces* was called 'the Cucamonga stink' in California's Central Valley, because although it was recognized as a spoilage problem, they did not know what caused it. The classic 'Brett' descriptors (see **10.4**) were consistently noticed in all these wines, though there was no method at the time to identify the ethyl phenols and other compounds responsible.

Unfortunately, winemakers were reticent to discuss their experiences with other winemakers, which delayed recognition of the widespread nature of the *Brettanomyces* problem for some years. 'Brett' was actually called 'the clap' by some wineries, because of winery-to-winery transmission, because the infection was easy to miss (a slogan for STD testing at the time was 'you can have it and not know'), and because of the stigma attached to the infection. The result was that *Brettanomyces* was almost a taboo subject in California for a number of years. A prominent winemaker, reached by phone in 1974 because 'Brett' character had been detected in one of his wines, thundered, 'Why the hell would you want to talk about a thing like that?' and slammed down the receiver. Few winemakers acknowledged publicly (or even to their own sales and marketing people) that their wineries were infected, although most who were producing red wine knew they had a *Bretttanomyces* infection in their cellars. To this day, a 'don't ask, don't tell' dynamic lingers on.

During the 1970s and 1980s, wines were intensely scrutinized for *Brettanomyces* infection in California at wineries and commercial labs, but formal research was seldom if ever funded until the 1990s. USA research funding still struggles mightily, though some very important studies have been performed already, and can continue if funded. More research is now carried out in other countries, because of funding considerations.

In some winegrowing countries, *Brettanomyces* was identified years earlier, but sometimes the awareness faded. For example, *Brettanomyces* was studied in South African wines in the late 1950s and early 1960s (van der Walt and van Kerken, 1960), and the genus *Dekkera* was named in 1964 (van der Walt, 1964). However, by 2004, many South African winemakers were unaware of *Brettanomyces* infections in their wineries (personal observation).

Winemakers in other countries also discovered relatively recently that *Brettanomyces* is the origin of certain sensory characteristics. Despite excellent Australian research on ethyl phenol production by *Brettanomyces* (Heresztyn, 1986), until the late 1990s it was assumed by many Australians that the so-called 'Hunter Valley stink' in Syrah was a regional or varietal characteristic. However, even in the 1970s, Californians visiting Australia recognized the telltale sensory characters of *Brettanomyces* in some of the wines, and some Australian winemakers were discussing the possibility that there was *Brettanomyces* in Australia (Carol Shelton, personal communication). Since 1999, when it was discovered that a large proportion of Australian reds had high ethyl phenol levels (Henschke *et al.*, 2004), the Australian Wine Research Institute have been extremely proactive in their research on *Brettanomyces* and in educating winemakers about the problem. By 2004, 1500 winemakers had attended their 'Institute Roadshows'.

While *Brettanomyces* was discovered in French wine in 1930 by researchers in Germany (Licker *et al.*, 1998), and studies continued in France over time (for example, Peynaud and Domercq, 1956), the link between *Brettanomyces* and 4-ethyl phenols in French wine was not observed until 1992 (Chatonnet *et al.*, 1992). Until then, *Brettanomyces* was assumed by French winemakers to produce only 'mousiness' (Ribereau-Gayon *et al.*, 2006). However, the other effects of *Brettanomyces* growth – horsey, barnyard, spicy descriptors – had been detected sensorily for many years in some French wines by winemakers and enologists in California, and *Brettanomyces* had been cultured from certain French wines since 1974 (personal observation). For example, a 1978 Bordeaux from a highly regarded château was cultured because of a complaint from a St Helena, CA, retailer, and was found to be heavily infected with *Brettanomyces*.

It is unfair to point to any one region as more thoroughly infected than another, however. Management practices usually make the difference in wines with 'Brett' character, not numbers of infected wineries. In the 1970s, Californians considered the Burgundy region in France to be a hotbed of *Brettanomyces* because of a miscommunication. Gerald Asher, then wine writer for *The San Francisco Chronicle*, asked Robert Mondavi why his 1971 Reserve Pinot Noir seemed so Burgundian, and he replied, 'Oh, that's from the touch of *Brettanomyces* we had in there.' In fact, that wine did not have *Brettanomyces*, it had *Pediococcus* (personal observation) which is very common in red Burgundies.

Does *Brettanomyces* come from the vineyard?

Whether or not *Brettanomyces/Dekkera bruxellensis* – the wine species – is commonly present in vineyards is very controversial. Some winemakers believe that *Brettanomyces* comes in on most grapes, therefore all red wines have *Brettanomyces*. Popular articles quoting wine writers and winemakers are filled with numerous references to *Brettanomyces* on grapes, but almost all of that evidence is anecdotal. A California winemaker declared in *Harpers Wine and Spirit Review* (Goode, 2003), 'Since Brett is largely ubiquitous, a rampant Brett infection is often more of a function of a large inoculum coming in on the grapes'.

Are these ideas about *Brettanomyces* substantiated by research? No. The opinion that *Brettanomyces* is part of the natural flora of all red wines is not borne out by testing of wines in hundreds of wineries over many years (personal observation). After reviewing formal research, independent lab results, and in-house studies performed by wineries, the inescapable conclusion is that not all wineries are infected with *Brettanomyces*, even when neighboring wineries are.

In fact, if *Brettanomyces bruxellensis* is present in vineyards, it is very difficult to find. A number of studies around the world over the past 35 years did not find the yeast on grapes (Fugelsang *et al.*, 1993; Oelofse *et al.*, 2008; also personal communication). In a 2007 French study, Renouf and Lonvaud-Funel (2007) developed a special liquid medium, EBB, to culture *Brettanomyces* yeasts from grapes, and then used polymerase chain reaction (PCR) to confirm the yeasts' identity. Similar studies in other regions would be helpful; a Napa Valley winery using EBB medium located no *Brettanomyces* on their grapes, but other yeast species (*Candida catenulata* and *Debaromyces hansenii*) grew. Those species also reacted with other tests presumed to be specific for *Brettanomyces* (personal communication, name withheld on request).

Many winemakers assume that *Brettanomyces* can populate vineyards if infected pomace is spread in the vineyard. This may indeed occur, but it has not yet been proven by controlled studies (a good topic for future research). The yeasts were found on airborne droplets within a winery (Connell *et al.*, 2002), and in insects in and around wineries by van der Walt and van Kirken (1960), though not on pomace. It has not been established that insects are actually vectors for infection of wine among different wineries. Species of *Brettanomyces* have been found in fruit orchards and in bees (Licker *et al.*, 1998), but it was not determined whether or not these belonged to *bruxellensis*; if not, they were unlikely to infect wine. All in all, *Brettanomyces* may be present at very low levels in some vineyards, but winemakers should not look to grapes as a primary source of *Brettanomyces* infection.

Brettanomyces in the winery

So how do the yeasts get into a winery? The most important vehicles of transmission of these yeasts from cellar to cellar are infected wines and wooden or plastic containers (Fugelsang *et al.*, 1993; Boulton *et al.*, 1996; also personal experience). Movement of bulk wine in a number of countries has spread

Brettanomyces yeasts internationally and domestically (Curtin *et al.*, 2007). When infected wine is purchased, unless the winery is prepared to deal with the infection before it enters the cellar, one wine can contaminate the entire cellar. Purchasing used red wine barrels from other wineries is economically very tempting, although *Brettanomyces* often comes along also (white wine barrels are much less frequently infected). There is no legal or moral requirement to disclose the *Brettanomyces* infection status of barrels for sale, even if known; it is definitely a 'buyer beware' situation. The practice of importing used barrels from Bordeaux wineries with the hope of capturing some of the 'gout de terroir' of that region has assisted in distributing *Brettanomyces* around the world. Used whiskey barrels have not been implicated in spreading *Brettanomyces*, presumably because the high alcohol content of whiskey prevents an infection from taking hold, even if somehow the yeasts were introduced.

Case histories abound of *Brettanomyces* entering a winery through wine or wood and taking up residence, infecting subsequent wines for many years thereafter. Three examples:

1. A new winery was built for the 2001 vintage. They were warned never to bring in anything from another winery, and had three vintages completely free of *Brettanomyces*, although the winery was in a valley full of infected wineries. Then, in 2004, the new winemaker bought 5000 liters of wine from another winery, but did not check it for *Brettanomyces* beforehand. The wine was infected, and as a result, the winery has been fighting *Brettanomyces* ever since (personal observation).
2. A winemaker whose small winery had been free of *Brettanomyces* unknowingly purchased an infected wine and used it to top all his red barrels, infecting the entire cellar.
3. The 1995 decision by a New Zealand winery to sell barrels after only three years sent *Brettanomyces* quickly around the country, into numerous wineries that previously were uninfected (personal observation).

Once in the winery, the yeasts are easily spread from wine to wine. Numerous studies have found infections in various places inside wineries, and as far back as 1960, it was recognized that these pools of *Brettanomyces* infection in the winery could lead to cross-contamination. 'The infection of wines and musts by *Brettanomyces* species is due to contamination spreading from latent foci of infection within the winery' (van der Walt and van Kirken, 1960).

Sanitizing measures such as SO_2, ozone, peroxycarbonate, and peracetic acid reduce the population in a barrel but do not kill all the yeasts, which can be 0.8 cm deep in the wood (Malfeito-Ferreira *et al.*, 2004). Shaving a barrel greatly reduces the infection (Pollnitz *et al.*, 2000) but does not always completely eliminate it. Asked why he was the only one in New Zealand with *Brettanomyces* in his Chardonnay, a winemaker immediately replied, 'That's because I use shaved red wine barrels for my Chardonnay' (personal communication, name withheld on request). The shaving got rid of the color but did not go down far enough to eliminate the infection deep in the staves.

Brettanomyces grows more readily in new barrels, but this is more likely to be related to the composition of new wood, rather than new barrels bringing in an infection. While some winemakers believe new barrels to be a primary source, this has not been confirmed by research (see **10.3**).

Avoiding *Brettanomyces* infection:

- Do not buy barrels, or wooden or plastic tanks, that were previously used for red wine in another cellar (white wine barrels could be infected but are much less likely to be).
- Do not buy bulk red wine unless it can be sterile-filtered before going into wood or plastic containers in your cellar.
- Do not buy equipment that cannot be sanitized with hot water or steam before use.

Acknowledgements

I would like to thank the following people for reviewing my question texts: Dr Charles Edwards, Washington State University; Dr Paul Grbin, University of Adelaide; Sarah Inkersell, Pacific Rim Oenology Services, New Zealand; C. M. Lucy Joseph, University of California at Davis, Zoran Ljepovic, Constellation Wines US, and Doug Manning, winemaking consultant, Napa, California.

Also thanks to Dr Ken Fugelsang, California State University, Fresno; Dr Trevor Phister, NCSU and Dr Bruce Zoecklein, Virginia Tech University for their helpful suggestions.

References

ARVIK, T. and HENICK-KLING, T., 2002, 'Overview: *Brettanomyces bruxellensis* occurrence, growth, and effect on wine flavor,' *Practical Winery and Vineyard*, May/June: 117–123.

BARNETT, J. A., PAYNE, R. W. and YARROW, D., 2000, *Yeasts: Characteristics and Identification*, 3rd edn, Cambridge, Cambridge University Press.

BOULTON, R., SINGLETON, V., BISSON, L. and KUNKEE, R., 1996. *Principles and Practices of Winemaking*, New York, Chapman and Hall.

CHATONNET, P., DUBOURDIEU, D. and BOIDRON, J. N., 1992, 'The origin of ethyl phenols in wines,' *J. Sci. Food Agric.* 68, 165–178.

CLAUSSEN, N. H., 1904, 'On a method for the application of Hansen's pure yeast system in the manufacturing of well-conditioned English stock beers,' *J. Inst. Brew.* 10, 308–331.

CONNELL, L., STENDER, H. and EDWARDS, C., 2002, 'Rapid detection and identification of *Brettanomyces* from winery air samples based on peptide nucleic acid analysis,' *Am. J. Enol. Vitic.* 53:4, 322–324.

CONTERNO, L., JOSEPH, C.M.L., ARVIK, T., HENICK-KLING, T. and BISSON, L., 2006, 'Genetic and physiological characterization of *Brettanomyces bruxellensis* strains isolated from wines,' *Am. J. Enol. Vitic.* 57:2, 139–147.

CURTIN, C., BELLON, J., COULTER, A., COWEY, G., ROBINSON, E., DE BARROS-LOPES, M., GODDEN, P., HENSCHKE, P., PRETORIUS, I., 2005, 'The six tribes of "Brett" in Australia –

distribution of genetically divergent *Dekkera bruxellensis* strains across Australian winemaking regions,' Aust. NZ Wine Ind. J. 20:6, 28–36.

CURTIN, C., BELLON, J., HENSCHKE, P., GODDEN, P. and BARROS-LOPEZ, M., 2007, 'Genetic diversity of *Dekkera/Brettanomyces bruxellensis* isolates from Australian wineries,' *FEMS Yeast Res.* 7, 471–481. doi: 10.1111/j.1567-1364.2006.00183.x.

DELANOE, D. and SUBERVILLE, N., 2005, 'Troubles et dépôts des boissons fermentées et des jus de fruits. Aspects pratiques du diagnostic', Chaintré France, Oenoplurimedia.

DIAS, L., DIAS, S., SANCHO, T., STENDER, H., QUEROL. A., MALFEITO-FERREIRA, M. and LOUREIRO, V., 2003, 'Identification of yeasts isolated from wine-related environments and capable of producing 4-ethylphenol,' *Food Microbiol.* 20, 567–574.

EGLI, M. and HENICK-KLING, T., 2001, 'Identification of *Brettanomyces/Dekkera* species based on polymorphism in the rRNA internal transcribed spacer region,' *Am. J. Enol. Vitic.* 52, 241–247.

FUGELSANG, K., OSBORN, M. and MULLER, C., 1993, '*Brettanomyces and Dekkera*, implications in winemaking', in *Beer and Wine Production, Analysis, Characterization, and Technological Advances*, Gump, B. (ed.), American Chemical Society, Washington, DC, 110–131.

GOODE, J., 2003, '*Brettanomyces*', *Harpers Wine and Spirit Weekly*, 18 April, 42–46.

HENSCHKE, P., BELLON, J., CAPONE, D., COULTER, A., COWEY, G., COZZOLINO, D., CURTIN, C., FIELD, J., GISHEN, M., GRAVES, P., LATEY, K., ROBINSON, E., FRANCIS, I.L., DE BARROS LOPES, M. and GODDEN, P., 2004, Incidence and control of *Brettanomyces*: the Australian perspective. *Abstr. Am. J. Enol. Vitic.* 55, 304A.

HERESZTYN, T., 1986, 'Metabolism of phenolic compounds from hydroxycinnamic acids by *Brettanomyces* yeasts,' *Arch. Microbiol.* 146, 96–98.

KUNKEE, R., 1996, 'Several decades of wine microbiology: have we changed or have the microbes?', Wine Spoilage Microbiology Conference, Toland, T. and Fugelsang, K. (eds). California State University, Fresno.

LICKER, J., ACREE, T. and HENICK-KLING, T., 1998, 'What is "Brett" (*Brettanomyces*) flavor?: a preliminary investigation', in *Chemistry of Wine Flavor*, Waterhouse, A. and Ebeler, S. (eds), Washington, DC, American Chemical Society Symposium series.

MALFEITO-FERREIRA, M., LAUREANO, P., BARATA, A., D'ANTUONO, I., STENDER, H. and LOUREIRO, V., 2004, 'Effect of different barrique sanitation procedures on yeasts isolated from the inner layers of wood,' *Abstr. Am. J. Enol. Vitic.* 55, 304A.

OELOFSE, A., PRETORIUS, I. and DU TOIT, M., 2008, 'Significance of *Brettanomyces* and *Dekkera* during winemaking: a synoptic review,' *S. Afr. J. Enol. Vitic.* 29:2, 128–144.

OELOFSE, A., LONVAUD-FUNEL, A. and DU TOIT, M., 2009, 'Molecular identification of *Brettanomyces bruxellensis* strains isolated from red wine and volatile phenol production,' *Food Microbiol.* 26, 377–385. doi: 10.1016/j.fm.2008.10.011.

PASTEUR, L., 1866, *Études sur le vin: ses maladies, causes qui les provoquent, procédés nouveaux pour le conserver et pour le vieillir*, Victor Masson et fils, Paris.

PEYNAUD, E. and DOMERCQ, S., 1956, 'Sur les *Brettanomyces* isolées de raisins et de vins', *Arch. Mikrobiol.* 24, 266–280.

POLLNITZ, A., PARDON, K. and SEFTON, M., 2000, 'Quantitative analysis of 4-ethylphenol and 4-ethylguauacol in red wine', *J. Chromatogr. A* 874, 101–109.

RENOUF, V. and LONVAUD-FUNEL, A., 2007, 'Development of an enrichment medium to detect *Dekkera/Brettanomyces bruxellensis*, a spoilage wine yeast, on the surface of grape berries,' *Microbiol. Res.* 162, 154–167.

RIBEREAU-GAYON, P., GLORIES, Y., MAUJEAN, A. and DUBOURDIEU, D., 2006, *Handbook of Enology Volume 2: The Chemistry of Wine Stabilization and Treatments*, John Wiley & Sons, Chichester.

SMITH, M. T., 1998, '*Brettanomyces Kufferath and van Laer*', in *The Yeasts*, 4th edn, Kurtman, C.P. and Fell, J.W. (eds), Elsevier, New York, 450–453.

SMITH, M. and VAN GRINSVEN, A., 1984, '*Dekkera anomala* sp. nov., the teleomorph of *Brettanomyces anomalus*, recovered from spoiled soft drinks,' *A. Van Leeuw.* 50:2, 143–148.

SMITH, M.T., YAMAZAKI, M. and POOT, G.A., 1990, '*Dekkera, Brettanomyces* and *Eeniella*: electrophoretic comparison of enzymes and DNA-DNA homology,' *Yeast* 6:4, 299–310.

VAN DER WALT, J. and VAN KERKEN, A., 1960, 'The wine yeasts of the Cape, Part V: Studies on the occurrence of *Brettanomyces intermedius* and *Brettanomyces schanderlii*,' *A. van Leeuw.* 26, 292.

VAN DER WALT, J., 1964, '*Dekkera*, a new genus of the *Saccharomycetaceae*,' *A. van Leeuw.* 30, 273–280.

10.3 How does *Brettanomyces* grow?

L. Van de Water

Brettanomyces can live on other microbes' nutritional leftovers, earning themselves the nickname 'junkyard dog of wine yeasts'. They are surprisingly versatile in their nutritive requirements. At least some strains of *Brettanomyces bruxellensis* can live in almost any red wine, and some whites, below around 15% alcohol or so, bottled or not, dry or sweet. They can use a number of carbon-containing substrates, some of which are always present in the wine. Low wine pH is not a limiting factor except as it relates to molecular SO_2 (Conterno *et al.*, 2006).

Red wines are most at risk for *Brettanomyces* infection and growth. Red wine pH is higher than most whites, so SO_2 protection is not as effective. Reds are often kept warmer during at least part of their cellaring, and they often are aged in infected cooperage for a much longer time than whites would be. However, white wines are not immune to *Brettanomyces*; infections have been found in white wines fermented or aged in used wood (even if shaved), and in sparkling wine.

It has often been noted that *Brettanomyces* yeasts grow more extensively in new barrels compared to old ones. It was therefore speculated that this meant that new barrels were contaminated with *Brettanomyces*, but this has not been substantiated. Rather, it has been shown that most strains can use the disaccharide cellobiose, a component of toasted wood, as a sole carbon source (McMahon and Zoecklein, 1999; Mansfield *et al.*, 2002). This is a very important aspect of *Brettanomyces* growth, especially in new barrels. Cellobiose is more abundant in new oak, promoting more active growth if new barrels are filled with infected wines (Boulton *et al.*, 1996; Fugelsang and Edwards, 2007).

At normal cellar temperatures (13°C/55°F) and conditions the yeasts grow slowly over some months or may even become temporarily dormant, though at a warm temperature (over 20°C/68°F) they may grow significantly in a few weeks (Chatonnet *et al.*, 1995). For example, warming stuck wines to encourage fermentation also encourages *Brettanomyces* growth, if the wine has already become infected. Inconveniently, exponential growth may be stimulated by a rise in storage temperature when unfiltered wine is sent to market.

Brettanomyces can grow throughout the wine, but the cells tend to settle in barrels or tanks, so the population is often higher near the bottom, sometimes much higher (Boulton *et al.*, 1996). In barrels with ill-fitting bungs or other containers with ill-fitting closures, there may also be a population near the top, taking advantage of dissolved oxygen to broaden their range of substrates. Some strains make biofilms that can adhere to surfaces in contact with the wine, such as tanks, hoses, and other equipment (Joseph *et al.*, 2007). Despite some reports, the yeast does not normally make a film on the surface of cellared wine (Fugelsang *et al.*, 1993). Occasionally a very thin film may form (which can extend up the sides of a small glass container for several cm) if an infected lab sample of wine is continually exposed to air for several months (personal observation).

Micro-oxygenation is contraindicated for *Brettanomyces*-infected wines, because research shows that they take up the oxygen more quickly than the phenolics do, stimulating Brett growth (du Toit *et al.*, 2005, 2006). While stimulated by oxygen, they do not need it; the yeasts also grow in unfiltered, bottled wine (Chatonnet *et al.*, 1992; Fugelsang and Zoecklein, 2003; Romano *et al.*, 2008; Coulon *et al.*, 2010; also personal observation).

Brettanomyces appear to be less sensitive to CO_2 than *Saccharomyces* because they have been isolated from bottled sparkling wine (Ciani and Ferraro, 1997; Licker *et al.*, 1998). One *methode champenoise* sparkling wine had viable *Brettanomyces* cells after having been *en tirage* (on yeast lees, with a large buildup of CO_2) for four years (personal observation).

Strain diversity
Winemakers everywhere – and even wine microbiologists – tend to think of *Brettanomyces/Dekkera* as one yeast, but actually we should speak of the members of this species as 'them' rather than 'it.' There are many strains of this yeast, with extremely variable genetics and characteristics. Even chromosome size and number varies among *bruxellensis* strains (Oelofse *et al.*, 2008). Thus, whenever one discusses *Brettanomyces*, much of what one can say about a particular strain can be contradicted by the behavior of another strain (Egli and Henick-Kling, 2001; Dias *et al.*, 2003; Conterno *et al.*, 2006; Barbin *et al.*, 2008). Strain differences can account for some (though not all) of the misconceptions about *Brettanomyces* that abound in the popular literature.

Everyone does agree that strain diversity and distribution in *Brettanomyces* are quite extensive. Curtin *et al.* (2007) tested 244 isolates in 31 regions in Australia and identified eight genotypes; 207 isolates from 28 regions were one genotype, a few isolates fit each of two other genotypes, and five genotypes had one isolate each. Most wineries had only one genotype, but a few had more. Barbin *et al.* (2008) found 23 strains in 24 isolates from two French wineries over three years. Martorell *et al.* (2006) found the same molecular pattern in strains from Portugal and from the USA.

Strain differences in growth and metabolism
All 35 strains in a study by Conterno *et al.* (2006) utilized arginine and proline, the two major amino acids in grapes, as a sole nitrogen source. Proline is not normally used by *Saccharomyces* so it is available after fermentation. Arginine is used during yeast fermentation, but some can be left over because of high levels in grapes, overenthusiastic nutrient supplementation, and yeast autolysis during lees aging.

Those 35 strains also had an absolute requirement for the vitamins thiamine and biotin, which are also used by other microbes. Competition for these and/or other nutrients may be relevant to observations in wineries. One winery noticed that barrels infected with *Brettanomyces* struggled with malolactic fermentation

(MLF) while uninfected barrels of the same wine did not, and barrels that had completed MLF did not grow *Brettanomyces* as readily as barrels of the same wine that had not yet undergone MLF (personal communication, name withheld on request). Romano et al. (2008) proposed that MLF might increase toxicity toward *Brettanomyces*.

Conterno et al. (2006) confirmed that different strains can metabolize different substrates, show different growth patterns and morphology, and have different tolerances to alcohol, SO_2, and temperature. Some strains did not produce the classic sensory hallmarks, 4-ethyl phenol (4EP) and 4-ethyl guaiacol (4EG), in the wine used, though the researchers postulated that those same strains may do so in other wines (Lucy Joseph, personal communication).

Sugars
As seen in Table 1, all of the 35 isolates could grow on the hexose sugars, glucose and fructose and the disaccharide sucrose, and most could use galactose, maltose, cellobiose, and trehalose; some strains could also use other sugars and sugar alcohols. Dry wines, with combined glucose and fructose levels by enzymatic analysis of 0.2 g/L or less, are quite able to support *Brettanomyces* growth (Chatonnet et al., 1995). *Brettanomyces* growth is too slow to compete successfully with *Saccharomyces* for sugar during yeast fermentation, but in stuck wines or bottled wines, *Brettanomyces* may help themselves to small amounts of various sugars along with other substrates.

Alcohol
Some strains (more European ones than US ones) can use ethyl alcohol as a sole carbon source in culture (Rodrigues et al., 2000, Conterno et al., 2006), and also in wine, reducing the alcohol level by a few tenths of a percent (personal observation). In aerobic conditions, many strains can produce alcohol and acetic acid from sugar. In anaerobic conditions acetic acid is not produced, and alcohol fermentation is at least temporarily inhibited, which is called the 'Custers Effect' (Ciani and Ferraro, 1997).

Variations in strain tolerance to alcohol have not been studied extensively, but this is definitely important. Barata et al. (2008) studied 29 strains, two of which tolerated almost 15% alcohol, but two did not even tolerate 14%. It may be significant the authors used mostly European strains, and included only two strains from California; many California wines 14% and over regularly grow *Brettanomyces*, and some strains manage alcohol levels at or over 15% (personal experience).

Growth patterns
There are also differences in growth patterns among strains. In a two-year study by Fugelsang and Zoecklein (2003), one strain reached a peak of cell growth, then declined (Fig. 1). Other strains grew, then declined so much that they were

Table 1 Summary of physiological characteristics of 35 *Brettanomyces* strains

Character tested	Frequency (%)	Isolates
Carbon source growth		
Arginine, cellulose, proline, tartrate	0	0
Adonitol	2	6
Arabinose, citrate, starch	3	9
Lactose, mannitol, raffinose	4	11
Ethanol	9	26
Glycerol	10	29
Lactate	12	34
Succinate	13	37
Malate	14	40
Galactose	28	80
Cellobiose, maltose	32	91
Trehalose	34	97
Sucrose	35	100
Nitrogen source growth		
Nitrate	25	71
Arginine, proline	35	100
Temperature growth		
at 37°C	13	37
at 10°C	11	31
Alcohol		
Tolerance >10%	35	100
Sulfite tolerance		
>30 mg/L at pH 3.4	17	49
pH growth		
at pH 2.0	33	94
4-EP and 4-EG (µg/L)		
High (>2000 4-EP; >1500 4-EG)	17	49
Medium (1000–2000 4-EP; 700–1500 4-EG)	6	17
Low (<50 4-EP; <60 4-EG)	7	20
None (<4.0 4-EP and 4-EG)	7	17

Source: Conterno *et al.* (2006).

not detectable by culturing, but later reached a second growth peak. Even when the culturable cell population had declined, the cells were still actively producing ethyl phenols. In another study (Barbin *et al.*, 2008), nine growth profiles of four types were found in 23 strains, including a two-stage growth by some strains. They wrote, 'One can assume that a compound necessary for growth was in default and that the yeast adjusted its metabolism before pursuing growth.' Whether or not the enzymes can continue to act in the wine after the cells have autolysed is not yet known. Much more research is needed on strain growth patterns to understand the behavior of these yeasts in wines.

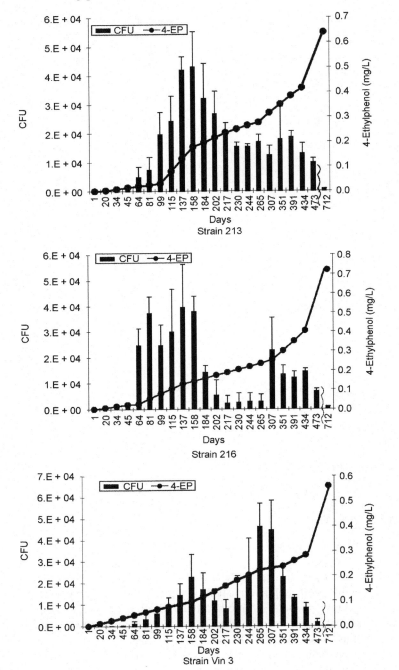

Fig. 1 Colony forming units/mL and concentration (mg/L) of 4-ethylphenol produced by three strains of *Brettanomyces bruxellensis* in Pinot noir wine. Source: Fugelsang and Zoecklein (2003).

References

BARATA, A., CALDIEIRA, J., BOTELHEIRO, R., PAGLIARA, D., MALFEITO-FERREIRA, M. and LOUREIRO, V., 2008, 'Survival patterns of *Dekkera bruxellensis* in wines and inhibitory effect of sulphur dioxide,' *Int. J. Food Microbiol.* 121:2, 201–207. doi: 10.1016/j.ifoodmicro.2007.11.020.

BARBIN, P., CHEVAL, J-L., GILIS, J-F., STREHAINANO, P. and TAILLANDIER, P., 2008, 'Diversity in spoilage yeast *Dekkera/Brettanomyces* isolated from French red wine: assessment during fermentation of synthetic wine medium,' *J. Inst. Brew* 114:1, 69–75.

BOULTON, R., SINGLETON, V., BISSON, L. and KUNKEE, R., 1996. *Principles and Practices of Winemaking*, New York, Chapman and Hall.

CHATONNET, P., DUBOURDIEU, D. and BOIDRON, J. N., 1992, 'The origin of ethyl phenols in wines,' *J. Sci. Food Agric.* 68, 165–178.

CHATONNET, P., DUBOURDIEU, D. and BOIDRON, J.N., 1995, 'The influence of *Brettanomyces/Dekkera* spp. yeasts and lactic acid bacteria on the ethyl phenol content of red wines,' *Am. J. Enol. Vitic.* 46, 463–468.

CIANI, M. and FERRARO, L., 1997, 'Role of oxygen on acetic acid production by *Brettanomyces/Dekkera* in winemaking,' *J. Sci. Food Agric.* 75:4, 489–495.

CONTERNO, L., JOSEPH, C.M.L., ARVIK, T., HENICK-KLING, T. and BISSON, L., 2006, 'Genetic and physiological characterization of *Brettanomyces bruxellensis* strains isolated from wines,' *Am. J. Enol. Vitic.* 57:2, 139–147.

COULON, J., PERELLO, M., LONVAUD-FUNEL, A., DE REVEL, G. and RENOUF, V., 2010, *Brettanomyces bruxellensis* evolution and volatile phenols production in red wines during storage in bottles,' *J. Appl. Microbiol.* 108, 1450–1455. doi: 10.1111/j.1365-2672.2009.04561.x.

CURTIN, C., BELLON, J., HENSCHKE, P., GODDEN, P. and BARROS-LOPEZ, M., 2007, 'Genetic diversity of *Dekkera/Brettanomyces bruxellensis* isolates from Australian wineries,' *FEMS Yeast Res.* 7, 471–481. doi: 10.1111/j.1567-1364.2006.00183.x.

DIAS, L., DIAS, S., SANCHO, T., STENDER, H., QUEROL. A., MALFEITO-FERREIRA, M. and LOUREIRO, V., 2003, 'Identification of yeasts isolated from wine-related environments and capable of producing 4-ethylphenol,' *Food Microbiol.* 20, 567–574.

DU TOIT, W., PRETORIUS, I. and LONVAUD-FUNEL, A., 2005, 'The effect of sulphur dioxide and oxygen on the viability and culturability of a strain of *Acetobacter pasteurianus* and a strain of *Brettanomyces bruxellensis* isolated from wine,' *J. Appl. Microbiol.* 98, 862–871.

DU TOIT, W.J., LISKAK, K., MARAIS, J. and DU TOIT, M., 2006, 'The effect of micro-oxygenation on the phenolic composition, quality and aerobic wine spoilage microorganisms of different South African red wines,' *S. Afr. J. Enol. Vitic.* 27:1, 57–67.

EGLI, M. and HENICK-KLING, T., 2001, 'Identification of *Brettanomyces/Dekkera* species based on polymorphism in the rRNA internal transcribed spacer region,' *Am. J. Enol. Vitic.* 52, 241–247.

FUGELSANG., K. and ZOECKLEIN, B., 2003, 'Population Dynamics and Effects of *Brettanomyces bruxellensis* Strains on Pinot noir (*Vitis vinifera* L.) Wines,' *Am. J. Enol. Vitic.* 54:4, 294–300.

FUGELSANG, K. and EDWARDS, C., 2007, *Wine Microbiology*, New York, Springer.

FUGELSANG, K., OSBORN, M. and MULLER, C., 1993, '*Brettanomyces and Dekkera*, implications in winemaking', in *Beer and Wine Production, Analysis,*

Characterization, and Technological Advances, Gump, B. (ed.), American Chemical Society, Washington, DC, 110–131.

JOSEPH, C. M.L., KUMAR, G., SU, E. and BISSON, L., 2007, 'Adhesion and biofilm production by wine isolates of *Brettanomyces bruxellensis*,' *Am. J. Enol. Vitic.* 58, 373–378.

LICKER, J., ACREE, T. and HENICK-KLING, T., 1998, 'What is "Brett" (*Brettanomyces*) flavor?: a preliminary investigation', in *Chemistry of Wine Flavor*, Waterhouse, A. and Ebeler, S. (eds), Washington, DC, American Chemical Society Symposium series.

MCMAHON, H. and ZOECKLEIN, B., 1999, 'Glycoside activity in *Brettanomyces bruxellensis* strains,' *J. Ind. Micro. Biochem.* 23, 198–203.

MANSFIELD, A., ZOECKLEIN, B. and WHITON, R., 2002, 'Quantification of glycosidase acitivity in selected strains of *Brettanomyces* and *Oenococcus oeni*,' *Am J. Enol. Vitc.* 53:4, 303–307.

MARTORELL, P., BARATA, A., MALFEITO-FERREIRA, M., FERNANDEZ-ESPINAR, M., LOUREIRO, V. and QUEROL, A., 2006, 'Molecular typing of the yeast species *Dekkera bruxellensis* and *Pichia guilliermondii* recovered from wine related sources,' *Int. J. Food Microbiol.* 106, 79–84. doi: 10.1016/j.ifoodmicro.2005.05.014.

OELOFSE, A., PRETORIUS, I. and DU TOIT, M., 2008, 'Significance of *Brettanomyces* and *Dekkera* during winemaking: a synoptic review,' *S. Afr. J. Enol. Vitic.* 29:2, 128–144.

RODRIGUES, N., GONÇALVES, G., PEREIRA DA SILVA, S., MALFEITO-FERREIRA, M. and LOUREIRO, V., 2000, 'Development and use of a new medium to detect yeasts of the genera *Dekkera/Brettanomyces*,' *J. Appl. Microbiol.* 90, 1–12.

ROMANO, A., PERELLO, M., DE REVEL, G. and LONVAUD-FUNEL, A., 2008, 'Growth and volatile compound production by *Brettanomyces/Dekkera bruxellensis* in red wine,' *J. Appl. Microbiol.* 104, 1577–1585. doi: 10.1111/j.1365-2672.2007.03693.x.

10.4 What do *Brettanomyces* do to wines?

L. Van de Water

Whether or not one appreciates the changes, it is obvious that *Brettanomyces/Dekkera* do affect the sensory characteristics of wine, sometimes only to a small degree, but sometimes profoundly. The more extensive the growth, and the longer the cells are present in the wine, the greater the sensory impact. However, it is essential to distinguish the sensory effects of *Brettanomyces* from other defects. For example, a 'barnyard' character may indeed be caused by *Brettanomyces*, but that descriptor is also used for H_2S or *Pediococcus*. Laboratory tests are necessary to distinguish different microbiological origins of sensory defects.

Production of ethyl phenols

Most *Brettanomyces/Dekkera bruxellensis* strains (perhaps all, in the right circumstances) produce the ethyl phenols 4-ethyl phenol (4EP), 4-ethyl guaiacol (4EG), and 4-ethyl catechol (4EC), by a two-step enzymatic process (cinnamate dehydrogenase followed by vinyl phenol reductase), from the hydroxycinnamic acids coumaric, ferulic, and caffeic, respectively (Hereszytn, 1986a; Chatonnet *et al.*, 1992, 1995, Hesford *et al.*, 2004; and many others).

Pichia guilliermondii, a yeast found on grapes and in wineries, can produce 4EP but has not been found to do so in wine (Stender *et al.*, 2001; Jensen *et al.*, 2009). Dias *et al.* (2003) remarked, '*D. bruxellensis* remain as the sole agents of "phenolic off-flavours" in wines.'

Production of ethyl phenols varies with *Brettanomyces* strain (Conterno *et al.*, 2006; Romano *et al.*, 2008; Oelofse *et al.*, 2009; Harris *et al.*, 2009; Coulon *et al.*, 2010), grape variety (Pollnitz *et al.*, 2000), and composition of the wine (lower alcohol leads to higher levels; higher residual sugar leads to higher levels), levels of precursors, temperature (more are produced at a higher temperature than lower), available oxygen (oxygen leads to higher ethyl phenols), and other variables (du Toit *et al.*, 2005, 2006; Suárez *et al.*, 2007; Romano *et al.*, 2008, Jensen *et al.*, 2009, Benito *et al.*, 2009), not all of them identified.

More 4EP usually is produced than 4EG, but ratios vary, causing pronounced sensory variations. Chatonnet *et al.* (1992) reported an 8:1 ratio for 4EP:4EG. In a survey of 61 Australian reds, Pollnitz *et al.* (2000) found average ratios of 4EP:4EG for Cabernet of 10:1, for Shiraz of 9:1 (highest 23:1) and for Pinot Noir 3.5:1 (lowest 1.6:1). A later Australian survey found 4EP:4EG ratios in Cabernet from 2:1 to 21:1 with higher ratios in cooler regions than in warm ones (Coulter *et al.*, 2003). An average ratio of 10:1 is often assumed (Romano *et al.*, 2009).

Sensory effects of ethyl phenols

The ethyl phenols 4EP and 4EC are usually described as having 'Band-Aid®' or 'plastic' odors, and 4EG smells 'medicinal', 'spicy' or like 'clove' or 'burnt beans.' The combination of these and other metabolites results in a character with descriptors such as 'sweaty horse saddle', 'wet dog/wet wool', 'barnyard', 'manure', 'gamey', 'tobacco', and others (Fugelsang et al., 1993; Boulton et al., 1996, Licker et al., 1998; Arvik and Henick-Kling, 2002). Fruitiness is reduced, the roundness and intensity in the mouth are diminished, and there is often a bitter, metallic finish.

Traditionally, levels of 640 parts per billion, (ppb, also written as micrograms/liter or nanograms/ml) or 425 ppb have been cited as the sensory threshold for 4EP in wine (Chatonnet et al., 1992; Godden et al., 2004; Ribereau-Gayon et al., 2006; Romano et al., 2009), but Peter Godden of the AWRI was quoted as saying, 'I've not been able to find an Aussie winemaker who doesn't find 100 micrograms/litre [4EP] negative' (Goode, 2003).

In fact, as many AWRI tastings have demonstrated, sensory detection levels for ethyl phenols vary greatly with wine composition, taster sensitivity, presence of other *Brettanomyces* metabolites, and even wine temperature (Curtin et al., 2008; Cliff and King, 2009). In one study, the average detection level for a 10:1 mixture of 4EP and 4EG in Bordeaux wine was 92 ppb, but the levels in samples judged 'not tainted' by experienced tasters ranged from 5 to 1370 ppb, with an average of 403 ppb (Romano et al., 2009). What is very important to understand is that the sensory perception of *Brettanomyces* character depends on many variables.

Fatty acids

Two fatty acids, isovaleric acid (IVA) and isobutyric acid (IBA) also contribute significantly to 'Bretty' odors (Fugelsang et al., 1993; Licker et al., 1998; Coulter et al., 2003; Oelofse et al., 2008; Romano et al., 2009). Their production has not been correlated with ethyl phenol production, but they may tend to mask ethyl phenol perception somewhat (Romano et al., 2009). The sensory effect of the combination, however, may be considered more unpleasant than ethyl phenols alone (Oelofse et al., 2008; also personal observation) as IVA adds a 'rancid', 'vomit', 'sweaty feet', 'wet goat' or 'stinky cheese' component to the wine aroma (it is also a female pheromone).

Color degradation

Brettanomyces growth has been associated with color degradation in some wines. It has been postulated that the glycosidase activity found in many strains (Mansfield et al., 2002) may lead to formation of colorless anthocyanins (Oelofse et al., 2008), but this hypothesis has not been confirmed.

'Mousiness'

At least some *Brettanomyces* strains are capable of producing the 'mousy taint' compounds acetyltetrahydropyridine and acetylpyrroline from the amino acids ornithine and lysine (Hereszytn, 1986b; Costello et al., 2001; Costello and Henschke, 2002; Grbin et al., 2007; Snowdon et al., 2006), which are liberated by yeasts after fermentation. The affected wine has a spectacularly offensive aftertaste of mouse urine or rancid macadamia nuts (more familiar to some people than mouse urine). Despite numerous references to an intense odor, these compounds are only slightly volatile at wine pH, so 'mousy' wines in fact do not usually smell strongly of mice, but rather have a 'popcorn' or 'jasmine rice' aroma (Heresztyn, 1986b; Arvik and Henick-Kling, 2002; Snowdon et al., 2006). 'Mousiness' is now so rare that many writers may not have experienced an actual 'mousy' wine themselves, or at least not for some years.

In the mouth, or if the wine is rubbed between the hands, the compounds become volatile at the higher pH of the body. So, when a 'mousy' wine is tasted, the nasty retronasal (palatal) aroma suddenly develops and is perceived as a horrible aftertaste after a few seconds' delay. The length of the delay differs for different people tasting the same wine (Paul Henschke, personal communication), and the varying ability of tasters to detect the off-flavor appears to be genetic, perhaps related to saliva pH (Snowdon et al., 2006).

While it is well established that *Brettanomyces* can produce the 'mousy' compounds, especially in the presence of oxygen, *Brettanomyces* yeasts are not the only source. Lactic acid bacteria also produce the 'mousy' heterocycles, perhaps even more so than *Brettanomyces* alone (Costello et al., 2001; Arvik and Henick-Kling, 2002). A 'mousy taint' can also occur after heating wine if yeast lees and/or lactic acid bacteria are present (personal observation).

Acetic acid

Most, If not all, *Brettanomyces* strains produce large amounts of acetic acid aerobically in culture (Freer, 2002), and in wine if sufficient oxygen is present (Ciani and Ferraro, 1997; du Toit et al., 2005, 2006). However, *Acetobacter* bacteria are nearly universal companions of *Brettanomyces* in cellared wine, so a rise in volatile acidity is not necessarily attributable to the yeasts. A survey of bottled wines by the Australian Wine Research Institute did not find a correlation between levels of ethyl phenols and volatile acidity (Godden et al., 2004).

Other effects

Brettanomyces can also produce biogenic amines (Caruso et al., 2002), though strain differences have not been established. Some European countries put voluntary or compulsory limits on biogenic amines, so this is a topic for further research.

Brettanomyces metabolites can reduce the perceived fruitiness of wines, especially reds, by masking the fruity esters (Fugelsang et al., 1993; Licker et al., 1998; Cliff and King, 2009) or perhaps by producing esterases that metabolize some of them. As wines age, and fresh fruitiness diminishes, ethyl phenols may also seem more intense. Varietal character often becomes much less prominent after *Brettanomyces* growth (Boulton et al., 1996), sometimes to the point that experienced tasters may not be able to identify the variety, even if they do not find the wine to be spoiled (personal observation).

Spoilage or complexity?
This is a controversial question. At lower levels of ethyl phenols, many people often consider the effects of *Brettanomyces* to be a 'complexity' rather than spoilage, though levels of perception and preferences vary widely. Wines that some tasters consider completely spoiled are judged to be acceptable or even pleasant by others (Boulton *et al.*, 1996; Arvik and Henick-Kling, 2002; Goode, 2003; Curtin *et al.*, 2008; Romano *et al.*, 2008).

Some wine writers actively appreciate some 'Brett' character, going so far as to call people worried about the effects of *Brettanomyces* 'Brett nerds' (Lynch, 2004). Winemakers do occasionally blend wines with a small amount of 'Brettiness' to improve ratings (personal communications, names withheld on request). Prepared *Brettanomyces* cultures are available for use in beermaking, and may perhaps have been used in wine. However, it is wise to remember the admonition, 'never trust a microbe.' A strain of *Brettanomyces* that produced just a slight complexity in one wine in one situation may completely spoil another wine. Some winemakers seek 'good' strains of *Brettanomyces* which would add complexity but not spoil wine; however, the very real possibility that a strain might behave well in one wine, but badly in another, tends to discourage this search. At higher levels of ethyl phenols, their flavor begins to predominate, which many tasters dislike. A 'metallic' aftertaste may also develop, which is often perceived negatively even by those who are not offended by the aromas.

We tend to enjoy flavors to which we have become accustomed. In the past 30 years or so, there has been so much increase in 'Bretty' wines that for some people, that character is normal for red wine. At a wine industry conference in 2003, most tasters thought that a wine with 4000 ppb 4EP was disgusting, but one said that to him it just smelled 'normal' (personal observation). Indeed, 'Brett' character is sometimes thought to be a 'gout de terroir', although it is hard to understand how a microbe found worldwide could be considered part of a particular terroir. 'The same *Brettanomyces* characters will be imparted to all varieties in all climates. *Brettanomyces* metabolites are dominant and mask other flavors.' (Boulton *et al.*, 1996). Commercial winemakers who do not find 'Brett' character offensive should remember that many consumers consider very 'Bretty' wines to be unpleasant to undrinkable. A smaller winery may be able to target consumers who appreciate the 'Brett' character, but larger wineries may not have this option.

Interestingly, more than one winery making expensive wines in more than one country has based their signature style on the effects of *Brettanomyces*. Fugelsang *et al.* (1993) noted that 'some (not all) internationally recognized, award-winning wines had perceivable "Brett character"'. However, other wineries have found less acceptance for their intensely 'Bretty' wines, and have had to withdraw some of them from the market at considerable expense (names withheld on request). Few wine microbiologists, and winemakers whose wines have spoiled from the effects of *Brettanomyces*, would agree with Marc Perrin of Château de Beaucastel, who said, 'There are certainly some *Brettanomyces* in every natural wine because *Brettanomyces* is not a spoilage yeast (as many people think) but one of the yeasts that exist in winemaking' (Goode, 2003).

For a discussion of *Brettanomyces* growth in bottled wines, see **10.9**.

References

ARVIK, T. and HENICK-KLING, T., 2002, 'Overview: *Brettanomyces bruxellensis* occurrence, growth, and effect on wine flavor,' *Practical Winery and Vineyard*, May/June: 117–123.

BENITO, S., PALOMERO, F., MORATA, A., CALDERÓN, F. and SUÁREZ-LEPE, J., 2009, 'Factors affecting the hydroxycinnamate decarboxylase/vinylphenol reductase activity of *Dekkera/Brettanomyces*: application for *Dekkera/Brettanomyces* control in winemaking,' *J. Food Sci.* 74:1, M16–M21.

BOULTON, R., SINGLETON, V., BISSON, L. and KUNKEE, R., 1996. *Principles and Practices of Winemaking*, New York, Chapman and Hall.

CARUSO, M., FIORE, C., CONTURSI, M., SALZANO, G., PAPARELLA, A. and ROMANO, P., 2002, 'Formation of biogenic amines as criteria for the selection of wine yeasts,' *World J. Microbiol. Biotechnol.* 18, 159–163.

CHATONNET, P., DUBOURDIEU, D. and BOIDRON, J. N., 1992, 'The origin of ethyl phenols in wines,' *J. Sci. Food Agric.* 68, 165–178.

CHATONNET, P., DUBOURDIEU, D. and BOIDRON, J. N., 1995, 'The influence of *Brettanomyces/Dekkera* spp. yeasts and lactic acid bacteria on the ethyl phenol content of red wines,' *Am. J. Enol. Vitic.* 46, 463–468.

CIANI, M. and FERRARO, L., 1997, 'Role of oxygen on acetic acid production by *Brettanomyces/Dekkera* in winemaking,' *J. Sci. Food Agric.* 75:4, 489–495.

CLIFF, M. and KING, M., 2009, 'Influence of serving temperature and wine type on perception of ethyl acetate and 4-ethyl phenol in wine,' *J. Wine Res.* 20, 45–52. doi: 10.1080/09571260902978535.

CONTERNO, L., JOSEPH, C.M.L., ARVIK, T., HENICK-KLING, T. and BISSON, L., 2006, 'Genetic and physiological characterization of *Brettanomyces bruxellensis* strains isolated from wines,' *Am. J. Enol. Vitic.* 57:2, 139–147.

COSTELLO, P., LEE, T. and HENSCHKE, P., 2001, 'Ability of lactic acid bacteria to produce N-heterocycles causing mousy off-flavour in wine,' *Aust. J. Grape and Wine Res.* 7, 160–167.

COSTELLO, P. and HENSCHKE, P., 2002, 'Mousy off-flavor of wine: precursors and biosynthesis of the causative N-heterocycles 2-ethyltetrahydropyridine, 2-acetyltetrahydropyridine, and 2-acetyl-1-pyrroline by *Lactobacillus hilgardii* DSM 20176,' *Agric. Food Chem.* 50, 7079–7087.

COULON, J., PERELLO, M., LONVAUD-FUNEL, A., DE REVEL, G. and RENOUF, V., 2010, *Brettanomyces bruxellensis* evolution and volatile phenols production in red wines during storage in bottles,' *J. Appl. Microbiol.* 108, 1450–1455. doi: 10.1111/j.1365-2672.2009.04561.x.

COULTER, A., ROBINSON, E., COWEY, G., FRANCIS, I. L., LATTEY, K., CAPONE, D., GISHEN, M. and GODDEN, P., 2003, '*Dekkera/Brettanomyces* yeast – an overview of recent AWRI investigations and some recommendations for its control,' Proceedings of a Seminar

Organized by the Australian Society for Viticulture and Oenology, *Grapegrowing on the Edge, Managing the Wine Business, Impacts on Wine Flavor*, Bell, S., de Garis, K., Dundon, C., Hamilton, R., Partridge S. and Wall, S. (eds), 41–50.

CURTIN, C., BRAMLEY, B., COWEY, G., HOLDSTOCK, M., KENNEDY, E., LATTEY, K., COULTER, A., HENSCHKE, P., FRANCIS, L. and GODDEN, P., 2008, 'Sensory perceptions of "Brett" and relationship to consumer preference', in *Proceedings of the Thirteenth Australian Wine Industry Technical Conference*, Adelaide, SA, 207–211.

DIAS, L., DIAS, S., SANCHO, T., STENDER, H., QUEROL. A., MALFEITO-FERREIRA, M. and LOUREIRO, V., 2003, 'Identification of yeasts isolated from wine-related environments and capable of producing 4-ethylphenol,' *Food Microbiol.* 20, 567–574.

DU TOIT, W., PRETORIUS, I. and LONVAUD-FUNEL, A., 2005, 'The effect of sulphur dioxide and oxygen on the viability and culturability of a strain of *Acetobacter pasteurianus* and a strain of *Brettanomyces bruxellensis* isolated from wine,' *J. Appl. Microbiol.* 98, 862–871.

DU TOIT, W.J., LISKAK, K., MARAIS, J. and DU TOIT, M., 2006, 'The effect of micro-oxygenation on the phenolic composition, quality and aerobic wine spoilage microorganisms of different South African red wines,' *S. Afr. J. Enol. Vitic.* 27:1, 57–67.

FREER, S. N., 2002, 'Acetic acid production by *Dekkera/Brettanomyces* yeasts,' *World J. Microbiol. Biotechnol.* 18, 271–275.

FUGELSANG, K., OSBORN, M. and MULLER, C., 1993, '*Brettanomyces* and *Dekkera*, implications in winemaking', in *Beer and Wine Production, Analysis, Characterization, and Technological Advances*, Gump, B. (ed.), American Chemical Society, Washington, DC, 110–131.

GODDEN, P., COULTER, A., CURTIN, C., COWEY, G. and ROBINSON, E., 2004, 12th AWITC, '*Brettanomyces* workshop: Latest research and control strategies,' Melbourne, Australia.

GOODE, J., 2003, '*Brettanomyces*', *Harpers Wine and Spirit Weekly*, 18 April, 42–46.

GRBIN, P. R., HERDERICH, M., MARKIDES, A., LEE, T. and HENSCHKE, P., 2007, 'The role of lysine amino nitrogen in the biosynthesis of mousy off-flavor compounds by *Dekkera anomala*,' *J. Agric. Food Chem.* 55:26, 10872–10879.

HARRIS, V., FORD, C., JIRANEK, V. and GRBIN, P., 2009, 'Survey of enzyme activity responsible for phenolic off-flavour production by *Dekkera/Brettanomyces* yeast', *Appl. Microbiol. Biotechnol.* 81: 1117–1127. doi: 10.1007/s00253-008-1708-7.

HERESZTYN, T., 1986a, 'Metabolism of phenolic compounds from hydroxycinnamic acids by *Brettanomyces* yeasts,' *Arch. Microbiol.* 146, 96–98.

HERESZTYN, T., 1986b, 'Formation of substituted tetrahydropyridines by species of *Brettanomyces* and *Lactobacillus* isolated from mousy wines,' *Am. J. Enol. Vitic.* 37, 127–132.

HESFORD, F., SCHNEIDER, K., PORRET, N. and GAFNER, J., 2004, 'Identification and analysis of 4-ethyl catechol in wine tainted by *Brettanomyces* off-flavor,' *Abstr. Am. J. Enol. Vitic.* 55, 304A.

JENSEN, S., UMIKER, N., ARNEBORG, N. and EDWARDS, C., 2009, 'Identification and characterization of *Dekkera bruxellensis, Candida pararugosa* and *Pichia guilliermondii* isolated from commercial red wines,' *Food Microbiol.* 26, 915–921. doi: 10.1016/j.fm.2009.06.010.

LICKER, J., ACREE, T. and HENICK-KLING, T., 1998, 'What is "Brett" (*Brettanomyces*) flavor?: a preliminary investigation', in *Chemistry of Wine Flavor*, Waterhouse, A. and Ebeler, S. (eds), Washington, DC, American Chemical Society Symposium series.

LYNCH, K., 2004, 'Attack of the Brett nerds,' in *Inspiring Thirst*, Ten Speed Press, Berkeley, CA.

MANSFIELD, A., ZOECKLEIN, B. and WHITON, R., 2002, 'Quantification of glycosidase acitivity in selected strains of *Brettanomyces* and *Oenococcus oeni*,' *Am J. Enol. Vitc.* 53:4, 303–307.

OELOFSE, A., PRETORIUS, I. and DU TOIT, M., 2008, 'Significance of *Brettanomyces* and *Dekkera* during winemaking: a synoptic review,' *S. Afr. J. Enol. Vitic.* 29:2, 128–144.

OELOFSE, A., LONVAUD-FUNEL, A. and DU TOIT, M., 2009, 'Molecular identification of *Brettanomyces bruxellensis* strains isolated from red wine and volatile phenol production,' *Food Microbiol.* 26, 377–385. doi: 10.1016/j.fm.2008.10.011.

POLLNITZ, A., PARDON, K. and SEFTON, M., 2000, 'Quantitative analysis of 4-ethylphenol and 4-ethylguauacol in red wine', *J. Chromatogr. A* 874, 101–109.

RIBEREAU-GAYON, P., GLORIES, Y., MAUJEAN, A. and DUBOURDIEU, D., 2006, *Handbook of Enology Volume 2: The Chemistry of Wine Stabilization and Treatments*, John Wiley & Sons, Chichester.

ROMANO, A., PERELLO, M., DE REVEL, G. and LONVAUD-FUNEL, A., 2008, 'Growth and volatile compound production by *Brettanomyces/Dekkera bruxellensis* in red wine,' *J. Appl. Microbiol.* 104, 1577–1585. doi: 10.1111/j.1365-2672.2007.03693.x.

ROMANO, A., PERELLO, M., LONVAUD-FUNEL, A., SICARD, G. and DE REVEL, G., 2009, 'Sensory and analytical re-evaluation of "Brett character",' *Food Chemistry* 114, 15–19. doi: 10.1016/j.foodchem.2008.09.06.

SNOWDON, E., M., BOWYER, M., GRBIN, P. and BOWYER, P., 2006, 'Mousy off-flavor: a review,' *J. Agric. Food Chem.* 54, 6465–6474. doi: 10.1021/jf05528613.

STENDER, H., KURTZMAN, C., HYLDIG-NIELSEN, J., SORENSEN, D., BROOMER, A., OLIVEIRA, K., PERRY-O'KEEFE, P., SAGE, A., YOUNG, B. and COULL, J., 2001, 'Identification of *Dekkera bruxellensis (Brettanomyces)* from wine by fluorescence *in situ* hybridization using peptide nucleic acid probes,' *Appl. Environ. Microbiol.* 67:2, 938–941. doi: 10.1128/AEM.67.2.938-941.2001.

SUÁREZ, R., SUÁREZ-LEPE, J., MORATA, A. and CALDERÓN, F., 2007, 'The production of ethylphenols in wine by yeasts of the genera *Brettanomyces* and *Dekkera*: a review,' *Food Chem.* 102, 10–21. doi: 10.1016/j.foodchem.2006.03.030.

10.5 How do I sample for *Brettanomyces* testing?

L. Van de Water

Record-keeping

When wine is tested for microbes, extensive records need to be kept or the results will not be very useful. Keeping the following information is highly recommended:

- wine identity
- container identity, including each individual barrel sampled
- sampling method and date
- date tested
- person or commercial lab doing the tests
- test methods
- test procedure if done in-house
- test results
- interpretation, and
- action taken.

Sampling procedures

Use new plastic or glass containers with new caps/corks, or resanitized containers and caps (autoclave, or run through a dishwasher with no soap, or rinse with 70% beverage alcohol; do not try to resanitize corks) for samples that are to be cultured. Do not rinse sampling devices or containers with unsterilized water. Sampling devices must be new or carefully resanitized between barrels to avoid contaminating the next sample, and also the barrel itself.

A recommended resanitizing procedure in the cellar is to rinse the device with water, then 70% beverage alcohol (not 95%, it is more effective if diluted to 70%), then with the next wine. If high-proof beverage alcohol is not available, 100-proof vodka will have to do. Do not use 'denatured' or 'reagent' alcohol, or 'meths'; these have additives that are poisonous to microbes as well as people. It is all right to put in an SO_2 rinse as well, but SO_2 is not a contact sterilant, so the alcohol rinse is important.

Except for chemical or sensory analysis, sampling is critical no matter which test method is used, because *Brettanomyces* cells tend to settle to the bottom of a barrel or other container (Barata *et al.*, 2008). Take a racking valve sample from a tank after letting at least a liter of wine run out of the valve (it can be returned to the top of the tank). Do not take top samples except for film evaluation. If a cell count that is representative of the wine is desired, the wine must be stirred or circulated. If the purpose is to find any *Brettanomyces* cells that may be present, a sample can be taken from the bottom valve of a tank, or from the bottom of a barrel with a sterilized hose. Another way to sample barrels is to use a device that takes a sample near but not at the bottom of a barrel, without a sucking action that disturbs lees.

Sampling new vs old barrels

New barrels bind free SO_2 at around twice the rate of older barrels (Ribereau-Gayon et al., 2006). Also, cellobiose (used by many strains of *Brettanomyces*, see **10.3**) is higher in new wood (Fugelsang et al., 1993), and there is also more trapped oxygen, so *Brettanomyces* tends to grow more quickly and extensively in new oak. If a wine is aging in a mixture of new and older barrels, the new barrels and old barrels should be sampled separately for SO_2 measurements and *Brettanomyces* tests.

For culturing *Brettanomyces*, it is extremely important not to take samples within one to three weeks of adding SO_2. Winemakers have observed for 30 or more years that at least some *Brettanomyces* strains respond to SO_2 by losing their ability to grow in culture (personal observation). The cells remain alive but temporarily do not grow on culture media (Millet and Lonvaud-Funel, 2000; du Toit et al., 2005), resulting in false negatives based on plating (no visible colonies on agar = no *Brettanomyces* present). This state is sometimes called VBNC, although there are questions about whether this is the correct terminology. At least in some cases, the cells apparently can recover from this state and may reactivate in the wine (Umiker and Edwards, 2007). If SO_2 has been added recently to a wine, it would be best to use PCR-based methods of detection.

Unfortunately, it is still not possible to take samples for *Brettanomyces* testing from a barrel without destroying the barrel (Kenneth Fugelsang, personal communication), because the yeasts can penetrate deep into the wood (Malfeito-Ferreira et al., 2004). Testing wine aged in a barrel is still the best way to test for *Brettanomyces* infection in that barrel.

References

BARATA, A., CALDIEIRA, J., BOTELHEIRO, R., PAGLIARA, D., MALFEITO-FERREIRA, M. and LOUREIRO, V., 2008, 'Survival patterns of *Dekkera bruxellensis* in wines and inhibitory effect of sulphur dioxide,' *Int. J. Food Microbiol.* 121:2, 201–207. doi: 10.1016/j.ifoodmicro.2007.11.020.

DU TOIT, W., PRETORIUS, I. and LONVAUD-FUNEL, A., 2005, 'The effect of sulphur dioxide and oxygen on the viability and culturability of a strain of *Acetobacter pasteurianus* and a strain of *Brettanomyces bruxellensis* isolated from wine,' *J. Appl. Microbiol.* 98, 862–871.

FUGELSANG, K., OSBORN, M. and MULLER, C., 1993, '*Brettanomyces and Dekkera*, implications in winemaking', in *Beer and Wine Production, Analysis, Characterization, and Technological Advances*, Gump, B. (ed.), American Chemical Society, Washington, DC, 110–131.

MALFEITO-FERREIRA, M., LAUREANO, P., BARATA, A., D'ANTUONO, I., STENDER, H. and LOUREIRO, V., 2004, 'Effect of different barrique sanitation procedures on yeasts isolated from the inner layers of wood,' *Abstr. Am. J. Enol. Vitic.* 55, 304A.

MILLET, V. and LONVAUD-FUNEL, A., 2000, 'The viable but non-culturable state of wine micro-organisms during storage', *Lett. Appl. Microbiol.* 30, 136–141.

RIBEREAU-GAYON, P., GLORIES, Y., MAUJEAN, A. and DUBOURDIEU, D., 2006, *Handbook of Enology Volume 2: The Chemistry of Wine Stabilization and Treatments*, John Wiley & Sons, Chichester.

UMIKER, N. and EDWARDS, C., 2007, 'Effects of SO_2 on *Brettanomyces*', presented at Santiago, Chile, July.

10.6 What methods do I have available to detect *Brettanomyces* infection?

L. Van de Water

There are several methods for detecting *Brettanomyces*, all helpful to the winemaker. Rather than selecting one method only, a combination of methods gives the most complete picture of the status of a *Brettanomyces* infection in a wine. Wineries are best served by utilizing all methods at their disposal. The key is to select the appropriate methods for the information you need.

Sensory detection

Sensory evaluation of wines should begin with tasting of fermentors every day, and continue every month or two until the wine is bottled. After bottling, wines should be tasted and analyzed every few months during the first two years, to detect any changes in the bottle.

With practice, winery workers can become proficient in recognizing the first signs of 'Brett' character developing in cellared wine. Even if no specific sensory character is perceived, if there is an unexplained difference between one container and others in the same batch, *Brettanomyces* or other microbial activity should be suspected. When the sensory effects of *Brettanomyces* growth are strong, the wine has been affected permanently, but even noticing it at this point is better than missing it entirely. Train everyone handling the wine, in the cellar, lab, and tasting room, to recognize 'Bretty' character. Use samples of 4-ethyl phenol (4EP) and 4-ethyl guaiacol (4EG), and other compounds such as isovaleric acid, plus a wine naturally spoiled by *Brettanomyces*.

- Levels of concern: any hint of 'Brett' character should be investigated by other detection methods.
- When to use sensory evaluation: any time that wine is handled, winery personnel should be alert to gassiness, cloudiness or odd aromas/flavors.

Chemical analysis

Routine chemical tests can indicate activity by a range of microbes, including *Brettanomyces*. Recommended tests include pH, free and total SO_2, volatile acidity, and glucose+fructose (separately, not added together). For a discussion of the meaning of the results, see **10.8**.

Measurement of the *Brettanomyces* metabolites 4EP and 4EG by gas chromatography-mass spectrometry confirms the sensory effects of *Brettanomyces*, and helps wineries correlate sensory observations with analytical data. As noted in **10.4**, sensory detection levels depend on wine composition, ratio of 4EP to 4EG, the individual taster, and many other variables, so measurement of ethyl phenols gives an objective reference point. Production of these ethyl phenols lags behind cell growth, and can continue to rise after cells have declined and are no longer detected in culture, so levels of ethyl phenols do not correlate well with cell growth patterns (Fugelsang and Zoecklein, 2003).

- Levels of concern for ethyl phenols: levels of 4EP over four ppb indicate *Brettanomyces* growth at some point. A level of around 100 ppb 4EP may be detectible by 250–500 ppb 4EP and 20–50 ppb 4EG, a sensory effect should be noticeable (Godden *et al.*, 2004; Guerra, 2008). At levels over 1000 ppb 4EP and 100 ppb 4EG the aroma and flavor of most wines would be seriously impacted.
- When to test 4EP and 4EG: during cellaring to confirm *Brettanomyces* growth, and to correlate with sensory observations.

Microscopic exam

Cells can be seen microscopically if there are at least 1000–2000 cells/ml, preferably more. This is a large population, but *Brettanomyces* cell counts sometimes reach a million per ml or more. Samples can be centrifuged to concentrate the cells at least 10 times, but particulate matter can make it difficult to see cells after centrifuging. Wine samples are prepared for direct examination as wet mounts by putting a small drop on a microscope slide and covering it with a cover glass of #1 thickness; these preparations last for half an hour or so before starting to dry up.

Wet mounts are examined with a very good phase-contrast microscope. Costs start around US$2000–3000 for one with suitable resolution with plan or planachromat objectives of $10\times$, $40\times$, and $100\times$ (oil). Eyepiece lenses are often $10\times$, but $15\times$ is greatly preferred because the microbes appear larger and thus are easier to find. Phase-contrast permits observation of microbes and particles directly without staining, because light going through the cell is retarded compared with light passing through the wine. This sets up a diffraction pattern that makes the cell membrane dark and the cell contents darker or lighter than the background.

If cells look clear but ghostly under phase contrast, without much distinction between them and the background, the phase rings in the condenser may be out of alignment, or an incorrect phase ring, which does not match the objective may have been selected. Causes of fuzzy images are oil on the $40\times$ objective (clean with lens cleaner and lens paper) or two cover glasses stuck together. Fingerprints on the cover glass and dried-up places on the slide can resemble yeasts or bizarre microbes. Look around for the edge of an air bubble; the microbes will be in the same plane of focus as the air bubble, and it will then be easy to tell a dried place from the wine sample.

Brightfield microscopes are not very useful for examining wine directly because the samples must be stained. White wines can be stained with Gram stain or 0.5% methylene blue and examined for yeast, but wine bacteria may resemble stain particles. Red wine pigment precipitates the stain, so red samples must be centrifuged and washed to remove the color, which may also remove some microbes. Phase-contrast microscopes are much preferable to brightfield for direct examination of wine.

Microscope operators must be trained specifically in wine microbe identification, not just general microbiology. Under the microscope, *Brettanomyces* cells are smaller than *Saccharomyces* cells, and are often apiculate (pointy). Some young cells may be ovoid (egg-shaped) but more will resemble olives or thin bowling pins. Bud scars tend to flatten the ends, so older cells begin to look like rowboats, barrels, or gothic arches ('ogive'), often with a bud on an angle at one of the edges. Identification cannot be confirmed by microscopic exam alone, so if any suspiciously apiculate yeasts are seen with the microscope, the wine should be checked for *Brettanomyces* by more definitive methods.

When growing in liquid (such as wine) *Brettanomyces* cells tend to make 'pseudomycelia', elongated structures resembling thin mold hyphae (Edwards, 2005). They range in length from 25 μm or so to 100 μm or much more. Small pseudomycelia can be seen in some cultures on Petri dishes, but they are much more extensive when growing in liquid.

- Levels of concern: any *Brettanomyces* cells (confirm by culturing or PCR) seen directly under the microscope indicate an extensive infection.
- When to examine wine microscopically for *Brettanomyces*: stuck wines (especially during reinoculation), wines that seem to be developing 'Bretty' characteristics, and any time that unexplained activity or sensory change (good or bad) occurs in one or more containers of the same batch. Topping wines should be examined for *Brettanomyces* and other microbes (*Acetobacter* especially) and cultured for *Brettanomyces*; microbiological disasters have happened because all the red wine in the winery was topped with wine infected with *Brettanomyces*.

Culturing
Culturing wines in Petri dishes is the traditional method of detecting *Brettanomyces*. This method is discussed in **10.7**.

Genetic methods: real-time PCR
In the past few years, tests based on polymerase chain reaction (PCR) for *Brettanomyces* and other wine microbes have become available (Phister and Mills, 2003; Cocolin *et al.*, 2004; Tessonnière *et al.*, 2009). The technology is very promising, representing a quantum leap from traditional methods. PCR methods take a few hours instead of days, and can often detect levels as low as 10 cells/ml, or possibly even lower if the wine is centrifuged to concentrate cells.

Real-time PCR is a molecular biology method, which amplifies DNA and links the amplification to fluorescence. In the first step, a small sequence of DNA (the 'target region') specific to the microbe of interest is isolated from the sample by lysing (breaking open) the cells, and this target is amplified using enzymes and primers, small pieces of DNA which bind to the selected target region. Then the primers and enzymes copy the target DNA billions of times using cycles of alternatively heating and cooling in an instrument called a thermocycler. The DNA is linked to fluorescence of a dye added along with the primer or in some cases attached to it. The amount of DNA originally present in the sample is estimated by comparing fluorescence in the sample to prepared standards of *Brettanomyces* (Phister and Mills, 2003; Phister, personal communication).

The methods are not without pitfalls. Care must be taken during extraction of the DNA to prevent contamination from extraneous DNA. Phenolic compounds in red wine interfere with PCR assays and must be removed by adding PVPP before cell lysis. It is difficult to prepare standards with known concentrations, which are essential to quantitative results, partly because of *Brettanomyces*' tendency to form pseudomycelia, and because standards must be freshly prepared, or the cell numbers may change. Tessonnière *et al.* (2009) also recommended using another microbe and its primer as an internal control to monitor the success of the procedure from DNA extraction through amplification and detection, to prevent false negatives.

As well as active cells, PCR detects VBNC cells and other cells that are temporarily unable to grow in culture (Millet and Lonvaud-Funel, 2000). Thus, PCR can detect cells which would be missed by culturing but which may still be producing ethyl phenols, and which may also reactivate and resume growth later on (Umiker and Edwards, 2007; Coulon *et al.*, 2010).

Detection of dead cells by PCR-based tests in wine is controversial. Forensic PCR tests were designed to detect long-dead cells, including a 68-million-year-old dinosaur bone tested successfully a few years ago, but the time between yeast cell death in wine and disintegration of the nucleus, releasing DNA, is not clear. Dr David Mills remarked, 'As long as the targeted sequence of DNA, typically only a few hundred base pairs long, is still intact, a PCR will likely amplify that DNA.' The dye ethidium monoazide is being investigated to bind with dead *Brettanomyces* cells and eliminate them from the DNA preparation before testing with PCR (Mills and Neeley, 2006; Phister, personal communication). This is a promising way to ensure that dead cells are not detected.

Depending on the DNA sequence selected, the method can be extremely specific for the target microbe species, or it may be designed for broader reactions. Specificity is the greatest strength of PCR-based tests, if specificity is assured. There are several primers for *Dekkera/Brettanomyces bruxellensis* in use at labs around the world; ideally, these primers would react with all strains of this species, but with no other species. However, primers usually are checked against certain other species, but not all species which could potentially react. There is an unpublished report that a primer currently in use for *B. bruxellensis* reacted with other yeasts, including *Candida catenulata* and *Debaryomyces hansenii*, though it did not react with *Dekkera anomala*. Another primer for *B. bruxellensis* did not react with those non-*bruxellensis* yeasts, or with certain other species that may also be confused with *B. bruxellensis* in culture (unpublished results). A thorough testing of more *Brettanomyces* primers against numerous potential cross-reactive species is underway and will be beneficial to prevent false positive results.

While very useful, PCR tests do not completely replace culturing to determine numbers of cells are alive and active. Both methods provide very useful information. For instance, if colonies grow quickly on a Petri dish (three to four days or so, depending on media used), they are actively growing in the wine, so more aggressive management is appropriate than if some small colonies struggle up after six or seven days or more. On the other hand, PCR tests can alert the winemaker to the presence of VBNC cells that may reactivate later.

- Levels of concern: same as for cultured wines (see **10.7**).
- When to test for *Brettanomyces* by PCR (if available and economically feasible): stuck wines during reinoculation, routinely on a schedule during cellaring (such as every three months), when wine is moved, if 'Brett' character is suspected, or if any unexpected activity is noticed during cellaring. Test blend components and also the final blend at least one week before bottling. Test bottled wines if activity or bottle variation is noticed.

Gene sequencing

Some labs offer identification of purified colonies by sequencing around 300–350 base pairs of yeast DNA, usually from the 26S ribosome. The yeast is identified as to genus and species, or the closest species in an extensive databank. Costs are around US$100 per colony. Results include closest match, other species by percentage match, and a phylogenetic tree.

- When to send for sequencing: if a complete identification by genus and species is desired on a microbe that has been isolated and purified.

Other methods

Z-Brett, an antibody-based test, detects 1000 cells/ml. It can be swamped by large numbers of other species (such as during fermentation), so this test is best for monitoring large increases in *Brettanomyces* populations during barrel-aging. The test cross-reacts with some other winery-related species, including certain *Candida* (surface film yeast).

Research-level tests such as epi-fluorescence, flow cytometry, peptide nucleic acid probes (PNA), and restriction fragment linear polymorphism (RFLP) are not widely available commercially at this time.

Interpreting test results

How much *Brettanomyces* is too much? The answer depends on the type and composition of the wine, the stage of processing of the wine, and on how the result fits with previous results. Other factors to take into account are the intended filtration before bottling (or lack thereof), and the history of similar wines in the winery. Ideally, there would be fewer than 25 cfu/ml during cellaring (see **10.7**), and zero before bottling, but this may not be achievable for all wines. It is wise to have wine periodically cultured for *Brettanomyces* or checked by PCR (preferably both) to establish a pattern for that particular wine.

Summary of tests
1. Sensory: detects 'Brett' character.
2. Microscopic exam: immediate result but at least 2000 cells/ml are required (less if sample is centrifuged).
3. Culturing: detects cells that are alive and not inhibited by SO_2. Direct culturing detects two to five cells/ml depending on sample amount used; membrane culture detects one cell in the amount filtered. Takes up to seven days in optimum conditions, longer in other conditions.
4. PCR-based tests: detects live cells, VBNC cells, and may or may not detect some kinds of dead cells. Can detect down to 10 cells/ml. Takes a few hours.
5. Sequencing: requires a purified culture. Yeasts are identified to genus and closest species in the databank.

References

COCOLIN, L., RANTSIOU, K., IACUMIN, L., ZIRONI, R. and COMI, G., 2004, 'Molecular detection and identification of *Brettanomyces/Dekkera bruxellensis* and *Brettanomyces/Dekkera anomalus* in spoiled wines,' *Appl. Environ. Microbiol.* 70:3: 1347–1355. doi: 10.1128/AEM.70.3.1347-1355.2004.

COULON, J., PERELLO, M., LONVAUD-FUNEL, A., DE REVEL, G. and RENOUF, V., 2010, *Brettanomyces bruxellensis* evolution and volatile phenols production in red wines during storage in bottles,' *J. Appl. Microbiol.* 108, 1450–1455. doi: 10.1111/j.1365-2672.2009.04561.x.

EDWARDS, C., 2005, *Illustrated Guide to Microbes and Sediments in Wine, Beer and Juice*. Pullman, VA, Wine Bugs LLC.

FUGELSANG., K. and ZOECKLEIN, B., 2003, 'Population Dynamics and Effects of *Brettanomyces bruxellensis* Strains on Pinot noir (*Vitis vinifera* L.) Wines,' *Am. J. Enol. Vitic.* 54:4, 294–300.

GODDEN, P., COULTER, A., CURTIN, C., COWEY, G. and ROBINSON, E., 2004, 12th AWITC, '*Brettanomyces* workshop: Latest research and control strategies,' Melbourne, Australia.

GUERRA, B., 2008, 'Research update: Is *Brettanomyces* sneaking into our wines? A review of the factors that favor Brettanomyces development during wine aging, including available means of control,' *Wine Business Monthly*, June.

MILLET, V. and LONVAUD-FUNEL, A., 2000, 'The viable but non-culturable state of wine micro-organisms during storage', *Lett. Appl. Microbiol.* 30, 136–141.

MILLS, D. and NEELEY, E., 2006, 'Molecular methods to characterize wine microorganisms', in *International Wine Microbiology Symposium Proceedings*, California State University, Fresno.

PHISTER, R. and MILLS, D., 2003, 'Real-time PCR assay for detection and enumeration of *Dekkera bruxellensis* in wine,' *Appl. Environ. Microbiol.* 7430–7434. doi: 10.1128/AEM.69.12.7430-7434.2003.

TESSONNIÈRE, H., VIDAL, S., BARNAVON, L., ALEXANDRE, H. and REMIZE, F., 2009, 'Design and performance testing of a real-time PCR assay for sensitive and reliable direct quantification of *Brettanomyces* in wine', *Int. J. Food Microbiol.* 129, 237–243. doi: 10.1016/j.ifoodmicro.2008.11.027.

UMIKER, N. and EDWARDS, C., 2007, 'Effects of SO_2 on *Brettanomyces*', presented at Santiago, Chile, July.

10.7 How can I culture the *Brettanomyces* strain in my wine?

L. Van de Water

Brettanomyces cells are grown by inoculating a Petri dish with wine and incubating it for a week or so, depending on conditions. Despite advice to the contrary, if the conditions are right, it is not difficult to culture *Brettanomyces/Dekkera*. Given optimum conditions, they can grow in seven days or fewer, and most colonies can be identified as *Brettanomyces* or not; ambiguous colonies can be sent to a wine analysis lab for confirmation.

Important: Do not culture wine for *Brettanomyces* within one to three weeks after adding SO_2 to the wine. The cells may become temporarily unculturable (Millet and Lonvaud-Funel, 2000; Umiker and Edwards, 2007), so false negatives could result (see **10.5**). Test these samples with PCR methods instead, or wait to take the sample.

Materials

Culture media can be purchased pre-sterilized (shelf life until opened is several years), or purchased dry and prepared in-house in an autoclave (Wisconsin Aluminum Foundry Electric Steroclave 50X has a large capacity and is cost-effective). Media can be sterilized in glass or narrow-mouth polypropylene bottles with caps. Also needed are a small 30°C incubator, some sterile disposable plastic pipettes and disposable plastic 60×15 or 100×15 mm Petri dishes. To test filtered or bottled samples, pre-sterilized or resterilizable filter holders and nonsterile receiving flask, pre-sterilized membrane filters, vacuum pump, reagent alcohol, alcohol lamp, and metal forceps are also required.

Media

The general yeast media Wallerstein Labs Nutrient Medium (WLN), YM Green, and YM agar are often used. The first two have bromocresol green dye which yeast colonies pick up to some degree, aiding greatly in identification. Use only WLN produced in the USA; for reasons unknown, some brands produced elsewhere do not always grow *Brettanomyces* reliably, often resulting in much slower growth or even false negatives (personal observation). One Australian winery said that while their wines tested positive for 4EP/4EG, they could not grow *Brettanomyces*; a visit to the winery determined that in fact, they were growing *Brettanomyces* but the colonies were so slow-growing and so tiny (<0.5 mm after more than a week) that they did not see them. Once they changed to a difference source of WLN, their cultures grew as expected. Culturing a sample known to have culturable *Brettanomyces* cells at the same time as the samples helps prevent false negatives.

Reducing the amount of dry WLN used from 80 g/L (the amount on the bottle) to 60 g/L promotes *Brettanomyces* growth (personal observation). Supplementing media with thiamine and biotin encourages growth of these yeasts. Adding p-coumaric acid, metabolized by *Dekkera/Brettanomyces* into 4EP, to the media will result in development of the characteristic 'phenolic' odor of 4EP, which aids in identification.

Cycloheximide

Brettanomyces strains that have been tested so far grow on agar containing 100 ppm or more of the antibiotic cycloheximide, formerly known by the brand name Actidione. It is added after autoclaving the media, to suppress *Saccharomyces*, *Zygosaccharomyces*, and most (though not all) other yeasts (Edwards, 2005). Unless purchased media already contain 50 ppm cycloheximide, add 1 ml of a sterile solution of 0.5% cycloheximide per 100 ml of agar after autoclaving. Media with lower levels of cycloheximide (such as WL Differential Medium which has only 4 ppm) allow growth of too many other cellar yeasts, including some species of *Candida*, *Pichia*, and other genera (personal observation). Cycloheximide is a carcinogen and teratogen so in its pure form, it should be handled with great care, using gloves, dust mask, and lab coat.

Chloramphenicol or other bactericides are often added to media to suppress growth of bacteria, which are not sensitive to cycloheximide at all. Otherwise, they could swamp the *Brettanomyces* colonies, and acetic acid sometimes (not always) produced by *Acetobacter* bacteria in culture may confuse *Brettanomyces* identification.

Other cycloheximide-resistant yeasts

Kloeckera apiculata (teleomorph *Hanseniaspora uvarum*), the most populous yeast on grapes (Fleet and Heard, 1993), and a number of other vineyard yeasts are cyclohexmide-resistant. *Kloeckera* somewhat resembles *Brettanomyces*, and is often mistaken for it. Fortunately, *Kloeckera* dies during yeast fermentation or shortly afterwards, so it does not cause confusion in cultures of cellared wines. If juice or must is cultured, *Kloeckera* makes green colonies in one to two days on WLN with cycloheximide, as opposed to later-appearing white colonies of *Brettanomyces*.

There are also some non-*Brettanomyces*, cycloheximide-resistant yeasts that have been found in wine cellars and as incidental contaminants in wines. These yeasts can cause false positive results for *Brettanomyces* unless the colonies are examined carefully, macroscopically and microscopically. Unlike *Brettanomyces*, these yeasts grow quickly and do not produce acetic acid on the Petri dish, but the colonies may closely resemble *Brettanomyces*. The microscopic morphology of these yeasts is usually (though not always) quite different from *Brettanomyces*. Some of these yeasts identified so far through C. M. Lucy Joseph, Culture Collection Curator at University of California, Davis, are *Candida boidinii*, *C. cantarelli* (very common), *C. catenulata*, *C. ishiwadae*, *Debaryomyces hansenii*, *Lodderomyces elongisporus*, and *Pichia guilliermondii* (unpublished results). Further studies continue, including PCR tests (see **10.6**).

Other media

Other media for *Brettanomyces* growth have been developed. While popular in Europem DBDM uses ethanol as a sole carbon source, which not all strains can utilize, especially Californian ones (Conterno *et al.*, 2006), so false negatives may occur. Certain non-*Brettanomyces* yeasts associated with wine cellars also grow on this medium (Dias *et al.*, 2003).

The liquid medium BSM ('*Brettanomyces* Specific Medium', not the same as Bacterial Standard Medium, also called BSM) from Millipore allows growth of *Kloeckera* (Iland *et al.*, 2007) and some other cycloheximide-resistant yeast species (Louriero and Malfeito-Ferreira, 2003; Romano *et al.*, 2008; personal observation). *Brettanomyces* colonies take 10 days or more to appear on BSM, and they are difficult to see against a white filter, being white, cream, or slightly pink from red wine color. Calcium carbonate agar is sometimes used to confirm acetic acid production by *Brettanomyces* because a clear space forms around the acid-producing colony.

'Easy Blue', a pre-poured, blue/green-colored agar for *Brettanomyces* testing (Lebrun Labs, www.lebrunlabs.com), is offered for wineries without microbiology lab facilities. Cycloheximide and a bactericide are added to suppress growth of most other microbes. An odor of 4EP develops if *Brettanomyces/Dekkera* grow. The instructions do say that colonies should appear after four days or more, but do not state clearly enough that earlier-appearing colonies cannot be *Brettanomyces*. Recently, a winemaker (name withheld on request) panicked when colonies appeared the next day, but of course they were not *Brettanomyces*. As with other media for *Brettanomyces* culturing, *Kloeckera* and other cycloheximide-resistant yeasts will also grow on Easy Blue, though most (except *Pichia guilliermondii*) do not produce 4EP. The colonies should be checked microscopically for confirmation, either in-house or by a commercial wine-oriented lab (not general microbiology labs, which report only total colony counts and cannot identify specific wine microbes).

'Sniff-Brett' cultures involve smelling 4EP production by *Brettanomyces* colonies on a special medium instead of examining colonies under a microscope (for wineries without microscopes). Depending on population, it can take two to seven days to develop the smell.

Culturing procedure

To melt pre-sterilized, solid agar, heat the agar bottle in a container of water in a microwave, one minute at a time (do not allow to boil) until completely melted. Before pouring the agar, on the bottom (not the top) of disposable plastic Petri dishes, write agar type, lot number, and date poured with a permanent marker. Pour Petri dishes half full. Poured Petri dishes should be poured no more than one week before inoculating with a sample, and should be kept upside down in a closed plastic container in a cool place. Write the sample identity and date cultured on the bottom of the Petri dish before inoculating with sample.

If there are 10 to 20 cfu (colony-forming units) per ml or more, the wine can be pipetted directly on a Petri dish containing an agar medium. Centrifuging samples can help concentrate cells. If there are more than around 500 cfu/ml, the result will be a solid 'lawn' or TNTC (too numerous to count). To set up a direct culture, shake the sample and immediately use a sterile pipette to dispense 0.2 to 0.5 ml of wine onto a 60 mm or 100 mm Petri dish containing agar. Do not tip the dish to move the wine around; the cells tend to go to the edges. To move the wine around the dish, use a Pasteur pipette bent in a flame to resemble a hockey stick.

If the wine has been filtered, or if the population is expected to be quite low, the sample is filtered through a sterile $0.45 \mu m$ (micron) membrane and the membrane is cultured on agar or on a broth-soaked pad. For a membrane filtration culture, set up a sterilzed filter holder on a receptacle (which itself need not be sterile) to catch the filtered wine. Put a pre-sterilized $0.45 \mu m$ 47 mm filter in the holder.

If there is a cork in the bottle, dip a corkscrew worm into reagent alcohol and set it alight; remove the bottle capsule, dip the neck of the bottle into alcohol and flame it also. Open the bottle without touching the corkscrew worm or the bottle neck. If there is a screw cap, flame the neck of the bottle to be tested, and break the seal using a swab or tissues dipped in alcohol to cover your hand. Then pour wine carefully into the filter holder for vacuum filtration. Our labs filter 250 ml for yeast cultures, 100 ml for *Brettanomyces* only, and 100 ml for bacteria.

Flame metal forceps with alcohol and transfer the filter aseptically to an agar Petri dish, or the filter and broth-soaked pad to an empty sterile Petri dish. Incubate Petri dishes upside down at 28–31°C. For direct cultures of wine, leave Petri dishes right side up overnight in an incubator, then turn upside down the next day. Using WLN and exactly the protocol described in this section, visible colonies appear in three to seven days. The faster the colonies grow, the more active the *Brettanomyces* cells are in the wine. If visible colonies appear in three or four days, different recommendations for handling the wine may be given than if small colonies struggle to grow after six or seven days. At lower temperatures, or if using other media, the colonies take longer to appear.

Identifying colonies

On WLN+C (cycloheximide) and other green media, *Brettanomyces* forms small, shiny, white colonies that turn olive green to light-green over time. The colonies produce acetic acid changing green media to yellow, which is helpful in identification. A slow-growing yeast making a strong acetic acid smell on a Petri dish of WLN+C is confirmed to be *Brettanomyces* or *Dekkera*.

After growing, some colonies will need to be checked microscopically, either in-house or sent to a wine analysis laboratory. Most *Brettanomyces* cells are smaller than *Saccharomyces*, and though some young *Brettanomyces* cells can resemble *Saccharomyces*, the majority will be more apiculate (pointed). A prominent feature of a *Brettanomyces* culture is extreme polymorphism (many different shapes), so the cells will be quite variable, even within the same culture

(C.M. Lucy Joseph, personal communication; Christian Butzke, personal communication; Zoran Ljepovic, personal communication; also personal observation).

Different strains of *Brettanomyces* cannot be distinguished by culturing. Although morphological differences can be observed among some strains, strain identity cannot be confirmed without extensive genetic and physiological research.

Even within one strain, cells vary greatly in size. Some are 3–4 μm in width to 5–8 μm in length (Millet and Lonvaud-Funel, 2000), but may be much longer. Some young cells may be ovoid (egg-shaped) but more may resemble olives. The first bud that a cell produces forms at one end, so the budding cell looks like a thin bowling pin, but it leaves a large bud scar, flattening the end somewhat (Fig. 1). Successive buds cannot form at the same place, so later buds are offset from the ends of the cell, and older cells begin to look like gothic arches ('ogive'), rowboats, watermelon seeds, or even barrels, with a bud sticking out from a corner of the flat end(s). Pseudomycelia (see **10.6**) may begin to form in culture but can become much more extensive in liquid, such as wine.

It is very important to note that identifying wine microbes microscopically, including *Brettanomyces*, takes some training. The very few texts about identifying wine microbes that are available are extremely useful (Edwards, 2005). Texts cannot replace hands-on training in identifying wine microbes, however, and attending a wine microbe identification session is recommended. A number of universities and commercial labs offer workshops on wine culturing techniques and wine microbe identification. These are best if one or more days are spent laying the foundation, then at least another day is spent working with the details of microbe identification (personal experience).

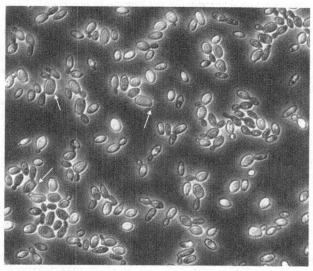

Fig. 1 *Brettanomyces* cells showing polymorphism. The three arrows indicate bud scars. Photo: © Lia Van de Water.

- Levels of concern: these vary depending on the situation and stage of production. In wineries believed to be free from *Brettanomyces*, even one colony is cause for alarm. In cellars with known infections, wines with 100 colony-forming units (cfu)/ml should be recultured frequently to monitor population increase; wines with 500 cfu/ml need better management; wines with more than 1000 cfu/ml require immediate attention to prevent spoilage, if the 'Brett' character is not already too intense.
- When to culture for *Brettanomyces*: routinely on a schedule (such as every three months) during cellaring, when wine is moved (before SO_2 is added!), if 'Brett' character is suspected, or if any unexpected sensory change occurs during cellaring. Culture blend components and also the final blend at least one week before bottling. Culture bottled wines if unexpected sensory change or bottle variation is noticed.

Summary of what winemakers should know about *Brettanomyces* culturing:

- Do not take samples within a few weeks after adding SO_2; false negatives can result.
- Following the exact procedures above, colonies should appear on WLN agar containing cycloheximide within three to seven days. On other media and under other conditions, colonies can take 10–14 days to appear.
- Culture wines periodically, not just once, to monitor patterns of growth in each wine over time.

References

CONTERNO, L., JOSEPH, C.M.L., ARVIK, T., HENICK-KLING, T. and BISSON, L., 2006, 'Genetic and physiological characterization of *Brettanomyces bruxellensis* strains isolated from wines,' *Am. J. Enol. Vitic.* 57:2, 139–147.

DIAS, L., DIAS, S., SANCHO, T., STENDER, H., QUEROL. A., MALFEITO-FERREIRA, M. and LOUREIRO, V., 2003, 'Identification of yeasts isolated from wine-related environments and capable of producing 4-ethylphenol,' *Food Microbiol.* 20, 567–574.

EDWARDS, C., 2005, *Illustrated Guide to Microbes and Sediments in Wine, Beer, and Juice*, Pullman, WA, Wine Bugs LLC.

FLEET, G. H. and HEARD, G. M., 1993, 'Yeasts-growth during fermentation,' in *Wine Microbiology and Biotechnology*, Fleet, G. H. (ed.), Switzerland, Harwood Academic Publishers, 27–55.

ILAND, P., GRBIN, P., GRINBERGS, M., SCHMIDTKE, L. and SODEN, A., 2007, *Microbiological Analysis of Grapes and Wine: Techniques and Concepts*, Patrick Iland Wine Promotions Pty Ltd, Campbelltown, Australia.

LOUREIRO, V. and MALFEITO-FERREIRA, M., 2003, 'Spoilage yeasts in the wine industry,' *Int. J. Food Microbiol.* 86, 23–50.

MILLET, V. and LONVAUD-FUNEL, A., 2000, 'The viable but non-culturable state of wine micro-organisms during storage', *Lett. Appl. Microbiol.* 30, 136–141.

ROMANO, A., PERELLO, M., DE REVEL, G. and LONVAUD-FUNEL, A., 2008, 'Growth and volatile compound production by *Brettanomyces/Dekkera bruxellensis* in red wine,' *J. Appl. Microbiol.* 104, 1577–1585. doi: 10.1111/j.1365-2672.2007.03693.x.

UMIKER, N. and EDWARDS, C., 2007, 'Effects of SO_2 on *Brettanomyces*', presented at Santiago, Chile, July.

10.8 How can I manage *Brettanomyces* in the cellar?

L. Van de Water

Don't bring it in
The very **best** way to deal with *Brettanomyces* is not to bring it into the cellar. Once it is established in a winery cellar, managing it, not eliminating it, is the goal (Fugelsang *et al.*, 1993; Boulton *et al.*, 1996). While there are many anecdotal references for *Brettanomyces* entering winery cellars by various means, and it is possible that the yeasts may occasionally come in on grapes, overwhelming evidence from studies in a number of countries points to wine or wooden containers as the major vectors for *Brettanomyces* contamination of cellars (Boulton *et al.*, 1996; Fugelsang *et al.*, 1993). Refraining from purchasing red wine or wood used for red wine from other cellars (no matter how extremely inconvenient it is to pass up used red wine barrels for sale!) has protected many wineries from *Brettanomyces* infection, until a mistake was made and *Brettanomyces* was brought in with used barrels or infected wine.

Quarantine possibly infected lots
Before bringing wine into a larger cellar, keep the wine in quarantine until it can be cultured (and tested by PCR if available), or simply sterile-filter all received wines on the way into the winery from the tanker truck (Fugelsang *et al.*, 1993). Even then, put newly arrived wine into stainless steel, not wood, until it can be checked for residual *Brettanomyces*. Smaller wineries (and especially amateur producers) would be well advised never to bring in wine or any other item from another cellar. Do not top cellared wines with wine from another cellar, even bottled commercial wines. A purchased wine, no matter how tasty, may still have *Brettanomyces*, or it may have substrates (sugar, etc.) that could feed a *Brettanomyces* population if one is already resident in the winery.

***Brettanomyces* growth in the cellar**
Danger signals for *Brettanomyces* growth in cellared wines include

- a wine that was quiet becoming active
- a wine that was clear becoming cloudy
- an unexpected change in aroma, flavor, or analysis results.

If any of these changes are noticed, take samples for analysis immediately, including pH, free and total SO_2, volatile acidity, and microbial testing. Examine directly under a phase-contrast microscope, but even if other microbes are seen microscopically, check for *Brettanomyces* by culturing or PCR, because they may also be present at levels too low to observe, but high enough to cause problems.

Wine composition influences *Brettanomyces* growth in cellared wines, especially nutrient depletion (sugar and nitrogen), pH and SO_2 management, and

330   Winemaking problems solved

turbidity. Cellar operations are also very important, including temperature control, minimizing oxygen contact, avoiding cross-contamination, population control, and sanitation.

Wine composition influencing *Brettanomyces* growth
Minimize substrates for growth (especially sugar and nitrogen)
Brettanomyces can grow on very small amounts of hexose sugars (0.2 g/L, well below what is considered 'dry'). However, they do avail themselves of more residual sugars if available. Stuck ferments are prime candidates for spoilage, especially from *Brettanomyces* and *Lactobacillus* (Godden et al., 2004). Procedures to encourage resuming stuck fermentations encourage *Brettanomyces* growth, if the wine is already infected. The most important must adjustment to avoid stuck ferments is to reduce Brix to 25–25.5 if it is higher; others include reducing pH to below 3.6 during fermentation and adding at least 40 ppm total SO_2 before fermentation.

Supplementation with complex nutrients, not just diammonium phosphate (DAP), in two or three portions at appropriate times and levels is very important in preventing stuck ferments and in avoiding leftover nitrogen. Amounts and timing of nutrient supplementation depend on tests of yeast-available nitrogen (YAN) on the juice. Nitrogen supplements should be calculated carefully and added during the first half to two-thirds of fermentation, because the cells are unable to bring in nitrogen later on, so late additions provide nitrogen that *Brettanomyces* can use (Coulter et al., 2003). Also, if too much DAP is added early in fermentation, utilization of amino acids can be inhibited, and some of these can be used by *Brettanomyces* as a sole nitrogen source (Conterno et al., 2006).

Appropriate yeast strain selection and proper handling of yeast during rehydration and inoculation also help to prevent stuck ferments. Especially if grapes are damaged or diseased, be sure to add a vigorous dry yeast; do not ferment rotten or damaged grapes without adding yeast. To discourage growth of non-*Saccharomyces* vineyard yeasts before fermentation, which deplete nutrients and can inhibit *Saccharomyces*, minimize the length of time the must spends between 10 and 15°C (50–60°F). During fermentation, maintain strict temperature control so musts do not get too hot or too cold for *Saccharomyces* yeasts.

SO_2 and pH management
Regular sensory evaluation and analysis of SO_2 help winemakers watch for signs of microbial growth. Test for *Brettanomyces* population and SO_2 levels in wines in new barrels separately from older ones holding the same batch of wine; new wood binds SO_2 much more quickly than old wood (Ribereau-Gayon et al., 2006).

The SO_2 in a wine exists as molecular, free, and bound. *Brettanomyces* yeasts are not sensitive to legal levels of bound SO_2 (Licker et al., 1998). When evaluating SO_2 effectiveness, it is very important to consider molecular SO_2, not just free SO_2! Over many years, certain levels of free SO_2 have been said to control *Brettanomyces* completely, but recommendations not based on molecular SO_2 are misleading. The free SO_2 level is meaningless unless it is taken with the wine pH to calculate molecular SO_2 (Margalit, 2004). For example, in *Handbook of Enology*, the authors say, 'a free SO_2 concentration of 30 mg/L SO_2 always results in the total elimination of all viable populations [of *Brettanomyces*] after 30 days', but then they clarify this statement as applying to 'red wines of normal pH levels (3.4–3.5)' and state that this level would be insufficient at pH 3.8 (Ribereau-Gayon *et al.*, 2006).

This brings up one of the most pervasive – and most consistently overlooked – problems with making recommendations about unfamiliar wines. The Bordeaux wines Dr Ribereau-Gayon refers to as 'normal' had a pH of 3.4 to 3.5, but a 'normal' wine in warmer climates – west coast USA, Chile, Argentina, South Africa, Australia, and others – may have a much higher pH unless it is strictly managed once the grapes are picked. Left to their own devices, some wines in those places may have a pH of 3.8 or higher. Thus, SO_2 management in Bordeaux, and in other cool climate regions as varied as (for example) New Zealand's South Island, Michigan, and California's Santa Cruz Mountains, is much easier because of naturally lower pH. Sadly, if winemakers do not understand the effects of climatic differences between their region and others, they may take advice (in vineyards as well as in wineries) that are appropriate for somewhere else, but inappropriate for their own grapes and wines.

In 1982 at the University of California at Davis, Clark Smith developed a formula and chart to assist winemakers in calculating molecular SO_2 from free SO_2 and pH (Fugelsang *et al.*, 1993). This discovery about wine pH was a revelation, explaining so much about the influence of pH that was not understood before. Indeed, in the 1970s, pH was often considered to be of little or no consequence to winemaking (personal observation). More than thirty years later, most commercial winemakers worldwide use the pH/SO_2 chart (Margalit, 2004), and those who do not, should do so.

For example, at 3.5 pH, 30 ppm free SO_2 calculates to 0.6 ppm molecular SO_2, sufficient to delay or inhibit growth of many *Brettanomyces* strains, though not all. But at 3.8 pH, 30 ppm is only 0.3 ppm molecular SO_2, too little to impact *Brettanomyces* growth. Note, however, that strain differences influence the response of *Brettanomyces* SO_2; a level of molecular SO_2 that will inhibit one strain may not inhibit another (Barata *et al.*, 2008; Umiker and Edwards, 2007).

Making larger SO_2 additions less frequently rather than smaller, more frequent ones, results in a higher percentage of free SO_2 rather than bound. Coulter *et al.* (2003) advise, 'Therefore it is beneficial to make a large addition of SO_2 to red wine after MLF, before aging in barrels, in order to reduce the populations of *Dekkera/Brettanomyces*.'

Brettanomyces is not especially sensitive to pH in the pH range of wine (Conterno *et al.*, 2006), but the lower the pH, the more molecular SO_2 is present at the same free SO_2 level, so the SO_2 is more effective. This is very important when managing a *Brettanomyces* infection. Even one tenth of a pH unit makes a difference to the percentage of molecular SO_2, and to the management of the wine. If the pH is over 3.65–3.70, winemakers can reduce the pH to 3.6 or less with tartaric acid during cellaring for greater SO_2 effectiveness, then bring the wines to the desired pH with potassium carbonate or bicarbonate before bottling. This makes *Brettanomyces* management much more successful, and it uses only ions that are naturally found in wine. Alternatively, Wine Secrets (www.winesecrets.com) perform electrophoresis to reduce acidity as the same time as reducing pH.

Lower turbidity
SO_2 binds more readily in cloudy wines, leaving less free SO_2. If at all possible, clarify the wine (even just racking off gross yeast lees) before adding SO_2. Cleaning up wine as soon as it is finished yeast fermentation and MLF can help prevent spoilage. Chatonnet *et al.* (1995) noted that delaying racking, and thus delaying SO_2 addition and removal from solids, could result in significantly higher ethyl phenol levels. Lees contact also increases substrates for *Brettanomyces* growth through yeast autolysis.

Cellar conditions influencing *Brettanomyces* growth
Temperature control
Keeping the wine below 15°C/60°F greatly reduces *Brettanomyces* growth rate. Some strains (mostly European) can continue to grow at that temperature (Conterno *et al.*, 2006), but their growth will be slowed; other strains will cease growth (though not die). Special vigilance is needed in cellars whose temperature rises during the summer, encouraging a *Brettanomyces* 'bloom'.

Minimizing oxygen contact
The top surface of wine in tanks should be checked and topped at least once a week, and gassed regularly with N_2, CO_2, or argon. If a film is seen, this indicates that the headspace protection is not sufficient to prevent growth of aerobic microbes. *Brettanomyces* yeasts do not need oxygen, but their growth is stimulated by oxygen, which also allows them to use a wider range of substrates. Micro-oxygenation should not be performed on wines already infected with *Brettanomyces* because the yeasts take up the oxygen more quickly than the target molecules, wine phenolics (du Toit *et al.*, 2005). Each time wine is racked, some oxygen pickup occurs, though delaying racking must be balanced against the increase in micronutrients that occurs during yeast autolysis.

Avoid cross-contamination
Clean hoses thoroughly to remove debris, stain, tartrate deposits, residual wine, etc. Sanitize hoses, topping equipment, sampling devices and receiving containers between batches of wine, unless you know that both batches are already infected with *Brettanomyces*. Trading bungs among barrels can easily spread

infection (note: SO_2 solution does not penetrate into the bung to kill sequestered microbes). Check the *Brettanomyces* infection status of topping wine before use, or simply sterile-filter topping wine.

Do not ferment wines in *Brettanomyces*-infected wood, do not top wines that are not dry with infected wine, and do not put stuck or sweet wine into infected barrels, to avoid infecting wine with *Brettanomyces* while residual sugar is present. Do not put infected wine into new barrels; try to keep new barrels free of *Brettanomyces* as long as possible. Winemakers are often advised to 'keep the problem children together', that is, to keep infected wine away from uninfected wine. Zoran Ljepovic (QA/QC, Constellation Wines US) advises, 'We should put infected wines into quarantine because we cannot predict what the yeasts will do' (personal communication).

Population control if needed (fining/filtering)
Keep track of *Brettanomyces* population by culturing (and also PCR if possible). If the population is increasing rapidly, reduce the cell population by depth filter, diatomaceous earth, or crossflow filtration. Continue monitoring to watch for regrowth. Even if cell growth declines, the dormant cells may still be producing ethyl phenols. Monitoring 4EP and 4EG can track this, though it may be more economical to filter the wine to remove most of the cells than to check ethyl phenols frequently. If filtration is impossible, egg white fining and racking can remove a large proportion of the cells (personal observation). Other fining agents used to reduce yeast population are bentonite (not advised for reds because it tends to strip flavor), gelatin and PVPP.

Cellar and barrel sanitation
Sanitation is extremely important in preventing any microbial spoilage, including *Brettanomyces*, from spreading throughout the cellar. Effective cellar sanitizers include quaternary ammonia, peracetic acid, and peroxycarbonate. Chlorine is no longer recommended for use in wineries because of the potential for trichloroanisole formation (TCA, TeCA), strong moldy-smelling taints which can enter the wine through the air in the cellar.

Alas, there is no reliable way that is available to remove all the microbes from a barrel all the time (Coulter *et al.*, 2003; Fugelsang *et al.*, 1993; also personal experience) except perhaps steaming, which removes oak flavor. You can kill some of the microbes all the time, and all the microbes some of the time, but not all the microbes all the time. More microbial toxicity is assumed than actualized. Burning sulfur wicks/rings in between wines, and whenever barrels are stored empty, has been shown to be more effective against *Brettanomyces* than SO_2 gas (Ribereau-Gayon *et al.*, 2006). While burning sulfur reduces *Brettanomyces* population, it does not eliminate the yeasts, which may penetrate 0.8 cm deep in the wood (Malfeito-Ferreira *et al.*, 2004). The same is true for ozone, discussed below, and any other current treatment for infected barrels. Boulton *et al.* (1996) explained, 'We know of no sure way to sterilize infected barrels.'

A procedure to greatly reduce *Brettanomyces* contamination in barrels using high-power ultrasound has been developed in Australia and offered by Cavitus (Yap *et al.*, 2008). It is expensive, but warrants evaluation by wineries.

Ozone treatment of barrels, when carried out correctly, is very effective in reducing the *Brettanomyces* population (Guerra, 2008), and has been considered a godsend for some wineries battling *Brettanomyces* in barreled wines. *Brettanomyces* may re-grow in the wine, and/or in wines subsequently aged in the same wood. Ozone has no residual sanitizing effect, so if infected wine is re-introduced into an ozonated barrel, the yeasts will grow as readily as if the barrel had not been treated.

Effective (and safe) use of ozone requires adherence to strict procedures (Suárez *et al.*, 2007; Doug Manning, personal communication). The barrels need to be prepared with a high-pressure hot water wash so that there are no tartrates or other particles in them; they must be completely cooled after hot water treatment (by waiting, or using cold water rinse), before using ozone. Ozone should be charged into very cold, tightly filtered water. An ozone meter is essential; measure the ozone in the water going into the barrel (should be 2.0 to 2.5 ppm) and coming out (should be 0.1 to 0.2 ppm). Rinse barrels with cold water before filling with wine. In most cellars, the procedure should be followed outside to reduce ozone exposure of winery workers, but wearing a tag to check cumulative ozone exposure is strongly recommended anyway. Many companies offer ozone treatment of barrels as a service.

What is important?
Brettanomyces management depends on principles that apply to managing other microbes. The most important aspect of management of *Brettanomyces* – or any other microbe – in the cellar is to understand that an integrated, holistic approach is essential (Coulter *et al.*, 2003). These authors advise, 'In situations where only some of the suggested control strategies are implemented it is likely that reductions in 4-ethyl phenol concentrations will be sporadic.' Dr Charles Edwards of Washington State University compares the process of *Brettanomyces* management to putting a number of hurdles in its path. One hurdle is not enough; to stop its progress, at least several 'hurdles' must be in the way, combining to prevent rampant growth of microbes (Edwards, 2007).

References
BARATA, A., CALDIEIRA, J., BOTELHEIRO, R., PAGLIARA, D., MALFEITO-FERREIRA, M. and LOUREIRO, V., 2008, 'Survival patterns of *Dekkera bruxellensis* in wines and inhibitory effect of sulphur dioxide,' *Int. J. Food Microbiol.* 121:2, 201–207. doi: 10.1016/j.ifoodmicro.2007.11.020.
BOULTON, R., SINGLETON, V., BISSON, L. and KUNKEE, R., 1996. *Principles and Practices of Winemaking*, New York, Chapman and Hall.
CHATONNET, P., DUBOURDIEU, D. and BOIDRON, J.N., 1995, 'The influence of *Brettanomyces/*

Dekkera spp. yeasts and lactic acid bacteria on the ethyl phenol content of red wines,' *Am. J. Enol. Vitic.* 46, 463–468.

CONTERNO, L., JOSEPH, C.M.L., ARVIK, T., HENICK-KLING, T. and BISSON, L., 2006, 'Genetic and physiological characterization of *Brettanomyces bruxellensis* strains isolated from wines,' *Am. J. Enol. Vitic.* 57:2, 139–147.

COULTER, A., ROBINSON, E., COWEY, G., FRANCIS, I. L., LATTEY, K., CAPONE, D., GISHEN, M. and GODDEN, P., 2003, '*Dekkera/Brettanomyces* yeast – an overview of recent AWRI investigations and some recommendations for its control,' Proceedings of a Seminar Organized by the Australian Society for Viticulture and Oenology, *Grapegrowing on the Edge, Managing the Wine Business, Impacts on Wine Flavor*, Bell, S., de Garis, K., Dundon, C., Hamilton, R., Partridge S. and Wall, S. (eds), 41–50.

DU TOIT, W., PRETORIUS, I. and LONVAUD-FUNEL, A., 2005, 'The effect of sulphur dioxide and oxygen on the viability and culturability of a strain of *Acetobacter pasteurianus* and a strain of *Brettanomyces bruxellensis* isolated from wine,' *J. Appl. Microbiol.* 98, 862–871.

EDWARDS, C., 2007, 'Microbe Management: Minimizing Risk of Spoilage', presented in Auckland, New Zealand, November.

FUGELSANG, K., OSBORN, M. and MULLER, C., 1993, '*Brettanomyces* and *Dekkera*, implications in winemaking', in *Beer and Wine Production, Analysis, Characterization, and Technological Advances*, Gump, B. (ed.), American Chemical Society, Washington, DC, 110–131.

GODDEN, P., COULTER, A., CURTIN, C., COWEY, G. and ROBINSON, E., 2004, 12th AWITC, '*Brettanomyces* workshop: Latest research and control strategies,' Melbourne, Australia.

GUERRA, B., 2008, 'Research update: Is *Brettanomyces* sneaking into our wines? A review of the factors that favor *Brettanomyces* development during wine aging, including available means of control,' *Wine Business Monthly*, June.

LICKER, J., ACREE, T. and HENICK-KLING, T., 1998, 'What is "Brett" (*Brettanomyces*) flavor?: a preliminary investigation', in *Chemistry of Wine Flavor*, Waterhouse, A. and Ebeler, S. (eds), Washington, DC, American Chemical Society Symposium series.

MALFEITO-FERREIRA, M., LAUREANO, P., BARATA, A., D'ANTUONO, I., STENDER, H. and LOUREIRO, V., 2004, 'Effect of different barrique sanitation procedures on yeasts isolated from the inner layers of wood,' *Abstr. Am. J. Enol. Vitic.* 55, 304A.

MARGALIT, Y., 2004, *Concepts in Wine Chemistry*, 2nd edn, Wine Appreciation Guild, South San Francisco, CA.

RIBEREAU-GAYON, P., GLORIES, Y., MAUJEAN, A. and DUBOURDIEU, D., 2006, *Handbook of Enology Volume 2: The Chemistry of Wine Stabilization and Treatments*, John Wiley & Sons, Chichester.

SUÁREZ, R., SUÁREZ-LEPE, J., MORATA, A. and CALDERÓN, F., 2007, 'The production of ethylphenols in wine by yeasts of the genera *Brettanomyces* and *Dekkera*: a review,' *Food Chem.* 102, 10–21. doi: 10.1016/j.foodchem.2006.03.030.

UMIKER, N. and EDWARDS, C., 2007, 'Effects of SO_2 on *Brettanomyces*', presented at Santiago, Chile, July.

YAP, A., SCHMID, F., JIRANEK, V., GRBIN, P. and BATES, D., 2008, 'Inactivation of *Brettanomyces*/Dekkera in wine barrels by high power ultrasound,' *Wine Business Monthly* (Australia), 23:5, 32–40.

10.9 Can I bottle my wine unfiltered if it is infected with *Brettanomyces*?

L. Van de Water

It is well-established that *Brettanomyces* can grow in bottled wines (Chatonnet *et al.*, 1992; Fugelsang and Zoecklein, 2003; Cocolin *et al.*, 2004; Renouf *et al.*, 2007; Coulon *et al.*, 2010; also personal observation). So, the safest answer is 'no.' However, here are some recommendations to assess the relative risk.

Evaluating potential instability in the bottle

First, consider aspects of marketing and liability. Is bottle variation acceptable to your target market, or do your consumers expect a stable, consistent product? If the wine is being made for another company, and they have requested that the wine not be sterile-filtered at bottling, they need to sign a release absolving the bottling winery from all responsibility for microbial instability. These questions and any others related to bottle instability should be considered far enough ahead of time before proceeding to bottle wine infected with *Brettanomyces* without sterile filtration.

Then, ask the four questions:

1. Does it have residual fermentable sugar?
2. Does it have residual malic acid (if pH 3.3–3.4 or higher)?
3. Does it have *Pediococcus*?
4. Does it have *Brettanomyces*?

Testing for instability in the bottle

Assuming that the answer to the last question is yes, the others must also be investigated to prevent in-bottle spoilage from other microbes. Analysis by methods requiring a UV-VIS spectrophotometer, or at least a colorimeter (Megaquant, www.megazyme.com), are needed to assess stability. A winery that does not have the ability to run such tests should send samples to a commercial wine analysis lab.

Enzymatic glucose and fructose (not added together, tested separately), pH, free and total SO_2, and malic acid should be tested. If the wine is microbially unstable for other reasons such as residual fermentable sugar at 0.5 g/L or more, or malic acid over 0.1 g/L (in wine that underwent incomplete MLF), *Brettanomyces* may become irrelevant, and sterile filtration may be needed anyway. But if the other questions are answered 'no' (or if lysozyme is added to kill *Pediococcus*), then the focus can be on *Brettanomyces* infection status.

For most microbes, tests of levels of potential substrates are needed to help decide whether a wine can be bottled without sterile filtration. However, *Brettanomyces* can use a variety of substrates, so it is best to assume that cells can grow in the bottle if they are present. Because *Brettanomyces* can grow at levels of hexoses (glucose and fructose) well below those considered 'dry'

(Chatonnet et al., 1992, 1995), the absence of significant residual fermentable sugar is not a limiting factor in *Brettanomyces* growth in the bottle. In the 1980s some wineries bottled wines unfiltered on the assumption that *Brettanomyces* could not grow in dry wine, but they discovered that this idea was incorrect.

Handling wines infected with *Brettanomyces* before bottling
White wines infected with *Brettanomyces* should be sterile-filtered at bottling with a membrane filter because even a small amount of yeast growth will render the wine unacceptably cloudy. Even for reds, the 'safe' answer is, of course, to filter out all the cells, so they cannot reactivate in the bottle. If they grow, they could potentially spoil the wine, or they could grow a little bit and then stop.

Many winemakers are reluctant to filter reds before bottling. However, they need to weigh the potential for *Brettanomyces* (and perhaps other microbes) to reactivate in the bottle against their preference to avoid pre-bottling filtration. This is not easy, no matter how much experience the winemaker or consultant has. Levels of *Brettanomyces* infection can range from almost undetectable to 10 million cells/ml (personal observation). While almost no one would bottle a wine that is visually cloudy with cells, it is difficult to set a lower limit on acceptable *Brettanomyces* populations in bottled wine. Sometimes, *Brettanomyces* does not grow in the bottle even though there are 100 cfu/ml or more, and sometimes they grow extensively from fewer than 1 cfu/ml. Culturing the wine can help in this assessment; healthy-looking colonies that grow sooner (in optimal culture conditions) indicate a more robust and active population of *Brettanomyces* than tiny colonies that appear later on (see **10.7**).

One way to evaluate a wine's spoilage potential is to hand-bottle three or four bottles straight from tank or barrel (preferably at a racking), without an SO_2 addition or filtration. Keep them in a warm place (in an office, or next to a hot water heater). Open one bottle a month, taste, and check the *Brettanomyces* population. If the 'Brett' character increases or the population rises, the wine is probably not a good candidate for bottling unfiltered.

***Brettanomyces* growth and survival in the bottle**
When *Brettanomyces* grow in the bottle, there will almost always be bottle variation, sometimes quite dramatic (Chatonnet et al., 1992; also personal observation). One bottle may have little or no effects, but another bottle in the same case may be completely spoiled. The lack of consistency means that repeat customers are likely to be disappointed, whether they prefer the 'Bretty' or the 'non-Bretty' bottles. When nutrients or vitamins are depleted, *Brettanomyces* becomes dormant but does not necessarily die. Live cells of *Brettanomyces* have been cultured from wines bottled for more than 30 years (personal observation).

4-ethyl phenol (4EP) in the bottle

If cells are present in the bottle, even if they do not grow extensively, they can continue to produce ethyl phenols after bottling (Fugelsang and Zoecklein, 2003; Coulon *et al.*, 2010). So, in a wine without recognizable 'Brett' character but bottled unfiltered, the ethyl phenols can continue to increase, sometimes dramatically (Fugelsang *et al.*, 1993).

In some wines that have been sterile filtered to remove all cells, the perception of 'Brett' character, particularly ethyl phenols, may increase quite dramatically even though analysis shows that the levels remained the same (personal observation). There is as yet no good explanation for this sensory phenomenon. Surely, as wine ages, its freshness and fruitiness diminishes, allowing other components to become more prominent, but the sometimes startling increase in perceived 'Brett' character without an increase in ethyl phenols so far remains a mystery.

Considerations about bottling unstable wine

Some winemakers decide not to sterile-filter their wine or use inhibitors even if it has a known *Brettanomyces* infection. While wine microbiologists are all too aware that *Brettanomyces* growth in the bottle can be a disaster, they also realize that winemakers sometimes choose the option of bottling unstable wine. It is hard for many winemakers to imagine that a wine which has been quiet for months in the cellar could become active once it is bottled, but this does indeed occur (personal observation).

The procedures recommended in **10.10** represent a conservative approach to *Brettanomyces* stability in the bottle. Every winemaker knows examples of wines that were bottled with viable *Brettanomyces* cells which did not grow in the bottle. Sometimes the yeasts' explosive growth completely spoils a wine, but sometimes they grow only a small amount or not at all. One of the most difficult and frustrating predictions to attempt to make pre-bottling is whether or not a small population of *Brettanomyces* will grow to a large one in the bottle. One never knows, but no matter how many instances a winemaker can cite when *Brettanomyces* yeasts failed to cause in-bottle spoilage, most wine microbiologists and consultants know of instances in which they did, and caused an economic calamity for the winery.

In one example, a Merlot from another country was featured at a fancy tasting in the Napa Valley. On its way to the USA, *Brettanomyces* grew extensively in the bottle, so the wine had developed a strong 'Brett' character. The distressed winemaker had no choice but to pour the wine, but as soon as he arrived back in his country he sought assistance to help prevent future problems in the bottle. The spoiled Merlot was already enough of a problem economically; they could afford no more microbial disasters.

It is the potential for disastrous activity in the bottle that leads wine microbiologists to lean heavily toward procedures to achieve stability rather than risk spoilage. Whatever negative effects filtration may have (and many winemakers believe that the effects are only temporary), extensive *Brettanomyces* growth in the bottle is worse.

See **10.10** for information about inhibition of *Brettanomyces* at bottling.

References

CHATONNET, P., DUBOURDIEU, D. and BOIDRON, J. N., 1992, 'The origin of ethyl phenols in wines,' *J. Sci. Food Agric.* 68, 165–178.

CHATONNET, P., DUBOURDIEU, D. and BOIDRON, J.N., 1995, 'The influence of *Brettanomyces/Dekkera* spp. yeasts and lactic acid bacteria on the ethyl phenol content of red wines,' *Am. J. Enol. Vitic.* 46, 463–468.

COCOLIN, L., RANTSIOU, K., IACUMIN, L., ZIRONI, R. and COMI, G., 2004, 'Molecular detection and identification of *Brettanomyces/Dekkera bruxellensis* and *Brettanomyces/Dekkera anomalus* in spoiled wines,' *Appl. Environ. Microbiol.* 70:3: 1347–1355. doi: 10.1128/AEM.70.3.1347-1355.2004.

COULON, J., PERELLO, M., LONVAUD-FUNEL, A., DE REVEL, G. and RENOUF, V., 2010, *Brettanomyces bruxellensis* evolution and volatile phenols production in red wines during storage in bottles,' *J. Appl. Microbiol.* 108, 1450–1455. doi: 10.1111/j.1365-2672.2009.04561.x.

FUGELSANG, K. and ZOECKLEIN, B., 2003, 'Population Dynamics and Effects of *Brettanomyces bruxellensis* Strains on Pinot noir (*Vitis vinifera* L.) Wines,' *Am. J. Enol. Vitic.* 54:4, 294–300.

FUGELSANG, K., OSBORN, M. and MULLER, C., 1993, '*Brettanomyces* and *Dekkera*, implications in winemaking', in *Beer and Wine Production, Analysis, Characterization, and Technological Advances*, Gump, B. (ed.), American Chemical Society, Washington, DC, 110–131.

RENOUF, V., PERELLO, M-C, DE REVEL, G. and LONVAUD-FUNEL, A., 2007, 'Survival of wine microorganisms in the bottle during storage,' *Am. J. Enol. Vitic.* 58:3, 379–386.

10.10 How should I prepare my wine for bottling if it has *Brettanomyces*?

L. Van de Water

Reducing ethyl phenols
Trials performed with yeast hulls, live yeast, and potassium caseinate/casein reduce ethyl phenols. Reverse osmosis can be used to remove some of the ethyl phenols, reducing the objectionable 'Brett' characters. Several companies provide this service.

Assessing stability potential
Test the level of *Brettanomyces* infection **before** making the final SO_2 addition, by culturing and also by PCR if available, which can detect VBNC cells. Culturing is also important because the shorter the growth time in culture, the more active the cells are likely to be in the wine.

Reducing population by fining and filtration
If the wine will be sterile-filtered through an absolute 0.65 to 0.45 μm membrane filter at bottling, then population reduction methods are recommended to reduce the cell load which the sterilizing filters may need to stop. The membrane filter should be preceded by a sterilizing depth filter or the sterilizing setting on cross-flow or ultrafiltration. The final membrane filters should be just a precaution; they should not actually have to remove any cells. Membrane filters are very expensive, and the more cells they have to remove, the faster they will plug.

Egg white fining can precipitate around 90–95% of the yeast cells (personal observation), resulting in a much more manageable population. Other fining agents that can reduce yeast population are casein/potassium caseinate, PVPP, and gelatin. Bentonite fining also reduces yeast population, but it is usually considered to be unsuitable for red wines.

Filtering with diatomaceous earth (DE), crossflow, or a pad fliter reduces the *Brettanomyces* population dramatically, making it much easier to put the wine through the final membrane filters. The preparatory filters remove most of the yeasts, preparing the wine for sterile filtration.

One winery follows this procedure for reducing the *Brettanomyces* population: settle the wine for at least one to two months in the barrel, undisturbed. Rack the top 90% into a tank; filter the rest into the tank through DE or a pad filter. Fine the tank with egg whites, and settle for at least two weeks. Rack the top 90% into another tank; filter the rest into the second tank. By this time, the *Brettanomyces* population is down to only a few cells/ml, which are removed by sterile filtration at bottling.

It is very important to note that depth pads and cartridges may sport 'nominal' numerical designations matching those of membrane filters, such as '0.45 μm (micron)'. Understandably, winemakers often believe that the two types of filters are equivalent. However, the 'nominal' micron designations refer only to an average retention size under certain circumstances, so they do not give the same protection as an 'absolute' membrane filter.

The difference between these two types of filters, both in cartridges, is critical. Depth filters depend on entrapment of particles (including microbes) in the matrix of the filter; particle size, particle load, pressure, and other variables determine what gets through and what does not. Wines with large amounts of small microbes are most likely to have microbes passing through depth filters. Depth filters remove a very large proportion of the *Brettanomyces* cells, and sometimes all of them, though not necessarily. A 0.45 μm membrane filter, however, is an absolute barrier unless its integrity is compromised (a hole in it), or unless cells are small enough to pass through the pores.

Inhibiting *Brettanomyces* growth
SO_2
A molecular SO_2 of 0.5–0.6 ppm is recommended for reds at bottling, to inhibit *Brettanomyces* growth temporarily, though it does not kill the cells. In most red wines, adding enough SO_2 to kill all the yeasts (at least 0.8 ppm molecular SO_2) would result in too high a level of total SO_2, because much of the SO_2 will bind to wine components.

If SO_2 is added just before bottling the 'Brett' cells may respond by going into a 'stunned' state which is sometimes called 'VBNC' ('viable-but-not-culturable'). The cells may shrink by as much as 30% (Charles Edwards, personal communication, 2007); some *Brettanomyces* cells are rather small anyway, around two to four microns by five to eight microns (Millet and Lonvand-Funel, 2000). These authors reported that starved lactic acid bacteria, acetic acid bacteria, and cycloheximide-resistant yeasts presumed to be *Brettanomyces* had passed a 0.45 μm membrane filter. In practice, however, for many years wineries have had extremely reliable success stopping *Brettanomyces* (and bacteria) with a membrane filter of 0.45 μm. Certainly, more work needs to be done on VBNC cells.

Sorbate/sorbic acid
The interaction of *Brettanomyces/Dekkera* and sorbate is not clear. There are quite a few references about resistance, at least in cultures (Chatonnet *et al.*, 1992; Licker *et al.*, 1998; Oelofse *et al.*, 2009); and a few references to sensitivity (Suárez *et al.*, 2007). More investigation of the response of different strains to sorbate in wine would be helpful.

Dimethyldicarbonate (DMDC, trade name Velcorin®)
Brettanomyces is sensitive to DMDC (Renouf *et al.*, 2008). Recent references suggest 100–150 ppm (Costa *et al.*, 2008) or 200 ppm (Oelofse *et al.*, 2009). Because DMDC degrades naturally within six to eight hours, there is no residual protection for wines that are reinfected after treatment. It is best used at bottling,

and a special (expensive) doser is needed to achieve the correct dose in each bottle. DMDC is a hazardous material, very harmful to skin, so it requires care to use safely. It is not permitted in all countries.

Other inhibitors
Chitosan, a polysaccharide derived from chitin, is reported to inhibit *Brettanomyces* at 3–6 g/L (Suárez *et al.*, 2007). It may be possible to use killer toxins made by the yeast strains *Pichia membranefaciens* (PMKT2) (Santos *et al.*, 2009), *Pichia anomala* and *Kluveromyces wickerhamii* to inhibit *Brettanomyces*. This competitive effect has been described as 'the big fish eats the little fish' (Guerra, 2008). More research is needed to determine whether these yeast toxins would work in wine being prepared for bottling. Also, inhibition of some *Brettanomyces* strains by peptides derived from bovine lactoferrin is under investigation.

Thermal inactivation
Of course, flash pasteurization is very effective at killing all microbes in the wine. It does require expensive equipment, and not all winemakers are happy to subject their wine to this procedure, especially their finer wines. At least one study found that simply heating wine to 35°C/94°F killed *Brettanomyces* (Couto *et al.*, 2005). However, heating could have serious effects on wine quality. Many years ago, a heating 'experiment' was carried out inadvertently by a Napa Valley Winery (Doug Manning, personal communication). Their 1978 Cabernet had won an important gold medal, but it had live *Brettanomyces* cells in the bottle; bottles were cultured every week to detect a 'bloom' if it began. But over the Fourth of July, the winery air conditioning failed and the wines became hot (no idea how hot); the bottles (miraculously) did not leak, but suddenly the *Brettanomyces* were dead. Many subsequent cultures grew no *Brettanomyces*, even 20 years later (personal experience).

Preparing the bottling line
After all the microbes have been filtered out of the wine, it is important that it does not become re-contaminated during bottling. The bottling line must be thoroughly sanitized to remove any resident microbes, including *Brettanomyces*, which could contaminate the wine. Sterilization, the complete removal of all microbes, can be achieved by running hot water (85–90°C/180–185°F, measured coming out of a filler spout) through the line for 20–30 minutes. Hot water may not physically penetrate far enough but the heat does, if it is applied for a long enough period. However, in bottling lines with blind spots and convoluted pathways, some microbes may be able to escape even being killed by the heat. Bottling lines can be sanitized with chemicals such as peracetic acid, quaternary ammonia, and ozone (a tricky procedure because of ozone's toxicity), though chemical sanitizers are not as effective as heat in killing microbes, especially

those sequestered in out-of-the-way places in the bottling line, including bends and T-joints.

Once the bottling line has been sanitized, it should be checked with a bioluminescence device, which detects ATP from cells (live or dead), to see if the procedure has eliminated all the microbes. If not, the bottling line should be re-sanitized before proceeding.

Without detracting attention from the extreme importance of bottling line sanitation, in most well-appointed commercial cellars *Brettanomyces* infections in the bottle nearly always come from failure to remove all the cells already present in the wine before bottling, rather than *Brettanomyces* contamination in the bottling line (however, bottling line contamination is very common for fermentative yeasts). Prevention of *Brettanomyces* growth in the bottle usually focuses on eliminating an existing infection.

After bottling

At least three bottles per bottling day (more for high-speed lines) should be checked by membrane filtration culture. PCR methods are not sensitive enough to detect the very low numbers of cells (a few cells per bottle, not per ml) required to test for fermentation stability of tightly filtered wine, though PCR could be used to check for *Brettanomyces* in wine that was not sterile-filtered.

Wines bottled without sterile filtration should also be cultured at 3, 6, and 12 months and tasted critically for any evidence of *Brettanomyces* (or other) growth. If any sign at all of activity, sensory change, or bottle variation is noticed, microbial growth should be suspected and the wine checked thoroughly.

References

CHATONNET, P., DUBOURDIEU, D. and BOIDRON, J. N., 1992, 'The origin of ethyl phenols in wines,' *J. Sci. Food Agric.* 68, 165–178.

COSTA, A., BARATA, A., MALFEITO-FERREIRA, M. and LOUREIRO, V., 2008, 'Evaluation of the inhibitory effect of dimethyl dicarbonate (DMDC) against wine microorganisms,' *Food Microbiol.* 25:2, 422–427. doi: 10.10.1016/j.fm.2007.10.003.

COUTO, J., NEVES, F., CAMPOS, F. and HOGG, T., 2005, Thermal inactivation of the wine spoilage yeasts *Dekkera/Brettanomyces*, *Int. J. Food Microbiol.* 104:3, 337–344. doi: 10.1016/j.ijfoodmicro.2005.03.014.

GUERRA, B., 2008, 'Research update: Is *Brettanomyces* sneaking into our wines? A review of the factors that favor *Brettanomyces* development during wine aging, including available means of control,' *Wine Business Monthly*, June.

LICKER, J., ACREE, T. and HENICK-KLING, T., 1998, 'What is "Brett" (*Brettanomyces*) flavor?: a preliminary investigation', in *Chemistry of Wine Flavor*, Waterhouse, A. and Ebeler, S. (eds), Washington, DC, American Chemical Society Symposium series.

MILLET, V. and LONVAUD-FUNEL, A., 2000, 'The viable but non-culturable state of wine micro-organisms during storage', *Lett. Appl. Microbiol.* 30, 136–141.

OELOFSE, A., LONVAUD-FUNEL, A. and DU TOIT, M., 2009, 'Molecular identification of

Brettanomyces bruxellensis strains isolated from red wine and volatile phenol production,' *Food Microbiol.* 26, 377–385. doi: 10.1016/j.fm.2008.10.011.

RENOUF, V., STREHAIANO, P. and LONVAUD-FUNEL, A., 2008, 'Effectiveness of dimethyldicarbonate to prevent *Brettanomyces bruxellensis* growth in wine,' *Food Control* 19, 208–216.

SANTOS, A., SAN MAURO, M., BRAVO, E. and MARQUINA, D., 2009, 'PMKT2, a new killer toxin from *Pichia membranifaciens*, and its promising biotechnological properties for control of the spoilage yeast *Brettanomyces bruxellensis*,' *Microbiology* 155, 624–624.

SUÁREZ, R., SUÁREZ-LEPE, J., MORATA, A. and CALDERÓN, F., 2007, 'The production of ethyl phenols in wine by yeasts of the genera *Brettanomyces* and *Dekkera*: a review,' *Food Chem.* 102, 10–21. doi: 10.1016/j.foodchem.2006.03.030.

11
Particular wine quality issues

11.1 How do I know my samples are representative?
T. J. Payette

Representative sampling
While in the lab the winemaking team spends considerable time on getting procedures and processes correct to run each particular test on a sample of wine or juice. Yet, equally as important, if not more, are we obtaining data that represent that particular batch or tank of wine? Has the sample, in any way, been compromised so that it does not reflect the tank or lot of wine intended? One would not want to take an aggressive winemaking action on a wine only later to find out the action was not necessary.

How often does a winemaker go into the cellar, select a sample of wine and then run a particular test, taste analysis or blending trial on the sample drawn? Are these test(s) run with confidence? Do these numbers reflect the tank's true contents? Do we need to sample each individual 60-gallon barrel and spend countless hours in the laboratory? Not necessarily unless the winemaker suspects a certain flaw to be identified with a particular vessel. If that is the case, it is best to 'quarantine' that particular lot of wine until the proper blending time making sure not to infect other batches.

The author wants to emphasize for the purpose of this article the lab test results are correct and the lab technicians are not at fault. The numbers reflect the sample given to the lab but the sample is not *representative* of the complete batch of wine in that tank. Take the opportunity to resample a wine that shows suspicious numbers in the lab. Remain open-minded and always quiz yourself to the possibilities of improper sampling.

Below are some pointers to help with the scope of sampling properly.

Tank sampling
Sample valve
Perhaps one of the easiest situations to monitor a large quantity of wine, yet, this process must be taken seriously with particular attention to the contents.

If taking a sample from a stainless tank, understand where the sample is coming from in reference to the contents of the tank. If a cloudy sample is taken from a sample valve near the bottom of the tank, understand it may not be cloudy throughout the whole tank and most likely is very clear at the top of the tank. Was the sample valve cleaned after its previous use? Was the valve flushed of its spoiled contents to bleed off any high VA or bacterially loaded wine, prior to sampling, that may have formed in the unsealed body of that valve? This flushing action of the sample valve, due to the positive pressure in the vessel, has little risk for cross-contamination and it is recommended in order to obtain a representative sample of the tank's contents.

From the top lid or manhole
Perhaps one of the best ways to sample in well managed cellars is if all the tanks are kept topped up and a catwalk exists to each man way. This gives the winemaker a chance to visually inspect the wine while taking the sample. Caution should be expressed not to sample the surface of the wine but to get the collection flask well under the surface of the wine to collect a representative sample. The surface of the wine may have a lower SO_2 reading and false numbers other than these that do not reflect the majority of the wine. Please keep in mind this action could be a large source of cross-contamination if not done properly because certain items may contact each wine as samples are collected.

Ball valves
These are known to be the largest cause of panic while reading a test result from the lab. Ball valves often have high spoilage counts, if not cleaned properly, lending toward off values most particularly with volatile acidity and sulfur dioxide to name just two. Other tests may give false reports in terms of the tank's actual contents. One should flush ball valves diligently when taking samples and one should be able to clean these valves well while not in use.

Butterfly valves
These offer an excellent source to sample a tank's contents if a sample valve is not present. Care should be taken to flush these valves. This flushing is more to remove solids and less to remove any potentially spoiled wine. The butterfly valve will often collect solids in them and deliver an unrepresentative sample unless flushed prior to sampling.

Barrel sampling

This is often the easiest yet most winemakers try and dodge this exercise making large decisions after tasting one barrel in a particular batch. One needs to isolate the vessels of a certain lot of juice or wine. With proper record keeping and a logical marking of each container this process is not too bad. One may sample an equal portion from each vessel of a particular lot (recommended for blending session exercises) or fifty percent of the vessels for that lot. Many times the winemaker can hedge this knowing a particular test or cellar action will be run in the future. One could use the fifty percent rule, or even less, if only addressing minor actions. Knowing the wine will be racked from barrel in 6 months may lend toward running further tests, double checking current data and making larger corrections at that time if needed.

Stratification
If taking a sample from the very top of a tank – does the sample represent the complete wine tank's contents? If taking a sample from near the bottom – does this represent the wine tank's content? Should mixing be used to make the tank contents uniform?

Mixing
It may not always be prudent to mix a tank of wine for sampling purposes. Much of the lees and solids may have settled to the bottom of the tank and mixing the tank would only re-suspend those solids. Certain times the winemaker may want to mix the tank prior to sampling may include: prior to bottling, just after a racking, just after and during additions and anytime a true representative sample is known to be needed for a particular winemaking decision.

Blending sessions

Getting prepared for a blending session is a time to make sure your samples are *very representative* and broken down into areas of distinction – perhaps even inside various lots. For example:

- Mountain Fruit Cabernet New French Oak
- Mountain fruit Cabernet Old French Oak
- Mountain Fruit Cabernet New American Oak and so on.

All of these samples in the example may be from the same raw material but the cooperage influence incorporated into them has made them very different. These differences make blending sessions a joyful challenge and yet offer the best chance to have the flexibility and control needed during a blending session. This sampling will give the blending session the greatest flexibility and control to the outcome of the final blend. After wine samples are taken from each vessel – make sure to mix the sample so the actual sample taken will be representative of the complete number of vessels sampled and that incomplete mixing will not adversely affect the blending session's outcome in the cellar.

348   Winemaking problems solved

Fining trials
This is an important time to have a representative sample. Mixing a reasonably clean wine, free of sediment, is desired to make sure this important refinement tool is employed properly prior to fining the tank's complete contents.

Sampling collection beakers, vessels and containers
Make sure to take samples in clean containers free from any debris or residues. An example may be the adverse reaction to a sample taken in a beaker that was recently used during a sulfur dioxide addition or used to dissolve meta-bisulfite. If residual sulfur dioxide remains in that container, it may adversely affect the lab test results and needed additions may be overlooked. The lab test result will show ample quantities, when in fact, the actual tank contents sample may have indicated otherwise.

Temperature measurement and stratification
Outside of the sampling topic, keep in mind when looking at a thermostat on a tank, where the thermocouple is inside that tank. If the reading is from a lower area in the tank, it may not be given a representative reading of that tank's true temperature. This is especially important when cold stabilizing wines. Mix the tank prior to seeding (if seeding is the practice used for cold stabilizing) the wine. With large capacity tanks, one may notice after mixing the temperature may rise. Another temperature stratification check, prior to mixing, is taking a temperature reading from the top of the tank's contents. Notice any difference? One will often see a difference in warm cellars with tall and large capacity tanks.

Miscellaneous
Representative sampling applies well beyond the wine cellar. This principal has huge applications in the vineyard when sampling the raw fruit to determine when to pick a certain variety or block. This concept is often reflected in grape berries sampling – a potential article in itself not to be dealt with in this article.

With above knowledge, keep in mind how the wines were sampled and how important that sampling technique may be when a particular decision is being made. When in doubt – resample and re-run the test in question. The winemaker is encouraged to make sure to think about the sample he or she has and to think what is actually inside the tank. Keep a keen sense of when tasting or when chemical data from a sample does not 'measure up' to what is expected. Be sure to investigate all angles before proceeding with drastic processes toward a tank of wine. It may just be the sample!

The data collected, whether blending, tasting or chemical analysis in the lab, can only be as effective as the sampling. The samples' content should be directly related to the tank and it should represent as closely as possible the contents. Always keep in the winemakers' mind how a sample represents the tank contents while tasting, testing and blending.

Know what the sample represents.

11.2 How do I use factors of ten to make blend trials in the lab easier?

T. J. Payette

Trials by ten (making it easy)
While in the wine business, struggles often start with speaking and working in the odd amounts that we often do. Trials and fining agents are riddled with unfamiliar language and terms. Grams per liter, pounds per thousand and milligrams per liter project off most winemasters' tongues as if the world is in tune. Let's review how to make this simpler and to understand.

When?
We should perform trials anytime a question is raised on how to improve or change a wine. If a wine has a problem – identify the solution in a laboratory first. Then apply the desired corrective action in the cellar.

Why?
The reason we do trials is to experiment with refinement and correction of a juice or wine. Always work in small quantities with a wine so one does not create a larger problem, in a tank, in need of potential further corrective action. These trials can be tasted and tested to see what the results would or will have been if the addition was made to the actual tank or vessel of juice/wine. This eliminates guesswork and unnecessarily 'shooting from the hip.'

Where?
One should do these trials in the laboratory where control, on a small-scale amount of wine, is essential. The opportunities of what one can discover in the lab are almost endless. Let's always make our mistakes on a small scale in the laboratory before stepping into the cellar for any actions that may change the flavor, aroma or chemistry of any juice/wine. This lab area should be designed for this feature of experimentation. The metric system will be used. Once this is attempted, one will not step back into some of the complicated aspects or other forms of measurement. These trials can be used for many things including but not limited to: sugar additions, acid additions, fining agents, concentrates, de-acidifications, etc.

Tools needed
- Scales that measure in grams preferably to a tenth of a gram
- 3–100 milliliter graduated cylinder (plastic preferred)
- 1–50 milliliter graduated cylinder (plastic preferred)
- 1–10 milliliter pipette (Class A volumetric)

350 Winemaking problems solved

- 2–5 milliliter pipettes with graduated markings at each milliliter to the tenth (plastic preferred)
- Small glass beakers 250 milliliters plus or minus
- Representative sample(s) of each wine to be sampled
- Clean wineglasses
- Glass watch glasses to cover each glass
- Spit cup
- Other testing equipment to answer questions at hand
- Magnetic stir plate with stir bars and retriever for the stir bars
- Sharpie™ pen or pencil for marking beakers
- 95% ethanol to remove Sharpie™ pen marks off glassware.

How?
Start with something simple where results can be easily determined with the wineglass to give the confidence needed to build upon the procedure. An example of this may be a tartaric addition trial for pH correction and/or palate modification. Let's go over one example:

1. Start with an ample quantity of wine to work with in the lab – perhaps an 800-milliliter representative sample from a wine vessel.
2. Weigh accurately 1.0 gram of tartaric acid and fully dissolve the acid in approximately 85 milliliters of the base wine with which you are working.
3. Once dissolved, place the full amount into a 100 milliliter graduated cylinder or as one becomes more experienced you may just make the solution in the 100 milliliter graduated cylinder.
4. Bring the amount up to volume in the 100 milliliter graduated cylinder up to 100 milliliters mark with additional base wine. (One should be clear they have made a solution of 1.0-gram tartaric acid dissolved into 100 milliliters of wine.)
5. In a clean graduated cylinder, pipette 10 milliliters of the newly prepared acid stock solution into the cylinder. Bring to the complete 100 milliliters volume mark with the base wine. This should represent a 1.0 gram per liter tartaric acid addition.
6. Pipette 20 milliliters from the stock acid solution made in step four into another graduated cylinder and bring to volume at the 100 milliliter mark to represent the next addition level of 2.0 grams per liter tartaric addition.
7. Continue to add to the number of samples you care to do the trial on in standard logical increments.

Set up the tasting trial
1. Pour about 50 milliliters or a quantity one desires to smell and taste, of the base wine, into a control glass and place it in the left hand glass in the tasting area. (One should always taste against a control.)

2. Pour the trials to be tasted, made in steps 5, 6 and 7 above, in increasing increments in each wineglass progressing from left to right. Mark their contents.
3. Add to this flight any wines from past vintages you may want to review or any other blind samples from other producers you may care to use as a benchmark. Mark their contents.
4. Taste and smell each wine several times. Go through the flight and detect what wine may best match or improve the desired style one is trying to achieve.
5. Select the match and leave the room for 1 to 2 hours.
6. Return and re-taste to confirm your decision.

Should chemistries play an important role to reviewing certain additions be certain to run a necessary panel of lab test to ascertain the proper numbers are also achieved. One may need to balance taste, flavor and chemistry to make some tough choices. Have all the data necessary and available to make those choices.

Calculation
Once the fear of the metric system is overcome and confidence is achieved, the calculations become very simplistic. Let's take the above as the example. If we dissolve 1.0 gram of tartaric acid into 100 milliliters of wine we now have 0.1 gram of tartaric acid in every 10 milliliters of wine. From this base if we blend 10.0 milliliters (one-tenth gram of tartaric) into 100 milliliters of fresh wine – this represents the equivalent of one gram per liter.

If we were to have used 20 milliliters that would represent two grams per liter in the small 100-milliliter blend. If we keep track of what we are tasting or testing and select the trial we prefer, one can mathematically calculate how much of the given addition is needed in a tank of known quantity of juice or wine. One can also extrapolate this out to larger volumes in the laboratory should that be desired to work beyond a 100-milliliter sample.

Spicing it up!
Once the first set of trials is mastered one may build on to the next step projecting out what one may want to do with the wine. This could eventually, and perhaps should, build out to treat large enough samples that one could cold and protein stabilize the wine in the lab, filter to the projected desired micron size and taste with a panel.

Double-checking the results
From experience, one can get so creative in a lab it can be difficult to trace exactly how one arrived at a certain desired concoction. Copious notes should be kept and most often one can trace their steps. When in doubt, however, re-

perform the steps with each addition to reestablish and confirm the same results. This extra time is well worth doing before stepping into the cellar.

Summary
Given time and experimentation with this system many blending trials with additions will become easy and systematic. Trials will often take less than ten minutes to prepare and one may taste at several points during the day or use extra time to perform lab test to confirm desired objectives.

Other helpful tips
- Keep in mind not to over scrutinize your accuracy in the laboratory. By this I mean make sure that if you measure something very tightly in the laboratory make sure this action will be able to be duplicated outside the lab. It is not uncommon, early on, for winemakers to get extremely exact in the lab only to step into the cellar with sloppy control over what they had just experimented with. Food for thought on the practical side!
- One can use other base solutes should that be desired. It does not always have to be wine.
- This system can be used for dosage formulation for sparking wines.
- If accurate scales may be an issue, the winemaker may always start by weighing larger quantities and dissolving into solution then breaking down that solution. Example: if a winemaker wants a 1.0 gram per liter solution and the scales are not accurate enough to weigh one gram, the winemaker may dissolve 10.0 grams into 100 milliliters and then measure out 10 milliliters of that solution and this should roughly equate to one gram.
- Makes sure all solids are dissolved and dispersed equally into any solution.
- One may also be able to blend two trials in 50% to 50% solutions to get an example of a trial in the middle without having to make one up specifically to match the amount desired.
- Always remember your palate may become desensitized while tasting and to step away from tasting for an hour or two and then return to taste one's preference. You may be shocked you had become used to certain levels because of tasting such extremes.

Further reading
AMERINE, M. A., BERG, H. W. and CRUESS, W. V. 1972. *The Technology of Wine Making*. AVI Publishing Company, Westport.
DHARMADHIKARI, M. R. and WILKER, K. L. 2001. *Micro Vinification*. Southwest Missouri State University, Springfield.
ZOECKLEIN, B. W., FUGELSANG, K. C., GUMP, B. H. and NURY, F. S. 1999. *Wine Analysis and Production*. Chapman & Hall, New York.

Based on a verbal discussion with Mr Jacques Boissenot, Mr Jacques Recht and Chris Johnson.

Dedication

Chris Johnson, a long-time colleague and friend who worked with me to develop this system many years ago, passed away in April of 2009. He was head of all red winemaking at Kendall Jackson and he had his own family winery label called 'Blair Estates'.

11.3 Why did my wine's pH and titratable acidity (TA) both drop significantly after fermentation?

C. Butzke

As one of the two main 'acidity' measurements in grape juice and wine, pH is primarily used to crudely assess the progression of ripening prior to harvest (e.g., Brix × pH^2) and to determine free SO_2 requirements after fermentation. The other basic winery lab analysis, TA (titratable acidity), serves as an indicator and dial for the sensory perception of acidity, *crispness, sourness, tartness* – whatever you call it. High TA and low pH mean high 'acidity' and while they are only indirectly related, both depend on the amount of free hydrogen ions (H^+, *protons*) from major acids or their salts in the must or wine. TA is also proportional to the amount of potassium (K^+) ions present, as well as the *ratio* between hydrogen and potassium ions. In addition, the pH depends on the ratio between tartaric and malic acid, the two main contributors to acidity in wine. As one would expect, any major loss of malic acid – either from respiration during ripening in warm climates or from (un)intentional malolactic fermentation during or after alcoholic fermentation – will reduce TA and increase pH.

Potassium is by far the major trace element that the grape vine picks up from the soil. The plant – for a variety of physiological reasons – exchanges K^+ against H^+ as part of the free tartaric and malic acids the berry produces from carbohydrates or directly from CO_2, respectively, during the growing season.

The 'extent of exchange' depends mainly on enzyme systems located in the roots, and viticultural practices such as rootstock choice, spacing or overcropping can influence the concentration in berries and subsequently the juice, must and wine. Potassium-poor, e.g. chalk/limestone soils also limit the amount of potassium uptake. Within the berry, potassium in located in relatively high concentrations in the skins, thus any skin contact, cold soak or extended maceration may extract additional potassium into the wine.

The salt of tartaric acid and potassium, potassium bitartrate (KH $C_4H_4O_6$, *cream of tartar*) by itself and at saturation concentration would create a pH buffer of 3.6 in water. Practically, i.e., considering the other ions and alcohol in wine, additional potassium will buffer a wine toward pH 4. This may explain some traditional winemaking practices, such as the *ripasso* of Northern Italy (fermenting *Valpolicella* under addition of *Amarone* pomace), which – among more obvious other things – provides additional extraction of potassium from the skins to balance normally acidic, cool climate red wines. In this context, note that bleeding (*saignée*) of red musts also changes the skin (potassium) to juice ratio! On the contrary, the 'plastering' (addition of $CaSO_4$, *gypsum*) of must, historically practiced in the hot viticultural areas of southern Spain, turns a portion of the potassium bitartrate back into free tartaric acid which in turn lowers the pH.

In most wines, any precipitation (removal) of potassium bitartrate at pHs below 4.0 will result in a slight drop in pH from the release of protons from previously undissociated tartaric acid molecules. Even though the pH may decrease, be aware that a drop of 3 g/L of potassium bitartrate will result in a *loss* of 1.2 g/L tartaric acid (−0.12% TA). This normally happens during cold stabilization but can already occur at the end of fermentation. The solubility of potassium bitartrate is about 5 g/L in freshly crushed, unfermented must at 20°C (68°F), but drops to 2 g/L in the corresponding wine – cooled down to cellar temperature after fermentation – at 14% alcohol and 15°C (60°F).

In the example of a long cool growing season, the vine produced lots of acid, respired relatively little of it, and had plenty of opportunity to exchange of hydrogen ions for potassium. This meant the initial juice TA was high, but so was the pH. Assuming it was a red fermentation, the juice picked up additional potassium from the skins, and it may have gone through a partial MLF, due to insufficient amounts of SO_2 at the crusher or residual yeast with MLF genes.

Consequently, the supersaturation with potassium bitartrate caused the same to precipitate, thereby releasing H^+ from the still abundant but non-ionized tartaric acid, and quickly dropping the pH. At the same time, the TA dropped as well with the loss of potassium bitartrate and the loss of malic acid due to the premature malolactic fermentation. In this scenario, an initial composition of a juice from a vineyard berry sample that measured pH 3.7 and a TA of 7.5 g/L, could have easily resulted in a pressed wine of pH 3.5 and TA of 5.5 g/L.

Due to the complexity of the interaction of different ions in wine, their solubility, ionization, recombination, etc., many seemingly illogical scenarios await the winemaker. Unfortunately, there are no quick and easy methods to determine the potassium content of a sample, or to measure 'total acidity' of a juice, both of which would make modeling of a juice's behavior much more predictable.

Further reading

AMERINE, M. *et al.*, 1967. *Technology of Winemaking*, 2nd edn. pp. 410–411, AVI.
BOULTON, R. B. *et al.*, 1996. In: *Principles and Practices of Winemaking*, pp. 523–537, Springer.
BUTZKE, C. E. and BOULTON, R. B., 1997. Acidity, pH and potassium for grapegrowers. *Practical Winery and Vineyard*. 09/10, 10–16.

11.4 How do I improve my wine color?

J. F. Harbertson

Improving wine color is a tricky proposition. It has been long believed that wine color is a direct translation of fruit color. There really aren't good data to support this despite what seems like such a logical conclusion (Downey 2005). Some basic aspects of the pigments that are responsible for wine coloration are required knowledge to understand why this isn't really the case. As it will become apparent wine color is hard to deal with because you have to extract and keep the pigments by transforming them during the winemaking process. Wine color is dependent upon the following factors (aside from the fruit composition):

1. pH
2. Copigmentation
3. Polymeric pigment formation.

Wine color is highly dependent on the amount of acid in the wine. At lower pHs more of the pigment is in its colored form. Thus more acidic wines are typically more brightly colored. Copigmentation is known to provide more color and to shift the hue of the wine towards blue (Boulton 2001). Copigmentation is limited by the amount of pigment itself and the concentration of colorless co-factors because they are found in grapes at low concentrations. Good examples of varieties that typically have a lot of copigmentation are Petite Syrah and Zinfandel. Unfortunately the color changes gained from copigmentation are transient because the pigments are highly reactive forming 'polymeric pigments' after combining with several other classes of compounds including tannins (Somers and Evans 1979). The polymeric pigments (which is a bit of misnomer because they are not all really polymers) are more resistant to sulfur dioxide decolorization and changes due to pH than the grape pigments (Somers 1971). Thus polymeric pigments are the stable color in red wine present long after the original grape pigments have left. They are more of a brick red color than the original pigments. The pigments in general are fairly stable but they decline rapidly during wine aging and can be as low as 50% of the peak value after 6 months (Somers 1971).

Knowing all of this, what is an effective strategy at getting good color in wine? As in many things in winemaking the work must begin in the vineyard. Viticultural areas that have cool nights and warm days will usually be able to make more color. Over-cropping has long been an issue with lower color in vines with excessive yield (Kliewer and Weaver 1971). Deficit irrigation applied early in the season has been demonstrated to also enhance color in the fruit and reduce the berry size (Roby *et al.* 2004). In any case having a strong relationship between the vineyard and winery is essential for success.

Pigment extraction occurs typically quite quickly during the winemaking process. From five to seven days have been reported in literature for peak extraction (Nagel and Wulf 1979). Tannins that can react with the pigments to form stabilized color are extracted more slowly (Cerpa-Calderon and Kennedy

2008). Depending on the type of wine you are trying to make, you can macerate longer to extract more tannin to help stabilize the color. If you have limited tank space, alternatives such as enzymes that quickly degrade cell walls in the skin and pulp are available that can help extract pigments and tannins (Ducruet et al. 1997). In most cases the color improvements are converted into polymeric pigments so you may not measure higher concentrations of free pigments. The only caveat to the enzymes is that you must experiment to use the right amount. Generally the enzyme use guidelines are purposely a bit vague on the package so as to avoid litigation. Use too little and no effect is observed, too much and you will have a cap that turns to oatmeal and your cellar crew will not be pleased with you. Enzyme reactions are highly dependent on temperature and pH. Enzymes are proteins after all and their structures are highly dependent on water, so ethanol will disrupt their activity. Just make sure your additions are early during the fermentation to avoid this problem.

A pre-fermentation cold soak where the must is held at low temperatures (10–15°C) prior to fermentation has been utilized for a number of years by winemakers as a means of extracting and retaining more color in wine. It is too bad that this technique has never been proven to actually work reliably (Heatherbell et al. 1996, Marais 2003, Reynolds et al. 2001, Zini et al. 2003, Álvarez et al. 2006). In most of these cases there is either no effect or very little improvement. The practice originated using the Pinot noir grape variety in Burgundy and is widely practiced in Oregon and California. It has been demonstrated to extract slightly more resveratrol than standard winemaking techniques (4.27 mg/L vs. 3.35 mg/L) but the significance of this difference has yet to be established (Clare et al. 2004). Many of the winemakers indicate that its use is not really just to enhance color extraction but to also avoid excessive seed tannin extraction because Pinot noir is an early ripening variety that often has green seeds that are rich in easily extractable tannins. The cold soak supposedly helps extract most of the pigments and then the wine can be pressed earlier before seed tannin extraction begins when the alcohol concentration has been raised in the ferment. In a commercial scale experiment with Pinot noir neither skin nor seed extraction of tannins was altered due to the use of a cold soak prior to fermentation (Peyrot des Gachons and Kennedy 2003). Thus evidence that cold soak does not help to extract unwanted tannins exists, but researchers have demonstrated that smaller tannins have less harsh astringent characters and since seeds primarily contain small tannins this has drawn this line of reasoning into question (Vidal et al. 2004). Further it is not clear why skin tannins are preferable considering that they are on an average three times larger than seed tannins.

Juice-run off prior to fermentation (bleeding or saignée) known is also a popular technique. In this technique after the fruit is crushed up to 20% of the juice is removed from the tank in an effort to simulate a smaller berry size and enhance the color and tannin concentration of the wine (Singleton et al. 1972). In many cases juice-run off works very effectively although in many cases the color enhancement does not persist (Gawel et al. 2001). Potentially this may occur in situations where not enough tannin and other phenolics are present to stabilize the color. Further in situations where the fruit is very high Brix and juice removal at the same rate as water addition is practiced no appreciable color enhancement has been observed (Harbertson et al. 2009).

Enological tannins that are derived from several sources (oak, grape, querbacho) have been touted as a means of stabilizing color in wine if added early when the pigments are being extracted. Although this seems like it should work there is very little evidence that it actually does. Experiments were tannins have been added pre-and post-fermentation have yielded some modest increases in color (Bautista-Ortín et al. 2007, Parker et al. 2007, Main and Morris 2007). Grape seed tannins added to Cynthiana (*Vitis aestivalis*) showed no significant increases in total phenolics but influenced the degree of browning in the wine and the amount of retained anthocyanins (Main and Morris 2007).

One of the main reasons that these experiments are being carried out with grape derived tannins is that the best evidence for the formation of stabilized pigments is with tannins that are derived from grapes (Somers et al. 1971, Remy et al. 2000). Oak tannins that are extracted into wine are primarily hydrolysable tannins where there is very little evidence to suggest that they combine with pigments to form stabilized pigment. Further the tannin extracts that are for sale are generally extracts of plant or wood tissues that contain not just tannins but other phenolic compounds. Even though the other compounds may be beneficial you are not adding as much tannin as you believe and this may explain why the color increases discussed earlier are very modest. A characterization of commercial tannins showed a 60-fold variation in the tannin concentration of 10 different available products (Obreque-Slíer et al. 2009). This points out the need to further investigate the different products to discover the major sources of variation and how to predict it. Wineries that attempt to use any tannin products are advised to use experimental controls and to maintain rigorous documentation to determine the usefulness of the new tannin products.

References and further reading

ÁLVAREZ, I., J. L. ALEIXANDRE, M. J. GARCÍA and V. LIZAMA. 2006. Impact of prefermentative maceration on the phenolic and volatile compounds in Monastrell red wines. *Analytica Chimica Acta* 563: 109–115.

BAUTISTA-ORTÍN, A. B., J. I. FERNÁNDEZ-FERNÁNDEZ, J. M. LÓPEZ-ROCA and E. GÓMEZ-PLAZA. 2007. The effects of enological practices in anthocyanins, phenolic compounds and wine colour and their dependence on grape characteristics. *J. Food Comp. Analysis* 20: 546–552.

BOULTON, R. 2001. The copigmentation of anthocyanins and its role in the color of red wine: a critical review. *Am. J. Enol. Vitic.* 52: 67–87.

CERPA-CALDERON, F. K. and J. A. KENNEDY. 2008. Berry integrity and extraction of skin and seed proanthocyanidins during red wine fermentation. *J. Agric. Food Chem.* 56: 9006–9014.

CLARE, S. S., G. SKURRAY and R. A. SHALLIKER. 2004. Effect of pomace-contacting method on the concentration of cis- and trans-resveratrol and resveratrol glucoside isomers in wine. *Am. J. Enol. Vitic.* 55: 401–406.

DOWNEY, M. 2005. Managing and manipulating winegrape tannins. Primary Industries Research Victoria Achievement Report 2004/05. (http://www.dpi.vic.gov.au/pirvic/achievementreport0405) p. 68.

DUCRUET, J., A. DONG, R. M. CANAL-LLAUBERES and Y. GLORIES. 1997. Influence des enzymes pectolytiques séléctionées pour l'oenologie sur la qualité et la composition des vins rouges. *Rev. Fr. Oenol.* 155: 16–19.

GAWEL, R., P. G. ILAND, P. A. LESKE and C. G. DUNN. 2001. Compositional and sensory differences in Syrah wines following juice runoff prior to fermentation. *J. Wine Res.* 12: 5–18.

HARBERTSON, J. F., M. MIRELES, E. HARWOOD, K. M. WELLER and C. F. ROSS. 2009. Chemical and sensory effects of saignée, water addition, and extended maceration on high brix must. *Am. Soc. Enol. Vitic.* 60: 450–460.

HEATHERBELL, D., M. DICEY, S. GOLDSWORTHY and L. VANHANEN. 1996. Effect of cold maceration on the composition, color and flavor of Pinot noir wine. In *Proceedings of the Fourth International Symposium on Cool Climate Enology and Viticulture.* T. Henick-Kling et al. (eds), pp. V1: 10–17. New York State Agricultural Experiment Station, Geneva.

KLIEWER, W. M. and R. J. WEAVER. 1971. Effect of crop level and leaf area on growth, composition, and coloration of 'Tokay' grapes. *Am. J. Enol. Vitic.* 22: 172–177.

MARAIS, J. 2003. Effect of different wine-making techniques on the composition and quality of Pinotage wine. I. Low-temperature skin contact prior to fermentation. *S. Afr. J. Enol. Vitic.* 24: 70–75.

MAIN, G. L. and J. R. MORRIS. 2007. Effect of macerating enzymes and postfermentation grape-seed tannin on the color of Cynthiana wines. *Am. J. Enol. Vitic.* 58: 365–372.

NAGEL, C. W. and L. W. WULF. 1979. Changes in the anthocyanins, flavonoids and hydroxycinnamic acid esters during fermentation and aging of Merlot and Cabernet Sauvignon. *Am. J. Enol. Vitic.* 30: 111–116.

OBREQUE-SLÍER, E., A. PEÑA-NEIRA, R. LÓPEZ-SOLÍS, C. RAMÍREZ-ESCUDERO and F. ZAMORA-MARÍN. 2009. Phenolic characterization of commercial enological tannins. *Eur. Food and Res. Tech.* 229: 859–866.

PARKER, M., P. A. SMITH, M. BIRSE, I. L., FRANCIS, M. J. KWIATKOWSKI, K. A. LATTEY, B. LIEBICH and M. J. HEDERICH. 2007. The effect of pre- and post-ferment additions of grape derived tannin on Shiraz wine sensory properties and phenolic composition. *Aus. J. Grape and Wine Res.* 13: 30–37.

PEYROT DES GACHONS, C. and J. A. KENNEDY. 2003. Direct method for determining seed and skin proanthocyanidin extraction into red wine. *J. Agric. Food Chem.* 51: 5877–5881.

REMY, S., H. FULCRAND, B. LABARBE, V. CHEYNIER and M. MOUTOUNET. 2000. First confirmation in red wine of products resulting from direct anthocyanin-tannin reactions. *J. Sci. Food Agric.* 80: 745–751.

REYNOLDS, A.G., M.A. CLIFF, B. GIRARD and T. KOPP. 2001. Influence of fermentation temperature on composition and sensory properties of Semillon and Shiraz wines. *Am. J. Enol. Vitic.* 52: 235–240.

ROBY G., J. F. HARBERTSON, D. O. ADAMS and M. MATTHEWS. 2004. Berry size and vine water deficits as factors in winegrape composition: Anthocyanins and tannins. *Aus. J. Grape and Wine Res.* 10: 100–107.

SACCHI, K. L., L. F. BISSON and D. O. ADAMS. 2005. Review of the effect of winemaking techniques on the phenolic extraction in red wines. *Am. J. Enol. Vitic.* 56: 197–206.

SINGLETON, V. 1972. Effects on red wine quality of removing juice before fermentation to simulate variation in berry size. *Am. J. Enol. Vitic.* 23: 106–113.

SOMERS, T. C. 1966. Wine Tannins-Isolation of condensed flavonoid pigments by gel-filtration. *Nature* 209: 368–370.

SOMERS, T. C. 1971. The polymeric nature of wine pigments. *Phytochemistry* 10: 2175–2186.

SOMERS, T. C. and M. E. EVANS. 1979. Grape pigment phenomena: interpretation of major colour losses during vinification. *J. Sci. Food Agric.* 30: 623–633.

VIDAL, S., L. FRANCIS, A. NOBLE, M. KWIATKOWSKI, V. CHEYNIER and E. WATERS. 2004. Taste and mouth-feel properties of different types of tannin-like polyphenolic compounds and anthocyanins in wine. *Anal. Chim. Acta* 513: 57–65.

ZINI, S., V. CANUTI, A. SILANI and M. BERTUCCIOLI. 2003. Criomacerzione in Toscana: Esperienze sul Sangiovese. *Ind. Bev.* 32: 16–23.

11.5 What is wine oxidation and can I limit it during wine transfer?

M. R. Dharmadhikari

Post fermentation processing of wine involves many operations such as clarification, stabilization, maturation and storage of wine. In performing these tasks a wine is often moved between the containers. During the wine transfer, if not well protected, a wine can be exposed to excess aeration (oxygen) resulting in wine oxidation. Wine oxidation can also occur when it is stored in partially filled containers and is not protected by an inert gas blanket. Once a wine is oxidized its quality is permanently impaired. Symptoms of wine oxidation include browning of color, loss of fruity varietal aroma and development of nutty, rancino, aldehydic flavor. Oxidation reaction can be enzymic as it commonly occurs in must or non-enzymic, or chemical auto-oxidation as it happens in wine. Phenolic compounds are the primary substrate for oxidization (Singleton 1989). The oxidizable phenols of wine include hydroxycinnamates, such as caftaric acid, catachins, gallic acid and anthocyanins. In chemical oxidation reaction (in wine), the phenolic oxidation generates hydrogen peroxide which oxidizes ethanol to acetaldehyde giving wine an aldehydic aroma.

Many factors influence wine oxidation. Important variables include types and amount of phenolic compounds, pH, temperature, amount of SO_2 and other antioxidants present in wine. Winemakers need to be especially cautious when handling a cold wine, such as during cold stabilization. Oxygen is more soluble at lower wine temperatures. However, the oxidation reaction speeds up when the temperature rises. As the cold wine warms up the greater amount of dissolved oxygen will contribute to serious wine oxidation. There is little information on the acceptable limits of oxidation. In the case of white table wines, some may improve with limited oxidation, but for most of the whites the tendency is to process them with minimum incidental oxidation. The red wines seem to improve with limited oxidation, but there is significant variation among red wines that show the beneficial effect of oxygen exposure. In order to minimize the adverse effects of oxidation during wine transfer the winemakers employ several techniques such as, using sulfur dioxide, using gentle pumps that minimize aeration, and checking hoses and fittings for leaks and flushing hoses and containers with inert gas prior to wine transfer.

In modern winemaking the inert gases are often used to minimize oxygen pickup in the head space of partially filled containers and also during wine transfers. The common inert gases used include; nitrogen, carbon dioxide, argon and a mixture of these gases in various proportions. For economic reasons the use of nitrogen and carbon dioxide seems to be more common. In order to provide an inert gas cover over the wine surface in a partially filled container, carbon dioxide or argons should be used. These gases are denser than the air and form an inert layer devoid of oxygen.

The danger of oxygen exposure is greater during the wine transfer. To minimize oxygen/air contact the system is purged with the inert gas. In the process of purging the inert gas is passed through the system such as hoses, transfer lines, equipment and the receiving tank to displace air. The wine is then transferred under an inert atmosphere.

Reference

SINGLETON, V. L. 1989. *Browning and oxidation of musts and wines*, Proceedings from the 4th Annual Midwest Regional Grape and Wine Conference, 4: 87–94.

11.6 What is the difference between oxidative and non-oxidative browning?

L. Bisson

Browning refers to the development of brown colors in foods and beverages. Browning can be either oxidative, requiring molecular oxygen exposure; or non-oxidative, occurring in the absence of oxygen. Oxidative browning involves the oxidation of phenolic substrates either spontaneously (non-enzymatic) or enzymatically due to the action of the grape enzyme polyphenol oxidase (PPO). Non-oxidative browning occurs when there is a specific interaction between amino acids and sugars usually catalyzed by heat. It is also called the Maillard reaction. This reaction is very important in some foods with high amino acid contents.

11.7 How do I identify and treat sulfur off-odors?

J. McAnnany

Identification

There are three ways to identify whether a wine contains sulfur off-odors; the penny/copper test, laboratory testing and sensory analysis.

The penny or copper test is a quick way to determine if your wine contains compounds associated with sulfur off-odors. Place a clean penny or copper sulfate solution in a glass of wine and swirl. If after a few seconds, the aroma improves then you have a sulfur issue.

The second option is to send a wine sample to an accredited wine laboratory for testing. The laboratory will either use a copper test (similar to the penny test) to confirm sulfur presence or will test for sulfur compounds using GCMS (gas chromatography-mass spectrometry). The GCMS method has had some issue lately with rumors of reporting false negatives, however, the method is still accepted in some circles.

The third option is to identify the presence of sulfur off-odors by sensory testing. This option should include more than the winemaker smelling the wine. For positive confirmation, it is important to send a wine sample to a professional sensory laboratory. Sensory professionals are trained to detect the nuances of sulfur off-odors and the descriptors that are associated with them.

Treatment

There is no one perfect cure for sulfur off-odors; however, many methods have been developed to reduce the sensory perception of off-odors. Depending on concentration and origin of sulfur compounds, some methods will work better than others.

The first and most popular method is treatment by copper. Copper additions are known to reduce sulfur off-odors very effectively. Although, copper does not react with oxidized sulfur compounds and tends to strip wine of aroma and flavor if added at high quantities. The legal limit for copper additions is 6ppm.

The second method was developed by Dominique Delteil, International Wine Consultant. His theory is to use copper in conjunction with an inactivated yeast product. The inactivated yeast will adsorb sulfur compounds and will release beneficial compounds (esters, fatty acids, etc.) that will limit the perception of copper. A recommended procedure can be found at http://www.vinquiry.com/pdf/07SumNwsltr.pdf.

Ascorbic acid is also brought up in regards to sulfur off-odor treatment. Ascorbic acid itself doesn't treat off-odors. It scavenges oxygen from the wine to convert dimethyl disulfide (which cannot be treated with copper) back to methyl mercaptans. Mercaptans can then be treated with copper.

There is some anecdotal evidence that enological tannins adsorb copper compounds thereby reducing sulfur off-odor perception. Contact your local tannin supplier for more information and recommendations.

Reference

DELTEIL, D., 2008. *Sulfur Off-Aromas: Guidelines for Red Wines*, available at http://www.vinquiry.com/pdf/SulfurOff-AromapaperforLodi2.pdf

11.8 What are some of the key steps to avoid sulfur off-odors?

J. McAnnany

The major compounds responsible for the scent known as sulfur off-odors are hydrogen sulfide (H_2S), ethanthiol, mehanthiol, dimethyl-sulfide, dimethyl-disulfide, among others. Each of these compounds occurs naturally during the winemaking process and each becomes noticeable at varying levels. Some of the sensory descriptors used to describe sulfur compounds include rotten egg, cabbage, burnt rubber, vegetal, onion, and garlic. It is important to take steps to avoid their development and keep levels under sensory threshold.

Pre-harvest

Starting with the vineyard, sulfur off-odors can be formed with the presence of certain pesticides and residues on wine grapes. This happens infrequently, but should be considered when encountering a sulfur issue. Winemakers should work with growers to avoid late season applications of elemental sulfur used for disease control.

Fermentation

Since many sulfur issues result from fermentation, it is important to follow good fermentation practices. First you must select a low sulfide producing yeast. It is also important to implement a proper nutrition program. Using a yeast stimulant to rehydrate the yeast and adding yeast nutrient after the first 6 hours and at 1/3 sugar depletion is a recommended strategy for keeping the yeast healthy and reducing the chance of producing sulfur off-odors. Aeration is another strategy for avoiding sulfur off-odors. Because yeast and yeast hulls are oxygen scavengers, adding oxygen when the cap forms and at 1/3 sugar depletion helps maintain healthy yeast cells (Lallemand, 2008).

Other strategies such as regulating temperature, mixing to avoid CO_2 toxicity, using lysozyme on native bacteria, and eliminating untoasted oak are all important to avoid sulfur odors.

Post fermentation and ageing

Ageing while on lees gives many benefits in regards to mouthfeel and other positive characteristics. However, if the lees contain yeast membranes that are stressed or if the lees are vegetal in aroma, then sulfur off-odors can develop. In these cases, it is important to rack wine off the lees. If *sur lie* ageing is still desired, suspicious lees can be replaced with a 'clean' lees product available from some companies (Delteil, 2008).

References

DELTEIL, D., 2008. *Sulfur Off-Aromas: Guidelines for Red Wines*, available at http://www.vinquiry.com/pdf/SulfurOff-AromapaperforLodi2.pdf

LALLEMAND, 2008. *Fermentation Product Catalog*, 37–38.

11.9 A sulfides analysis run by our wine lab shows the presence of disulfides in my wine. Should I treat it with ascorbic acid, sulfur dioxide (SO_2) and copper?

C. Butzke

Disulfides are the reaction product of two *mercaptans* in the presence of oxygen. They tend to smell like cooked corn, onions or cabbage, and in the opinion of very few winemakers contribute to the desirable aroma of wine. They form mostly when wine already had a problem with mercaptan formation. Hydrogen sulfide and the related mercaptans ('sulfur alcohols', *thiols*) form during the alcoholic fermentation as part of the yeast's nitrogen metabolism and due to nutrient deficiencies, particularly the amino acid *arginine*, stress conditions, or high concentrations of either *sulfate* or elemental *sulfur* from the vineyard. Winemakers sometimes try to strip the highly volatile H_2S gas from their wines by splashing the wine inside a partially filled container or aerating it during a tank transfer. However, H_2S is not only very volatile but it also readily oxidizes to elemental sulfur, thus remaining as source for reductive aromas if the solid sulfur particles settle into the lees.

While aeration is an acceptable practice during a pump-over of a still actively fermenting red must, *after* fermentation it is highly problematic. Aeration – and this includes micro-oxygenation – may lead to the oxidation of any mercaptans that have formed and the resulting development of disulfides. The problem here lies in our much higher sensory threshold for disulfides. Their temporary formation may make the mercaptan problem 'disappear', as we cannot smell the disulfides at the same concentration as the thiols. The odor detection threshold for H_2S and some mercaptans is about 1 to 2 $\mu g/L$, those for disulfides between 10 and 60 $\mu g/L$!

As a remedy against disulfides, the addition of the anti-oxidant ascorbic acid (Vitamin C) and sulfur dioxide had sometimes been suggested. The concept of this treatment had been the breakup and reduction of the disulfides back to their corresponding mercaptans by SO_2, under protection from oxygen by ascorbic acid, and followed by a binding of the mercaptans to copper ions from a subsequent dose of copper sulfate.

Unfortunately, the cleaving reaction is extremely slow at wine pH and would take months to complete, i.e. it is commercially impractical. In addition, treatment of wine with ascorbic acid is in general not recommended, including the treatment of 'untypical aging' notes. Ascorbate is not only very bitter but it reacts with oxygen or oxidized phenols (*quinones*) in wine to dehydro-ascorbic acid. This oxidized form (*dehydro-ascorbate*) is a potent source for the release of peroxide and subsequent browning and formation of acetaldehyde (leading to loss of free SO_2), once a treatment addition of the reduced form (*ascorbate*) is used up, e.g. during bottle aging.

On the other hand, a reductive environment after bottling, e.g. when closures are used that allow minimal oxygen permeation (such as screw caps or long corks) and in the presence of recommended amounts of free SO_2, the disulfides may split up in a period of several months and release the *reduced* smell of mercaptans.

Further reading

http://www.foodsci.purdue.edu/research/labs/enology/Coulter%20Unified%20Symp%20Reduction%20The%20Closure%20Issues.pdf

BOULTON, R. B. *et al.* 1996, in: *Principles and practices of winemaking*. Springer, New York, p. 289.

11.10 Can I use silver (alloys) instead of copper to bind reduced sulfur compounds such as hydrogen sulphide (H_2S) and mercaptans?

C. Butzke

The metal copper has a long history of use in both viticulture and enology. Toxic to downey mildew, it has been used as the active ingredient in the so-called Bordeaux Mixture since 1885. It had been present historically in the winery in its pure malleable form, e.g. as barrel topping cans or as part of an alloy such as brass in fermentor or hose fittings, etc. Both sources of copper, from the vineyard and the winery, have indirectly solved wine problems related to the formation of *reduced* sulfur components, most prominently hydrogen sulfide and its alcohol derivatives, the mercaptans (aka thiols). H_2S smells of flatus and sewage, mercaptans of burnt rubber, garlic and skunk.

As stainless steel tanks and fittings replaced contact with brass during the winemaking process, the addition of known quantities of copper *sulfate* (recently also *citrate*) to combat reduction in wine became a widely acceptable practice. Copper is an essential trace element in the human diet, and common multivitamin supplements contain about 2 mg of copper per daily dose. In the major wine-producing countries, the limit for copper content of finished wine is between 1.0 mg/L (OIV) and 0.5 mg/L (USA). To reach the copper intake equivalent to a vitamin pill, one would have to drink three to five bottles of wine per day. The US EPA national secondary maximum contaminant level for copper in drinking water is 1.0 mg/L.

However, concentrations as little as 0.6 mg/L of copper in wine can cause hazes and catalyze oxidative reactions, and once the extremely bitter copper sulfate has been added to wine, there are no safe fining agents on the market today that would allow for its removal. In the past, preparations that contained a sequestered complex of potassium ferrocyanide and ferrous sulfate were available, as an alternative to the direct 'blue-fining' with potassium ferrocyanide which is illegal in the US. Alternatively, winemakers have used fresh – metal-binding – yeast lees to remove excess copper from wine. In recent years, filter pads with resins that can bind copper, or that have bound copper imbedded, have been introduced. There is a caveat with such applications as it often remains unclear what other – possibly desirable – components are removed from the wine.

Looking for alternative metals to bind odorous sulfides, winemakers in some regions have experimented with soluble *silver* salts, such a silver chloride. Silver and its alloys are readily used in odor-eliminating kitchen gadgets and as nanoparticles imbedded in washing machines and clothing. Silver sulfide is even less soluble in wine than copper sulfide and both are easily settled and filtered out of wine. However, it takes significant additions of silver salts (more than 0.1 mg/L) to introduce enough silver ions to the wine that can quantitatively bind up sulfides and mercaptans. Silver can also react with tartaric acid in wine potentially forming a silver tartrate haze.

While silver is not an essential trace element, it is found in most human tissues, but appears to have no known physiologic function. The US EPA national secondary maximum contaminant level for silver in drinking water is 0.1 mg/L, only 10% that of copper. Adding any quantities of silver to wine is an illegal practice in the USA, so is a treatment with gold. For members of the OIV, 10 mg/L silver can be added for treatment of a wine, not more that 0.1 mg/L may remain in the finished product (Resolution OIV/OENO 145/2009).

It is important to remember that neither copper nor silver can bind disulfides (smelling of cabbage, onions, cooked corn, etc.), the oxidation product of two mercaptans. Once they have formed, they cannot be removed. It is there not recommended to splash and thereby aerate a stinky wine, unless only H_2S but no mercaptans are present which is impossible to tell by smell alone. It is recommended to have a complete reduced sulfides analysis panel run promptly by a commercial wine laboratory to determine the extent of the reduction issues of a particular wine. **11.9** discusses the futile application of ascorbic acid and SO_2 to break up disulfides.

Further reading

http://pmep.cce.cornell.edu/profiles/extoxnet/carbaryl-dicrotophos/copper-sulfate-ext.html

http://www.epa.gov/safewater/contaminants/index.html#sec

BOULTON, R. B. *et al.* 1996, in: *Principles and practices of winemaking.* Springer, New York, p. 289.

http://cira.ornl.gov/documents/SILVER.pdf

11.11 My wine smells like old fish what should I do?

J. F. Harbertson

With the problem of getting taint aromas in wine similar in nature to those normally associated with cork, a number of interesting problems have arisen. Since the taint aroma compound comes from the combination of chlorine or other halogens, many wineries have stopped using chlorinated sanitizing agents for any number of winery uses, especially on surfaces of unfinished wood where microorganisms can survive. Non-chlorinated sanitizing agents have become commonplace and are not without their own problems.

In one case we became aware of a winery that attempted to clean its filter with a non-chlorinated sanitizing agent. When a batch of wine was put through the filter the wine began to show bubbles on the surface, very similar to the head on a beer. The wine also had a distinct 'fishy' aroma to it and was very unappealing. About a barrel worth of wine had been filtered before the operator of the filtration realized that something was wrong and stopped the filtration. It was not clear at first what had caused the problem. So the winery had to investigate the composition of the cleaning agent.

Material safety data sheets (MSDSs) are required for chemicals that are used in the workplace and provide personnel with proper procedures for handling and working with the substances. If you purchase the compound from most companies, MSDSs come with it, if not there are several free internet-based databases available (MSDS online sources). In this case it was necessary to find out what the composition of the filter-sanitizing agent was so the MSDS was a good starting point. After evaluating the MSDSs we discovered them to comprise potassium hydroxide, alkali amines and some proprietary detergents. The use of scientific database systems that are available online is very helpful to evaluate compounds that are unfamiliar to you (online scientific search engines).

This provided all of the information we needed to determine what the problem was. Apparently the filter was not rinsed thoroughly all the way and a residue was left on the filter. The detergent caused the 'suds' on the surface of the wine, while the unpleasant 'fishy' aroma was caused by the amines. The pH of the wine was re-evaluated and we found no difference, which really isn't surprising given the buffering capacity of wines in general. The difficulty is that it is unlikely that the detergent would completely leave the filter.

The recommended ways of fixing the problem are actually not terribly difficult. Typically small-scale experiments carried out in the lab will give an idea of what type of treatment will improve the wine. In cases like these remember to be realistic and consider it successful if you can remove the offending aroma. Fining agents typically are indiscriminant and will remove both pleasant and unpleasant aromas. Several fining agents can be added to wine that can remove aroma components. The harshest is activated carbon, which resembles closely charcoal briquettes used in outdoor barbeques. Carbon should be used as a last ditch effort though. Bentonite is usually used to remove protein from wine but has some proven effects on the removal of aroma compounds on white wines (Ponzo-Bayón *et al.* 2003). We found reasonable success using butterfat. Half and half, which contains about 12.5% butterfat can be added straight to the barrel. Heavy creams tend to float on the surface and not settle out so the use of half and half is recommended. The amount added should be determined doing a small-scale experiment with different amounts of fining agent in the laboratory. Bearing in mind that the rate of the reaction is dependent upon the temperature experiments carried out in the lab will take less time, so for more accurate results just put your treated samples in the winery (Weiss *et al.* 2001).

The best way to avoid this type of problem is obviously not to get the cleaning agent in the wine to begin with. This is far easier stated than put into practice. Development of diagnostic tools to assess the presence of the cleaning agent can be developed internally based on either the appearance of soapy bubbles in the effluent of the filter, or perhaps even a pH difference. An experiment can be conducted to determine the minimum amount of time to remove the sanitizing agent at a given flow rate through the filter. This can be used as a guideline for cleaning the filter for its use in the winery.

MSDS online sources
The MSDS FAQ: http://www.ilpi.com/MSDS/faq/
MSDSSEARCH – The National MSDS Repository: http://www.msdssearch.com/
MSDS Solutions: http://www.msds.com
MSDS Sheets: http://www.msdsonline.com

Online scientific information sources
Scirus: http://www.scirus.com/
Online Journals Search Engine: http://www.ojose.com/
Google Scholar: http://scholar.google.com
Scitopia: http://www.scitopia.org/

References

POZO-BAYÓN, M. A., E. PUEYO, P. J. MARTÍN-ÁLVAREZ, A. J. MARTÍNEZ-RODRÍGUEZ and M. CARMEN POLO. 2003. Influence of yeast strain, bentonite addition, and aging time on volatile compounds of sparkling wines. *Am. J. Enol. Vitic.* 54: 273–278.

WEISS, K. C., L. W. LANGE and L. F. BISSON. 2001. Small-scale fining trials: effect of method of addition on efficiency of bentonite fining. *Am. J. Enol. Vitic.* 52: 275–279.

11.12 My wine is too astringent what do I do?

J. F. Harbertson

Not being able to control the amount of tannins in red wine or even in white wine (although more rare) is a problem that many winemakers have. The main things that are easy to do, but not ideal, are to blend the wine with something less tannic, or fine the wine with any number of proteins. Since the interaction between proteins and tannins that you add is identical to the way that tannins interact with salivary proteins, the treatment is fairly effective. Different proteins are available: gelatin, casein, isinglass, ovalbumin/conalbumin (egg whites) (Boulton *et al.* 1996). Gelatins often leave a residue because ethanol helps to solubilize the otherwise precipitated protein tannin complexes (Boulton *et al.* 1996). Aside from this problem gelatins are by far the most effective at removing tannins (Maury *et al.* 2001) as they share structural similarities to saliva proteins that some believe are only made for the sole purpose of protecting the gut from tannins (Lu and Bennick 1998). Casein (a milk protein) is very effective but must be solubilized in alkaline conditions prior to use. Egg whites are very popular because they are a natural product and can be purchased easily at a supermarket.

It has been generally thought that proteins bind primarily large tannins (Sarni-Manchado *et al.* 1999) but some recent work showed that each of the different proteins (gelatin, ovalbumin, isinglass, casein) and different size fractions of the same protein class interact differentially with different sizes of tannins (Cosme *et al.* 2009), but unfortunately no sensory was performed on the wines so it is difficult to know what the results actually mean. Regardless, allergen labeling may make fining wine with any of the animal derived products impractical although some effort has been made to evaluate plant-derived proteins (Maury *et al.* 2003). Recent studies of wine astringency demonstrated that tannins must be different by two-fold in order for a trained panel to be able to successfully differentiate the wines (Landon *et al.* 2008). Further since some of the polymeric pigments (the color that persists in wine) are actually capable of precipitating with protein there is a danger of losing stable color (Harbertson *et al.* 2003). All of this is pointing to controlling tannins as they are extracted during the winemaking process so you are not forced to blend away your problems, or add things to your wine.

Trusting your own palate to determine when the wine is astringent enough is a tricky proposition. Especially given that the ferment contains a lot of sugar, which interferes with your ability to perceive astringency (Valentová *et al.* 2002). Many winemakers have honed their skills so they are able to work through this problem. It has been recognized that there is quite a bit of variability between people and their ability to perceive astringency. A recent publication shed some light on this demonstrating that there was a fundamental difference between the compositions of the salivary proteins in the individuals (Sarni-Manchado *et al.* 2008). Those extremely sensitive to astringency had highly glycosylated proteins that didn't precipitate the tannins as effectively whereas people with less glycosylated proteins that are highly effective at precipitating tannins didn't perceive astringency as dramatically. Thus it is becoming more apparent that an independent means of assessing the tannin concentration of wine is necessary. Several commercial laboratories offer tannin measurements if you are not equipped with a laboratory and there are a couple of methods simple enough to do in your own laboratory and have been reviewed here (Harbertson and Spayd 2006, Herderich and Smith 2005). The methods based on precipitation of tannins with either protein or polysaccharides have been shown to have strong correlations with perceived astringency (Kennedy *et al.* 2006, Mercurio and Smith 2008). For wineries that are overworked as it is, models have been developed based on spectral data in the visible and ultra violet range, and the infra red range to measure tannins without having to do the reference method (Skogerson *et al.* 2007, Fernandez and Agosin 2007). The only necessary equipment is a spectrophotometer capable of measuring spectra in the previously mentioned range or a near-infra red spectrophotometer (very expensive, unfortunately).

Some practical advice when assessing wines for tannin concentrations is to first find wines you like and measure them. It is important to remember that many things go into the perception of quality and astringency. Astringency is known to be harsher when wines are more acidic while ethanol tends to make them less astringent (Gawel 1998). So measuring these components can also be helpful. This will allow you to determine at least a range of tannin concentrations for wines that you find acceptable. It is a good idea to keep in mind the age of the wine, as wines tend to get less astringent as they age. Basic research evaluating the relationship between tannin size and astringency showed that smaller tannins had less harsh astringent descriptors than larger tannins (Vidal *et al.* 2003). Experiments have been carried out to evaluate why this is (Haslam 1980, Vidal *et al.* 2002) and the most recent one concluded that tannins get smaller over time and reasoned this why they are less astringent (Vidal *et al.* 2002).

The tannin concentrations of red wines have been found to be highly variable at least an order of magnitude greater than that of the fruit (Harbertson *et al.* 2008). This suggests that winemaking plays a large role in determining the final tannin concentration of the wine. It has also been found that on average there were differences between commercial wines made from different grape varieties. Of the varieties tested the order was: Cabernet Sauvignon $\geq$ Zinfandel > Merlot > Syrah > Pinot noir. Although this is not surprising to anyone who has been making wine, it actually puts values to where previously there had only been conjecture.

The ability to troubleshoot the winemaking process has been one of the best uses of measuring tannins in the winery. In several instances we have been able to determine which step was causing the problem. The first instance was a winemaker who was upset about his wine being too tannic, but immediately had put down the problem to the grapes because the vineyard had a reputation for being tannic. But when we measured all of his wines they turned out to be on the higher end of the tannin concentration range. After assessing his technique he found that his built-in must-pump on his new crusher was shredding the skins and seeds and allowing the extraction to occur far too quickly. In another instance we assessed a cold-soak experiment and discovered high tannin concentrations when we were expecting high concentrations of color. Learning from previous experience we evaluated the must-pump and found the same problem and the pump was replaced with one with a more delicate mechanism and the problem abated.

References

BOULTON, R. B., V. L. SINGLETON, L. F. BISSON and R. E. KUNKEE. 1996. *Principles and Practices of Winemaking*. Chapman & Hall, New York

COSME, F., J. M. RICARDO DA SILVA and O. LAUREANO. 2009. Effect of various proteins on different molecular weight proanthocyanidins fractions of red wine during wine fining. *Am. J. Enol. Vitic.* 60: 74–81.

FERNANDEZ, K. and E. AGOSIN. 2007. Quantitative analysis of red wine tannins using Fourier-transform mid-infrared spectrometry. *J. Agric. Food Chem.* 55: 7294–7300.

GAWEL, R. 1998. Red wine astringency: a review. *Aust. J. Grape Wine Res.* 4: 74–95.

HARBERTSON, J. F. and SPAYD, S. E. 2006. Measuring phenolics in the winery. *Am. J. Enol. Vitic.* 57: 280–288.

HARBERTSON, J. F., E. A. PICCIOTTO and D. O. ADAMS. 2003. Measurement of polymeric pigments in grape berry extracts and wines using a protein precipitation assay combined with bisulfite bleaching. *Am. J. Enol. Vitic.* 54: 301–306.

HARBERTSON, J. F., R. HODGINS, L. THURSTON, L. SCHAFFER, M. REID, J. LANDON, C. F. ROSS and D. O. ADAMS. 2008. Variation in the tannin concentration of red wines. *Am. J. Enol. Vitic.* 59: 210–214.

HASLAM, E. 1980. In vino veritas: oligomeric procyanidins and the aging of red wines. *Phytochemistry* 19: 2577–2582.

HERDERICH, M. J. and P. SMITH. 2005. Analysis of grape and wine tannins: methods, applications, and challenges. *Aust. J. Grape Wine Res.* 11: 205–214.

KENNEDY, J. A., J. FERRIER, J. F. HARBERTSON and C. PEYROT DES GACHONS. 2006. Analysis of tannins in red wine using multiple methods: correlation with perceived astringency. *Am. J. Enol. Vitic.* 57: 481–485.

LANDON, J. L., K. WELLER, J. F. HARBERTSON and C. F. ROSS. 2008. Chemical and sensory evaluation of astringency in Washington state red wines. *Am. J. Enol. Vitic.* 59: 153–157.

LU, Y. and A. BENNICK. 1998. Interaction of tannin with human salivary proline-rich proteins. *Arch. Oral Biol.* 43: 717–728.

MAURY, C., P. SARNI-MANCHADO, S. LEFEBVRE, V. CHENYIER and M. MOUTONET. 2001. Influence of fining with different molecular weight gelatins on proanthocyanidin composition and perception of wines. *Am. J. Enol. Vitic.* 52: 140–145.

MAURY, C., P. SARNI-MANCHADO, S. LEFEBVRE, V. CHENYIER and M. MOUTONET. 2003. Influence of fining with plant proteins on proanthocyanidin composition of red wines. *Am. J. Enol. Vitic.* 54: 105–111.

MERCURIO, M. and P. A. SMITH. 2008. Tannin quantification in red grapes and wine: comparison of polysaccharide- and protein-based tannin precipitation techniques and their ability to model wine astringency. *J. Agric. Food Chem.* 56: 5528–5537.

SARNI-MANCHADO, P., A. DELERIS, S. AVALLONE, V. CHEYNIER and M. MOUTOUNET. 1999.

Analysis and characterization of wine condensed tannins precipitated by proteins used as fining agent in enology. *Am. J. Enol. Vitic.* 50: 81–87.

SARNI-MANCHADO, P., J.-M. CANALS-BOSCH, G. MAZEROLLES and V. CHEYNIER. 2008. Influence of the glycosylation of human salivary proline-rich proteins on their interactions with condensed tannins. *J. Agric. Food Chem.* 56: 9563–9569.

SKOGERSON, K., M. O. DOWNEY, M. MAZZA and R. BOULTON. 2007. Rapid determination of phenolic components in red wines from UV-visible spectra and the method of partial least squares. *Am. J. Enol. Vitic.* 58: 318–325.

VALENTOVÁ, H., S. SKROVÁNKOVÁ, Z. PANOVSKÁ and J. POKORNÝ. 2002. Time-intensity studies of astringent taste. *Food Chem.* 78: 29–37.

VIDAL, S., D. CARTALADE, J-M. SOUQUET, H. FULCRAND and V. CHEYNIER. 2002. Changes in proanthocyanidins chain length in winelike model solutions. *J. Agric. Food Chem.* 50: 2261–2266.

VIDAL, S., L. FRANCIS, S. GUYOT, N. MARNET, M. KWIATKOWSKI, R. GAWEL, V. CHEYNIER and E. WATERS. 2003. The mouthfeel properties of grape and apple proanthocyanidins in a wine-like medium. *J. Sci. Food Agric.* 83: 564–573.

11.13 I have a lactic acid bacteria problem in my winery. How do I control it to avoid high volatile acidity?

J. McAnnany

Lactic acid bacteria produce acetic acid from the sugars found in wine, glucose and fructose. Acetic acid is the main component of volatile acidity which gives a vinegar-like quality to wine. It is important to control lactic acid bacteria prior to fermentation, during stuck fermentations, or in sweet wines where there is a high amount of sugar present. It is also important to think about the timing of actions taken for controlling bacteria, since it is better to take care of the problem prior to alcoholic fermentation than dealing with the consequences later in the winemaking process. Knowing the type of bacteria is also important because certain strains require greater treatment than others. The following are some guidelines on how to control lactic acid bacteria to prevent the formation of acetic acid:

1. The use of lysozyme can be an important tool in controlling lactic acid bacteria. Lysozyme is a natural enzyme used to inhibit or suppress gram+ lactic acid bacteria, such as *Lactobacillus*, *Pediococcus* and *Oenococcus*. There are legal limits for the use of lysozyme and in white wines, it should be removed by bentonite fining before bottling (Dharmadhikari, 2008).
2. Sulfur dioxide, SO_2, is an effective compound in controlling the growth of lactic acid bacteria. Depending on wine pH, which influences the effectiveness of SO_2 on bacteria, relatively small amounts of SO_2 are needed to suppress bacterial growth.
3. Acid additions to lower wine pH can be effective in controlling lactic acid bacteria's ability to produce volatile acidity. At higher pH, lactic acid bacteria's conversion takes place at a much faster rate. Also, at higher pH, lactic acid bacteria can decompose sugars, citric acid and tartaric acid which lead to acetic acid production (Dharmadhikari, 2008).

Going forward, good sanitation practices are key to keeping lactic acid bacteria populations at a minimum. Sanitation also limits the formation of bacteria biofilms which are very difficult to eradicate. Storage of wine must take place in clean sanitized containers. All wine containers and equipment must be cleaned and sanitized before and after use. Sterile filter problem wines before bottling.

Reference

DHARMADHIKARI, M., 2008. *Lactic Acid Bacteria and Wine Spoilage*, available at http://www.extension.iastate.edu/Wine/Resources/lacticacidbacteriaandwinespoilage.htm

11.14 What are the types of hazes that can form in a wine?

L. Bisson

Over time cloudiness may develop in a wine that detracts from wine clarity. This is often a visual defect, but it may be accompanied by the production of off characters. Wine is initially clear but over time a visible haze or precipitate can form. The causes of haze problems in wines can be chemical, macromolecular or microbial. Chemical instabilities include metal ions, tartrate, polymerized phenols and oxidation products. Over time these components can engage in precipitate formation and appear as a visual cloudiness in the wine or as crystalline structures or lacquers coating the surface of a bottle. Macromolecular instabilities are caused by the formation of insoluble complexes between proteins and polysaccharides, and may also involve phenolic compounds and tannins. Proteins denature over time in wine exposing hydrophobic groups that tend to form associations or aggregates with other hydrophobic compounds or regions on other proteins. Eventually these aggregates become visible. Polysaccharides are soluble in water solutions but the displacement of water by ethanol can lead to the formation of aggregates as well that can become visible. Some hazes are mixtures of protein and polysaccharide. Finally if the wine contains sufficient nutrients such as sugar and nitrogen containing compounds at the end of fermentation microbial growth in the bottle can occur. If the malolactic fermentation has not completed prior to bottling, the lactic acid bacteria may grow in the bottle giving a visible haze. This is often accompanied by the appearance of carbon dioxide.

11.15 How much residual sugar will cause visible yeast growth and/or carbonation in the bottle?

C. Butzke

The recognition threshold for sweetness from residual sugar (R.S.) in a dry wine of average acidity is about 5 g/L (0.5%), while concentrations of 1 to 4 g/L can smoothen a wine without making it taste *off-dry*. In very acidic wine styles these numbers are significantly higher and a *méthode traditionelle* sparkling wine may taste *brut* even at 15 g/L R.S.

However, sensory perception of dryness is quite different from microbial stability of a particular wine, and cannot be used as a gauge for re-fermentation potential. After fermentation, a wine is commercially considered *dry* when its combined *Saccharomyces*-fermentable sugars (glucose and fructose) are below 1 g/L (0.1%). If the winery laboratory tests for R.S. by measuring *all* reducing sugars, including pentoses, then 2 g/L (0.2%) is commonly used.

Those numbers practically assure that the wine is safe from noticeable re-fermentation, i.e. cell growth and carbonation by *Saccharomyces* yeast strains. Nonetheless, any R.S., even below 2 g/L, can still serve as a substrate for our wine microbes such as *Brettanomyces* yeast or lactic acid bacteria.

Glucose and fructose are converted into roughly half ethanol and half CO_2, i.e., 1000 mg/L residual fructose can produce almost 500 mg/L CO_2 gas. As mentioned above, the solubility of CO_2 in wine is very high compared to other gases such as nitrogen or oxygen. A carbonation that is strong enough to push corks would occur beyond the saturation concentration of about 1400 mg/L at room/bottling temperature (20°C) and is strongly influenced by bottle headspace volume and closure type. A perceivable *spritz* may be tasted at 800 mg/L CO_2 which would require 1.6 g/L re-fermentable fructose or glucose.

Saccharomyces may grow if the recommended doses of sorbate and SO_2 are not met, or the sterile filtration prior to bottling was compromised (use bubble test!). It has been reported that even 100 mg/L residual pentoses can lead to a visible *Brettanomyces* yeast haze if proper SO_2 management and filter integrity tests are not implemented. A visible haze due to the growth of *Saccharomyces* must certainly be expected above 1000 mg/L (1g/L) R.S.

Further reading

http://www.vinquiry.com/pdf/05FallNwsltr.pdf

ZOECKLEIN, B. *et al.*, 1995. In: *Wine Analysis and Production*. Springer, New York, pp. 91–92.

11.16 What causes films to form on the surface of a wine?

L. Bisson

Films can form on the surface of a wine due to the growth of microorganisms. Both yeast and bacteria can be responsible for the film and sometimes the film is a mixture of both. *Acetobacter* is a common bacterial film former, and one that can convert the wine to vinegar. Sherry production relies on the metabolic activities of a film-forming yeast. Spoilage yeast such as *Pichia* and *Candida* can also make a film on the surface of a wine. The films may be a thin coating of the wine or may have the appearance of crinkled tissue paper. In severe cases multiple layers of cells can build up resulting in a very thick film on the wine. The organisms that form films like to grow at the air/wine surface interface which is why they coat the surface of the wine. They may be neutral, that is only causing a visual defect or they may cause significant deterioration of the wine, even converting the wine to vinegar.

11.17 A thin, dry, white, filamentous film has formed on the surface of my wine. What is it and how do I get rid of it?

J. P. Osborne

A microbial spoilage problem that can occur in wine during storage is the development of microbial films on the surface of the wine. A thin, dry, white film on the surface of wine is usually caused by the yeast *Candida vini* (formerly *Candida mycoderma*) and used to be referred to as 'mycoderma'. The formation of this film often starts off as small spots often referred to as 'wine flowers' that if left can rapidly spread over the entire surface of the wine. *Candida vini* is the primary yeast responsible for the film but other yeast such as *Pichia membranefaciens* may contribute and acetic acid bacteria are also often associated with wine films. The yeasts involved are oxidative and therefore require oxygen resulting in their growth only on the surface of the wine. This also explains the presence of acetic acid bacteria as they are also oxidative microorganisms. If wine conditions allow growth of oxidative yeast then they will also allow growth of oxidative bacteria.

Candida vini can produce sensorially undesirable products such as acetaldehyde and ethyl acetate via the oxidation of ethanol. In small quantities, acetaldehyde gives wine a distinctive 'nut like' aroma. However, *C. vini* can produce high concentrations of acetaldehyde to a point where it causes an objectionable in wine. Ethyl acetate is an ester of acetic acid and contributes a 'nail-polish remover' smell to the wine. *C. vini* can also oxidize some organic acids, producing a decrease in acidity. Overall, spoilage by *C. vini* leads to oxidized wine that seems flat and dominated by the smell of acetaldehyde.

Because *C. vini* is an oxidative organism, it is crucial to protect your wine from oxygen exposure. Formation of films often occurs in poorly topped barrels or tanks and wines with low alcohol, high pHs and low SO_2 levels are particularly susceptible. Therefore, maintaining topped tanks and barrels and keeping free SO_2 levels sufficiently high (see section regarding free and molecular SO_2) will help prevent spoilage by *C. vini*. It should be noted however, that one of the major products formed by *C. vini*, acetaldehyde, effectively binds free SO_2 and decreases its antimicrobial properties. Regular monitoring of free SO_2 levels after SO_2 additions are therefore recommended to ensure that adequate free SO_2 remains. Cellar temperature can also play a role as low temperatures (<15°C/60°F) can slow the growth of the film yeast. If small 'wine flowers' are noted early appropriate actions can be taken to prevent further microbial growth. However, if left unchecked growth of the film yeast will quickly spoil the wine.

If the film is widespread on the wine surface it may be useful to try and remove the film before treating the wine with SO_2. Care should be taken, however, when removing the film as the film is fragile and will disintegrate easily, sinking into the wine. Some winemakers have had success placing a few grams of potassium metabisulfite onto a small Petri dish that is allowed to float on wine in a barrel. Others have noted that spraying the surface of the wine with a strong SO_2 solution (10%) has helped control growth of film yeast. These methodologies rely on the fact that the film yeasts are only growing on the surface of the wine and so applying SO_2 to the surface will be more effective than mixing it in with the wine, thus lowering its concentration. The production of high concentrations of acetaldehyde, acetic acid, and ethyl acetate will make the wine unpalatable and difficult to recover. Fining (such as with gelatin) and close filtration may prove useful in some cases but often the wine is beyond repair. Prevention and early detection of the problem are key to minimizing the damage film yeast can cause to your wine.

In summary, a thin, dry, white, filamentous film that forms on the surface of the wine is caused primarily by the oxidative yeast *Candida vini*. This yeast oxidizes ethanol and can produce acetaldehyde, acetic acid, and ethyl acetate. The concentration of these products produced by the yeast can quickly render the wine spoiled. Preventative measures include maintaining topped barrels and tanks, adequate free SO_2, low pH, and low cellar temperatures. Treatment usually includes removal of the film (if possible) followed by treatment with SO_2 and topping.

Further reading

BALDWIN, G. 1993. 'Treatment and prevention of spoilage films on wine', *Aust. Grape Wine*, 378, 83–94.

ZOEKLEIN, B. W., FUGELSANG, K. C., GUMP, B. H. and NURY, F. S. 1995. *Wine Analysis and Production*, New York, Chapman and Hall.

11.18 I have a whitish film yeast growing on top of my wine in the tank (or barrel) that is resistant to sulfur dioxide (SO$_2$). What should I do?

C. Butzke

Film yeasts are particularly common in viticultural areas that use native American grape varietals or French-American crosses, both often rich in nutrients. These often bubbly surface biofilms appear to be synergistic layers of multiples species including strains of 'wild' wine-specific yeasts such as *Brettanomyces*, *Candida*, etc. They can appear so predictably during every vintage that the off-aroma they generate has frequently been mistaken for varietal aroma. It is particularly disconcerting as they appear resistant against the recommended doses of sulfur dioxide even at 0.85 mg/L molecular SO$_2$. The organisms may employ detoxification mechanisms similar to those used by the surface ('flor') yeasts in the production of Spanish Sherry or to *Saccharomyces cerevisiae*'s unique sulfite resistance in general. The winemaker's options are limited. The addition of sorbate may generate the feared *geranium* off-odor if malolactic bacteria growth is possible; i.e. at low free *and* bound SO$_2$. The use of antimicrobial agents such as allyl isothiocyanate disks floating on the wine surface is prohibited and will impart unpleasant off-odors. Historical remedies such as dispensing a layer of olive oil or paraffin on top of the wine will create other problems and is best left to home winemakers.

The most important aspect of prevention is the proper topping of tanks and barrels as these yeasts depend on the presence of oxygen. If tanks can only be kept partially full, frequent (bi-weekly) sparging with nitrogen or argon is advised to keep oxygen out of the headspace. Especially in warm areas with high relative humidity, the suspected organisms are part of the airborne microflora and will descend onto the wine's surface whenever possible. Therefore, it is not recommended to constantly re-open wine containers, e.g. for the ever-popular barrel or tank tastings with winery visitors. Opening bungs and topping barrels *too* frequently (more than once every two months) is also discouraged. In variable capacity tanks, organisms can often find a home around the inflatable and oxygen-permeable gaskets. In fixed capacity tanks, the air that replaces the wine sampled via the top gas vent is likely to be contaminated, even more so with the often neglected fluid inside the fermentation lock that acts as the barrier between cellar air and wine headspace.

Winemakers have had good success with UV radiation-based air filtration systems that keep the cellar air free of undesirable contaminants. Reducing the number of insects in the cellar especially bacteria-carrying fruit flies is also crucial to prevent microbial infection of the wine. The practice of spitting wine into the cellar drains after tasting it is unsanitary and will lead to growth of microbes, especially vinegar bacteria across drains and catch basins.

Excessive yeast nutrient additions to juices and musts – to prevent sluggish or stuck fermentations – is strongly discouraged as the left-overs will act as food for spoilage microorganisms. Exact measurement of the juice nitrogen status prior to fermentation is necessary in order to avoid microbial instabilities later on, in the tank or even in the bottle. Use of nutritional supplements that contain undefined amounts of yeast available nitrogen, phosphate, sterols, vitamins and other growth factors is unwise.

Further reading

PARK, H. and A.T. BAKALINSKY. 2000. SSU1 mediates sulfite efflux in *Saccharomyces cerevisiae. Yeast* 16, 881–888.

11.19 I am a great fan of Burgundian bâtonnage. How often should I stir my lees to release the most mannoproteins?

C. Butzke

Much as with red must cap punch-downs, the more frequent is not necessarily better or effective. Traditionally, in the cold cellars of Burgundy, the lees would be stirred every one to four weeks after fermentation, and not at all if the wine is cloudy or after malolactic fermentation has finished. There is no evidence that lees stirring (fr. *bâtonnage*) more than every other week has any benefit regarding the mouthfeel of the wine. In fact, constantly opening the barrel invites air, insects and microbes into your wine. Nor is more vigorous stirring recommended: the goal of bâtonnage is to gentle stir up the yeast lees that have settled to the bottom of the barrel or tank, not to make a yeast smoothie.

The desirable components that are extracted during aging on the lees are aromas of grilled nuts, toasted bread and roasted meats as well as 'mannoproteins', macromolecular combinations of sugars and proteins that reside in the yeast cell walls and contribute stylistically to the 'fat' mouthfeel of certain white wines. The mannoproteins can also bind tannins from new oak barrels and reduce astringency, yellowing and pinking. Their colloidal nature may enhance tartrate stability as well. It shall be noted though, that some winemakers recommend against the use of lees stirring altogether as it may diminish a wine's *finesse*.

Unfortunately, it turns out that the optimal conditions to extract mannoproteins from yeast cell walls are pH 7 and 38°C (100°F), far from, e.g., a Sauvignon Blanc aging *sur lie*. This does mean, however, that warmer cellar temperatures and higher pHs do enhance the extraction.

Much more important than the stirring regimen and the size of the winemaker's *dodine* is the actual amount of lees in the fermentor. The amount of yeast that grows during the lag phase prior to start of the alcoholic fermentation is directly related to the amount of nitrogen that the juice can provide. Lots of nitrogen means lots of yeast biomass, i.e. lots of extractable mannoproteins.

Some varieties such as Chardonnay are naturally rich in yeast available nitrogen (YAN), but individual juices should be tested for YAN using the NOPA method (see above) to assess potential for stuck fermentations as well as mannoproteins. Nutrient additions may enhance the formation of yeast biomass but may not be completely converted, leaving food for unwanted microorganism later on during the aging process.

By the way, the settling rate for yeast cells in wine or juice is independent of fermentor shape and size. Yeast cells, about 10 μm in diameter, take roughly 24 hours to fall 1 meter in juice, about 17 hours in the less viscous wine. In a standard barrel, about 66 cm sideways, all yeasts stirred in or up, will settle with 16 hours in juice or 11 hours in wine. In a 100 000 L tank that is 10 m high, the process will take 10 days or one week, respectively.

This has enormous impact on the uptake of nutrients prior to fermentation. Yeast cells that settle on top of each other in a short barrel or an experimental carboy compete for nutrients and may not get enough to grow or finish fermentation. Stirred small-scale fermentors are preferred to static ones, as they simulate the free-falling of yeast in a large fermentor surrounded by juice and its nutrients. Very cold juices (<15°C) or juices that had more than 100 mg/L SO_2 added may result in an extended lag phase during which the yeast may settle out completely even in an larger tank.

The stirring also aerates the wine slightly as oxygen is introduced through the bung hole. The wine remains protected by the reductive yeast lees and residual dissolved CO_2 from the fermentation, but since no SO_2 can be added if the wine is supposed to go through ML, it remains somewhat vulnerable to oxidative damage. In general, newer barrels tend to provide more oxygen to the wine via their stave's ellagitannins which get depleted with each subsequent use.

Compacted lees are a source for reduced sulfur off-odors, so stirring them up regularly, e.g., avoids the enzymatic reduction of sulfur-containing compounds such as yeast-derived SO_2 or elemental sulfur that may have formed during the racking and aeration of a H_2S-ridden wine.

Residual enzymatic activity by the settled yeast also prohibits the addition of SO_2 in the first few weeks of aging of a white wine on the lees, as it would be reduced to H_2S. After MLF has finished, the wine can be treated with SO_2 and left on the lees, as they should have no reductase activity left.

The management of reductive notes and oxidization is a fine balance as many of the grilled nuts or roasted meats aromas are related to traces of reduced sulfur components in the wine, specifically the acidic acid esters of certain mercaptans. During lees stirring and the introduction of small amounts of oxygen on the other hand, excess mercaptans can in turn be adsorbed by mannoproteins (as permanently bound disulfides!) and removed from the wine.

Actual autolysis of the yeast cell and a release of its internal amino acids and other components does not occur within the usual six months to one year window of white still wine making and aging. Only in late-disgorged sparkling wines that may have been *en tirage* for 2+ years, yeast autolysis will make a significant contribution to the wine aroma, often adding fruity rather than nutty/toasty notes. Yeast autolysis, however, may lead to the release of excess urea from yeast cells into the wine, causing the formation of undesirable ethyl carbamate.

Further reading

Recherches sur les composés soufrés volatils formés par la levure au cours de la vinification et de l'élevage des vins blancs secs. V. Lavigne-Cruege (w/D. Dubourdieu), PhD thesis, Université de Bordeaux 2, Bordeaux, France, 1996.

Identification et dosage de composés soufrés intervenant dans l'arôme 'grillé' des vins. Lavigne, V., R. Henry, D. Dubourdieu, Sciences des Aliments, 1998.

http://www.foodsafety.gov/~frf/ecaction.html

Incidence des conditions de fermentation et d'élevage des vins blancs secs en barriques sur leur composition en substances cédées par le bois de chêne. Chatonnet P., D. Dubourdieu, J.N. Boidron. *Sci. Aliments* 12: 665–685, 1992.

Index

α-iron, 242
acetaldehyde, 45, 172
acetic acid, 64, 309
Acetobacter, 96, 380
Acetobacter aceti, 268
Acidex, 21
acidity, 10, 101
acids, 9
acidulated sulphur dioxide, 260
active dried wine yeast, 71
aeration, 367
air pumps, 255
 care for winery pump, 214
AiroCide, 223
AISI 304, 243
AISI 316, 243
alcohol, 231, 302
alcohol level, 85
alcoholic fermentation, 271
aluminium oxide abrasives, 250
amino acids, 102
anodic reaction, 237
anodising, 242
ANSI/ASQC Z 1.4-1993 sampling plan, 178
anthocyanins, 10, 45
arginine, 98
aromatic maturity, 13
ascorbic acid, 176–7, 184, 367–8
Aspergillus niger, 28
ASTM A380, 249
astringency, 373–5
ATF Ruling 78-4, 50
ATP assay, 288
austenite, 242
austenitic 300 series stainless steel, 243
austenitic stainless steel, 243–4
autenitic-ferritic, 244

autochthonous yeast, 60

B-Brite, 274
β-phenylethanol, 94
back-flushing, 147
bag in box packaging, 184–5
Balling's scale, 22
barrel, 223–31
 cleaning procedure, 223–4
 dixie cup, styrofoam or bung, 227
 freshly emptied barrels management, 226
 management, 225–31
 monthly management, 228–30
  humidity, 230
  needed tools, 229–30
  rain, 229
  tartrate removal, 229
  when empty, 228
  when full, 228
 new barrels management, 225–6
  full fill, 226
  head swell, 225–6
  no treatment, 225
  quick rinse, 225
 spicing it up, 230
 storing of empty barrels, 226–7
  liquid sulphur dioxide, 226–7
  wicks and disks, 227
 types of rinse
  high pressure, 228
  hot water vs cold, 228
  ozone, 228
  water, 228
 wrapping it up, 230–1
barrel sampling
 mixing, 347
 stratification, 347

Index

base metals, 238
Basic oxygen steelmaking, 241
bâtonnage, 385–6
bentonite, 98, 109, 371
  heat-stabilising white wines, 102–4
  rehydration, 105–6
  slurry additions for bench trials, 106
  sodium or calcium bentonite, 105–7
berry sampling, 4
Bessemer process, 241
biofilms, 155, 262
biogenic amines, 309–10
bioluminescence, 288
bleach, 272
bleeding *see* saignée
blending sessions
  fining trials, 348
  sampling collection beakers, vessels and containers, 348
  temperature measurement and stratification, 348
Bordeaux, 188
Bordeaux Mixture, 369
botrytis, 17–18
Botrytis cinerea, 29, 33, 50
bottle bouquet, 195
bottling equipment, 255
bottling line
  disinfection, 154–5
  steaming, 156–9
  sterilisation, 150–2
Brettanomyces, 64, 121, 138, 166, 290–343
  culture in wine, 323–8
    colonies identification, 326–8
    culturing procedure, 325–6
    cycloheximide, 324
    materials, 323
    media, 323–4
    other cycloheximide-resistant yeasts, 324
    other media, 325
    polymorphism, 327
  effect on wines, 307–11
    acetic acid, 309
    colour degradation, 308
    ethyl phenol production, 307
    fatty acids, 308
    mousy taint, 308–9
    other effects, 309–10
    sensory effect of ethyl phenol, 307–8
    spoilage, 310
  growth, 300–4
    alcohol, 302
    CFU and concentration of 4-ethylphenol produced in Pinot noir wine, 304
    differences in growth and metabolism, 301–2
    patterns of growth, 302–3
    physiological characteristics of 35 strains, 303
    strain diversity, 301
    sugars, 302
  management, 329–34
    cellular conditions affecting growth, 332–4
    growth in the cellar, 329–30

    quarantine, 329
    SO_2 and pH management, 330–2
    wine composition influencing growth, 330
  methods of infection detection, 316–21
    chemical analysis, 316–17
    culturing, 318
    gene sequencing, 320
    microscopic exam, 317–18
    other methods, 320
    real-time PCR, 318–20
    sensory detection, 316
    test result interpretation, 320–1
  sample testing, 314–15
    new vs old barrels, 315
    record keeping, 314
    sampling procedures, 314
  source
    history and distribution, 292–3
    recent identification history, 293–4
    taxonomy, 292
    vineyard, 295
    winery, 295–7
  wine bottling, 336–9
    4-ethyl phenol in the bottle, 338
    bottling considerations, 338–9
    growth and survival in the bottle, 337
    instability testing, 336–7
    potential instability evaluation, 336
    wine handling prior to bottling, 337
  wine bottling preparation, 340–3
    after bottling, 343
    bottling line, 342–3
    ethyl phenol reduction, 340
    fining and filtration, 340–1
    growth inhibition, 341–2
    stability potential assessment, 340
    thermal inactivation, 342
  see also Dekkera
Brettanomyces bruxellensis, 262, 268
Brettanomyces infection, 290–343
Brettanomyces Specific Medium, 325
brightfield light microscope, 271
Brix, 9, 22, 47, 50
  adjusted YAN concentrations, 63
bromoanisoles, 273
browning, 363
bubble point membrane filter integrity testing, 146, 161, 218–22
  advantages, 222
  definition and principle, 218–19
  in-process bubble point test illustration, 221
  procedure, 220–2
bug catcher, 130
built detergents, 259
bungs, 227
butterfat, 372
BVS bottle finish, 181, 190

C6-alcohols, 42–3
Cabernet Sauvignon, 10, 63
Candida, 61, 380
Candida cantarelli, 324
Candida catenulata, 324
Candida ishiwadae, 324

Candida stellata, 268
Candida vini, 381
carbodoser, 174
carbon, 371
carbon dioxide, 26, 173, 258
carbonate, 173
carbonic maceration, 76
cartridge filtration, 136
casein, 109, 373
cast iron, 241
cathodic reaction, 237
caustic potash, 258
caustic soda, 258
    see also sodium hydroxide
cavitation, 213
cementite, 242
centrifugal pumps, 213–14
centrifugal stemmer/crusher, 199–200
centrifuge filtration, 137
ceramic, 242
Chardonnay, 16
chemical analysis, 316–17
chemical descaling, 249
chemical regeneration, 147
chitosan, 342
chloramphenicol, 324
chlorinated cleaners, 272
chlorine, 246, 272
chlorine dioxide, 273
chloroanisoles, 273, 276
chromium, 242
chromium (III) oxide, 242
cinder, 241
CIP system *see* cleaning-in-place system
citric acid, 115, 223, 251, 281, 282, 285
citric acid solutions, 220, 275
cleaning-in-place system, 275, 277
Clinitest method, 168, 169
cluster sampling, 4–5
cold soaking, 75
cold stabilising, 100–1
cold sterile bottling, 160
colloids, 141, 144
concentration-driven diffusion, 186
Concord, 41
condensed tannins, 45
copper, 367–8
copper sulphate test, 251
cork taint, 178
corrosion, 237–40
    galvanic corrosion, 239–40
    galvanic series of metals and alloys, 239
    metal and galvanic series resistance, 238–9
    metal corrosion, 237
    passivated materials, 246–8
        crevice corrosion, 247–8
        fretting corrosion, 246–7
        microbial corrosion, 248
crevice corrosion, 248
cross-contamination, 263–7
cross-flow filtration, 132–4, 137
Cucamonga stink *see Brettanomyces*
culturing, 318, 325–6

cupric casse, 193
Cutlery Grade stainless steel, 245
cycloheximide, 324

DAP *see* diammonium phosphate
Darcy's Law, 186
Debaryomyces hansenii, 324
DEDC *see* diethyldicarbonate
deformables, 142
Dekkera, 166, 290
depth filters, 139, 160
depth filtration, 130–1
diacetyl, 86
diammonium phosphate, 64, 67, 91, 330
diaphragm pumps, 207
    strengths and weaknesses, 210
diatomaceous earth, 130–1, 136, 137
diatomite, 130–1, 136, 137
diethyldicarbonate, 167–8
differential pressure, 145
dimethyldicarbonate, 167–8
disaccharide cellobiose, 300
disinfection, 154–5
disulphides, 367–8
dixie cups, 227
DMDC *see* dimethyldicarbonate
Dornfelder red wine, 125
Duplex alloys, 244

Easy Blue, 325
EC1118, 82
eddy-effect, 133
edible oils, 254
electric motor, 202–5
electrolyte, 238–9
electropolishing, 250
EN 1.4301, 243
EN 1.4401, 243
enological tannins, 45–6
environmental TCA, 276
epifluorescence microscopy, 268
ethanol, 93, 111, 121
ethidium monoazide, 319
ethyl acetate, 64, 381
4-ethyl guaiacol, 316–17
ethyl phenol, 340
    production, 307
    sensory effects, 307–8
4-ethyl phenol, 316–17, 338
Études sur le vin, 292

fatty acids, 89, 94, 308
Federal Register Title 27 CFR, 128
fermentors, 52
    effect on fermentation and wine storage, 55–6
    materials used in construction, 55–6
        cement/concrete tanks, 56
        plastic tanks, 55–6
        stainless steel tanks, 55
        wooden containers, 55
    red and white wines, 52–4
        material of construction, 52
        shape and size, 52–3

types and use in red and white wines
  other factors to consider, 53–4
  type of wine to be made, 53
ferric casse, 193
ferrite, 242
ferritic stainless steel, 243
fibreglass, 236
fibreglass resins, 236
Fick's Law of Diffusion, 186, 218
films, 380
  formation and removal, 381–2
filter, 136
filter porosity, 127
filtration, 138, 340–1
  see also wine filtration
fining, 340
first in-first out warehousing principles, 190
flexible impeller pumps, 207
  strengths and weaknesses, 208–9
foaming, 232–3
food grade lubricants, 253–4
fool's gold, 241
fortification, 47–8
fortified wines, 47
Fraction Defective Sampling Plan, 179
frozen grapes, 16, 50–1
fungicide residues, 89

γ-iron, 242
gallic tannins, 33
galling, 247
Gamay, 293
gamma butyric acid, 98–9
gasket, 180
gelatines, 373
gene sequencing, 320
geranium tone, 111
glass bead blasting, 250
glass head barrels, 230
glucanase enzymes, 233
glucans, 94, 233
glucosidases, 107
glutamate dehydrogenase, 65
glycerine, 253
GoFerm, 71
gold, 238
Gram stain, 270
grape analysis
  berry ripeness measurement and equipment used, 9–11
    acidity, 10
    phenolic composition, 10–11
    sugars, 9
  sampling, 1–2, 4–5
    berry sampling, 4
    cluster sampling, 4–5
    random sampling, 5
    systematic sampling, 5
  sensory ripeness assessment, 12–14
    aromatic maturity, 13
    grape pulp and structure maturity, 13
    skin maturity, 14
    tannin maturity, 13
  storage, transport and processing, 6–8

manual crushing, 6
preparing berries manually, 6–7
processing grape samples, 6
separating juice from berry solids, 7
storage and transport, 6
using mortar and pestle, 7
using muslin or cheese cloth to separate juice solids, 8
using sieve to separate juice and solids, 7
winemaking, 1–14
  importance of grape sampling and analysis, 1–2
grape berry
  compositional changes, 2
  development, 1–2
  measuring ripeness, 9–11
    acidity, 10
    phenolic composition, 10–11
    sugars, 9
  samples
    manual preparation, 6–7
    processing, 6
    separating juice from solids, 7
    storing and transporting, 6
    using mortal and pestle, 7
  sampling, 4
  sensory ripeness assessment, 12–14
  Victoria grape PPO thermal stability, 42
grape quality, 2
grape sampling, 1–2
  berry sampling, 4
  cluster sampling, 4–5
  random sampling, 5
  systematic sampling, 5
gravity fill bottling equipment, 232

HACCP *see* Hazard Analysis and Critical Control Points
halogen salts, 246
halogenated cleaners, 246
halogens, 246
Hanseniaspora, 61
Hanseniaspora uvarum, 75
harvest blocks, 2
Hazard Analysis and Critical Control Points, 278, 283
hazes, 378
hematite, 241
Henry's Law, 186
high speed modified centrifugal pumps
  strengths and weaknesses, 209
high-strength low-alloy steel, 236
high temperature-short time treatment, 102, 144
hot water, 260, 274
HSLA steel *see* high-strength low-alloy steel
HTST treatment *see* high temperature-short time treatment
Hunter Valley stink *see* Brettanomyces
hydrogen chloride, 123
hydrogen-evolution reaction, 238
hydrogen peroxide, 259
hydrolysable tannins, 45
hydrometer, 47

hyperfiltration, 128–9
hyperoxidation, 35
hyperoxygenation, 172
hypochlorite, 273

ice wine, 16, 50–1
inconsistent filling, 232–3
inline sparger, 173
integrity test, 146, 161–2
iodine, 274
iodophors, 259
iron alloy, 241
iron carbide, 242
iron (III) oxide, 237
iron oxide, 241
iron phosphate, 241
iron silicate, 241
isothermal membrane distillation, 135

juice clarification, 29
juice concentrate, 24
juice preparation, 15–51
  achieving desired sugar and alcohol concentrations, 47–8
  botrytis or rot management, 17–18
    red grapes, 17
    white grapes, 17–18
  calculation and addition of grape concentrate, 24–5
  cold settling of white juices, 31–2
  fortification, 47–8
    calculating volume of spirit to add, 48
    measurable Brix calculation, 47–8
    Pearson's square calculation, 49
  ideal temperature to press ice wine grapes, 50–1
  must clarification, 29–30
  pH and titratable acidity adjustment before fermentation, 21
  saignée and effect on red wine, 39–40
  skin contact on white wine style, 26–7
  sugar content nomenclature, 22–3
  thermovinification, 41–3
  wine press size, 15–16

K-sorbate see potassium sorbate
KBMS see potassium metabisulphite
kieselguhr/kieselgur see diatomaceous earth
Kieselsol, 107
Kloeckera, 61
Kloeckera apiculata, 67, 75, 324
KY jelly, 253

L-lactic acid, 95
L-malic acid, 95
LAB see lactic acid bacteria
laccase, 33, 102
lactic acid bacteria, 77, 85, 91, 96, 111, 144, 377
Lactobacillus, 81, 96, 124, 271
Lactobacillus casei, 262
Lactobacillus plantarum, 268
Lambic beer, 292
Lane-Eynon procedures, 168

lees, 88
lees press filtration, 137
lees stirring see bâtonnage
lemonade strength water, 284
Leuconostoc oenos, 91
liner, 180
loose RO, 128
low speed centrifugal pump, 209
lubrication
  destemmer. press and corker jaws, 253–6
    food grade lubricants, 253–4
    lubricants to be used, 254
  lubricating winery equipment, 254–6
    bottling equipment, 255
    pumps, 255
    sorting/vibrating tables, inclines, belted fruit elevators, 255
    stemmer/crusher, 254–5
    wine press, 255
  winery preventative maintenance, 235–6
  winery pump, 215
lye, 258
lysozyme, 88–9, 106–7, 124–5, 377

macrofiltration, 127
Madeirization, 196
magnesium, 236
magnetite, 241
malic acid, 89, 95, 97
malolactic bacteria, 81, 89, 94, 97, 121
malolactic fermentation, 77–97, 271, 301
  advantages and risk of inoculating with ML starter cultures, 83–4
    wine pH is greater than 3.5, 83
    wine pH is lower than 3.5, 83–4
  impact of temperature, 85
  impact of yeast used for alcohol fermentation, 93–4
    simulation of malolactic bacteria by yeasts, 94
  influencing factors, 88–9
    lees compaction, 88
    residual lysozyme activity, 88–9
    selected yeast strain, 88
    tannins, 88
  inoculation, 81–2
  malolactic bacterial growth/carbonation caused by residual malic acid, 97
  malolactic starter culture preparations, 77–8
    direct inoculation freeze-dried malolactic starter cultures, 78
    frozen malolactic bacteria starter cultures, 77
    liquid malolactic cultures, 77
    quick 1-step build-up cultures, 78
    traditional freeze-dried starter cultures, 78
  managing diacetyl levels in wine, 86
  monitoring, 95–6
    chemistry, 95
    microbiology, 96
    organic acid analytical methods comparison, 95
  nutrient addition, 91–2
    MLB nutritional requirements, 91

nutrient availability, 91
nutrient deficiencies, 92
other influencing factors
  excessive amounts of oxygen, 89
  fungicide residues, 89
  initial malic acid concentrations, 89
  yeast derived fatty acids and peptides, 89
storage and rehydration of ML starter cultures, 79–80
  direct inoculation starter cultures, 79
  frozen ML starter cultures, 79
  liquid ML starter culture suspensions, 79
  quick build-up starter cultures, 79–80
  traditional freeze-dried standard starter cultures, 80
malolactic starter cultures
  advantages and risk of inoculation, 83–4
    wine pH < 3.5, 83
    wine pH > 3.5, 83–4
  preparations, 77–8
    direct inoculation freeze-dried malolactic starter cultures, 78
    frozen malolactic bacteria starter cultures, 77
    liquid malolactic cultures, 77
    quick 1-step build-up cultures, 78
    traditional freeze-dried starter cultures, 78
  storage and rehydration, 79–80
    direct inoculation starter cultures, 79
    frozen ML starter cultures, 79
    liquid malolactic cultures, 77
    liquid ML starter culture suspensions, 79
    quick build-up starter cultures, 79–80
    traditional freeze-dried standard starter cultures, 80
manganese, 243
mannoproteins, 94, 102, 107, 385
marine grade stainless steel, 244
martensitic stainless steel, 244
Maser, 196
material safety data sheets, 215, 289, 371–2
  wine tanks, 278
  winery transfer hoses cleaning, 283
MBR, 79
mechanical cleaning, 274
mechanical cleanliness, 285
media migration, 131
medium-chain fatty acids, 94
Megaquant, 336
membrane filters, 160–1
membrane filtration, 96, 131–2
membrane porosity, 139
membrane storage, 148
methylene blue, 270
Metschnikowia, 61
micro-oxygenation, 172, 301
micropore clarification, 133
microscope, 270–1
microscopy, 317–18
mid-infrared spectrometry, 65
MIL-STD-105E, 178
milk, 109
MIR *see* mid-infrared spectrometry
MLB *see* malolactic bacteria

MLF *see* malolactic fermentation
molecular weight cut-off, 127–8, 133
molybdenum, 242
must clarification, 29–30
must preparation, 15–51
  2008 limit for DAP additions, 67
  avoiding oxygen exposure with white must, 35
  botrytis or rot management, 17–18
    red grapes, 17
    white grapes, 17–18
  calculation and addition of grape concentrate, 24–5
  consequences of water addition, 38
  enological tannins, 45–6
  facilitating fermentation of red musts, 37–8
  ideal temperature to press ice wine grapes, 50–1
  must clarification, 29–30
  pH adjustments, 19–20
    ability to process under hygienic and cool conditions, 20
    difference between actual and desired pH, 20
    fruit quality, 20
    wine style, 19
  pros and cons of using pectinase, 28
  saignée and effect on red wine, 39–40
  skin contact on white wine style, 26–7
  sugar content nomenclature, 22–3
  thermovinification, 41–3
  treating must from white grapes containing laccase, 33
  wine press size, 15–16
MWCO *see* molecular weight cut-off

nanofiltration, 128, 133
nanoseparation, 128
native flora fermentation, 60, 61
Natstep, 71
negative Pasteur effect, 302
nephelometric turbidity meter, 103
nickel, 242, 243
nitric acid passivation, 251
nitric and hydrofluoric acid pickling, 250
Nitrile/Buna-N, 216, 236
nitrogen, 98–9, 173, 220, 242
nitrogen by o-phthaldialdehyde, 65, 385
noble metals, 238
nominal filtration, 128
nominal rating, 160
non-deformables, 142
non-oxidative browning, 363
non-*Saccharomyces* yeast, 271
NOPA *see* nitrogen by o-phthaldialdehyde
Nutrient Medium, 323
nutrients, 91–2
nylon, 236

O-rings, 214, 215–16
1-STEP Kit, 79
Oenococcus, 124
Oenococcus oeni, 77, 81, 82, 84, 85, 86, 93, 96, 97, 262

off-dry taste, 379
osmosis, 135
osmotic distillation, 135
Oxiclean Free, 274
oxidation, 35, 189
oxidation-reduction reaction, 238
oxidative browning, 363
oxygen, 89, 258
  air and oxygen contact of wine, 170
  in bottles at filling, 176–7
  control of uptake at bottling, 172–5
    benefits of limited oxygen, 172
    elimination at bottling, 173–5
    elimination prior to bottling, 172–3
    inline oxygen sparger, 173
  pick up during bottling, 170–1
  solubility and uptake during processing and ageing, 170
  uptake into wine at bottling, 171
oxygen exposure, 35
ozonated water, 275
ozone, 236, 260, 277, 334

PAO *see* poly alpha olefins
paper chromatography, 97
particulates, 142
passivation, 242, 249
PCA *see* principal component analysis
Pearson's Square, 24, 37, 47, 49
pectinase enzymes, 141
pectinases, 28, 107
pectins, 141
Pediococcus, 96, 124
Pediococcus pentosaceus, 91
pentachlorophenol, 276
peptides, 89, 93–4
peracetic acid, 259–60
peripheral turbulent flow, 133
peristaltic pumps, 210
peroxides, 259–60
per(oxy)acetic acid, 259–60, 274
per(oxy)carbonate-based products, 274
perpendicular flow filtration, 130–4, 142
personal protective equipment, 289
Petite Syrah, 356
petrolatum, 253
pH, 10, 19–20, 21, 83–4, 106, 121, 330–2, 354–5
phase contrast microscope, 270, 271
phases, 241
phenolic composition, 10–11
phenolic ripeness, 9
Pichia, 61, 380
Pichia guilliermondii, 307, 324
Pichia membranefaciens, 381
pickling, 249
pig iron, 241
Pinot blanc, 125
Pinot gris, 125
Pinot Noir, 75, 125
piqûre lactique, 81, 83
piston pumps, 206
  strengths and weaknesses, 209
pitting, 248

plastering, 354
poly alpha olefins, 254
polyamides, 236
polyphenol oxidase, 42
polysaccharides, 141, 142
polyvinylpolypyrrolidone, 109
porosity, 127–8
Port-style wines, 47
positive displacement pumps, 206–7
  care for winery pump, 213
  diaphragm pump, 207
  flexible impeller pump, 207
  progressive cavity pump, 206
  reciprocating piston pump, 206
  rotary vane or lobe pump, 206–7
potassium, 354
potassium bicarbonate, 100–1
potassium bitartrate, 100, 167, 191, 193
potassium caseinate, 109
potassium metabisulphite, 100–1, 119–20, 171, 224
potassium metabisulphite solutions, 274
potassium sorbate, 100–1, 113, 115, 166–7
  structural formula, 114
potassium tartrate, 100, 223
Pouorbaix diagrams, 249
powder filtration, 131
PPE *see* personal protective equipment
PPO *see* polyphenol oxidase
pre-filter, 160
prefermentative cold maceration *see* cold soaking
PressPro 50, 15
pressure-driven permeation, 186
pressure hold testing, 146
principal component analysis, 58
proanthocyanidin tannins, 33
progressive cavity pumps, 206
  strengths and weaknesses, 210
proline, 98
protein stability, 103, 106–7
proxy compounds, 259–60
Proxycarb, 259
pump over, 53
pumps, 206–16
  care for winery pump, 213–17
    air pumps, 214
    centrifugal pumps, 213–14
    electrical considerations, 216
    lubrication, 215
    positive displacement pumps, 213
    sanitation issues, 216
    seals and O-rings, 215–16
  common problems, 211–12
    failure to prime, 211–12
    slowing down, losing pressure, 212
  different types for wine and/or must transfer, 206–7
    centrifugal pump, 207
    positive displacement pump, 206–7
  lubrication, 255
  strengths and weaknesses, 208–10
    diaphragm, 210
    flexible impeller, 208–9

394   Index

high speed modified centrifugal, 209
low speed centrifugal, 209
peristaltic, 210
piston pump, 209
progressive cavity, 210
rotary lobe, 209–10
see also specific pump
PVPP *see* polyvinylpolypyrrolidone
pyrite, 241

Quality Points Program, 257
quaternary ammonium compounds, 259, 275
QUATS, 259
quorum, 155

random sampling, 5
real-time PCR, 318–20
Rebelein, 168
reciprocating piston pumps, 206
red grapes, 37, 40
  managing botrytis or rot, 17
red musts, 69–70
  consequences of water addition, 38
  enological tannin addition before fermentation, 45–6
  facilitating fermentation, 37–8
red wine, 10
  effect of saignée, 39–40
  fermentors, 53
representative sampling, 345–6
residual sugar, 379
resveratrol, 357
reverse osmosis, 133, 149
reverse plating, 250
Riesling, 83, 125, 188
ripasso, 354
ripeness, 9–11
roller and wiper type stemmer/crusher, 200
Rosé, 39
rot, 17–18
rotary lobe pumps, 206–7, 255
  strengths and weaknesses, 209–10
rubber-based elastomers, 216

Saccharomyces, 31, 58, 61, 71, 73, 74, 93, 144, 163, 379
Saccharomyces bayanus, 51, 115
Saccharomyces cerevisiae, 57, 60, 63, 93, 115, 166, 262, 268
SAE 140, 255
safe, 258
saignée, 357–8
  effect on red wine, 39–40
  rates to be used with grapes of varying status, 39
5 percent salt spray test, 251
sampling procedures, 314
  new vs old barrels, 315
sand blasting, 250
sanitisation, 148
Sauvignon, 74
Sauvignon Blanc, 35, 66, 188
scalping, 182–3
screen filtration, 130

screw caps, 180–1
seals, 215–16
sensory analysis, 12
sensory detection, 316
sensory ripeness assessment, 12–14
Sequential Probability Ratio Test, 179
Shiraz, 10
silver, 369–70
skin contact, 16, 26–7
skin maturity, 14
slag, 241
smelting, 241
Sniff-Brett, 325
soda ash, 223, 258
sodium carbonate, 258
sodium hydroxide, 251
sodium *meta*-silicate, 258
sodium *ortho*-silicate, 258
sodium percarbonate, 259
sodium peroxycarbonate, 223
softening, 195
soluble oils, 254
sorbate, 111–12, 115
  advantages, 112
  disadvantages, 112
  health note, 112
  use in wine, 112, 113–14
    dose, 114
    potassium sorbate, 113
    sorbic acid, 113
sorbate/sorbic acid, 341
sorbic acid, 111, 113, 114, 115, 166–7
  concentration and rates of addition, 167
  disadvantages, 166–7
Spanish Sherry, 196
spoilage, 310–11
sponge ball, 284, 285, 286
spread plating, 96
stainless steel, 241–5
  classification, 243–4
    austenitic, 243–4
    autenitic-ferritic, 244
    ferritic, 243
    martensitic, 244
  cleaning or protection, 249–52
    methods of pickling and cleaning, 249–50
    passivation and pickling, 249
    passivation considerations, 251
    passivation verification, 251
  grades, 244–5
    200 series, 244
    300 series, 244–5
    400 series, 245
    500 series, 245
    600 series, 245
  iron and steel, 241–2
stainless steel wire brushing, 250
steam, 260
steaming crayons, 157
steel shot, 250
stemming/crushing operation, 199–201
  centrifugal stemmer/crusher, 199–200
  common problems, 200–1

roller and wiper type stemmer/crusher, 200
sterile bottling system, 136, 156, 162
sterile pads, 139
styrofoam cups, 227
succinic acid, 94
sugar ripeness, 9
sugars, 9, 302
sulphite ion, 119
sulphites, 116
sulphur dioxide, 27, 85, 93, 117–18, 119, 164–6, 176, 184, 220, 330–2, 341, 367–8, 377, 384–5
 adjusting SO_2 at bottling, 164–6
 disadvantages, 164
 free SO_2 required at wine pH, 122
 levels of free SO_2 to achieve molecular SO_2, 118
 prevention of malolactic bacteria or Brett growth, 121–3
 required free SO_2 levels, 165
 variation in SO_2 additions, 120
sulphur off-odours, 364–6
 identification, 364
 prevention
  fermentation, 366
  post fermentation and ageing, 366
  pre-harvest, 366
 treatment, 364
supercritical CO_2 injection, 144
synthetic closures, 182–3
systematic sampling, 5

T-Drains, 276
tangential-flow filtration, 132–4
tank sampling
 ball valves, 346
 butterfly valves, 346
 sample valve, 346
 top lid, 346
tannin maturity, 13
tannins, 45–6, 88, 102, 356–7, 373–4
tartaric acid, 19, 97
tartrate, 171, 185
TBA *see* 2,4,6-tribromoanisole
TCA *see* 2,4,6-trichloroanisole
Teflon, 246
temperature, 85
tetrachlorophenol, 276
the clap *see Brettanomyces*
thermal inactivation, 342
thermocycler, 319
thermovinification, 41–3
 advantages and disadvantages, 43
 Victoria grape PPO thermal stability, 42
tight RO, 128–9
titratable acidity, 10, 21, 23, 354–5
2,4,6-tribromoanisole, 276
tribromophenol, 276
2,4,6-trichloroanisole, 178, 246, 250, 273
trisodium phosphate, 258, 275
TSP *see* trisodium phosphate
turbidimeter, 29
turbulence, 232–3

ullage, 186
ultrafiltration, 127, 133
ultraviolet light, 261
UV-VIS spectrophotometer, 336

vaccenic acid, 95
vacuum-assisted bottling equipment, 232
vacuum drum filtration, 137
VBNC *see* viable but non-culturable organisms
Velcorin, 144, 168, 341–2
viable but non-culturable organisms, 268–9, 290
Viognier, 74
Vitis vinifera, 105
vortex, 280, 281

wad, 180
walnut shell blasting, 250
water regeneration, 147
Waukesha pumps *see* rotary lobe pumps
WD-40, 253
white grapes, 33
 managing botrytis or rot, 17
white must, 28
white wine
 case studies and applied response in nitrogen management, 68–9
 effect of skin contact, 26–7
  carbon dioxide, 26
  duration, 26
  fruit condition, 26
  sulphur dioxide, 27
  temperature, 26
 fermentors, 53
whole berry fermentation, 76
wine
 clarification, stabilisation and preservation, 98–125
  amorphous white precipitate formation, 115
  casein treatment, 109
  cold stability, 100–1
  'contains sulphites' vs 'no added sulphites,' 116
  heat-stabilising white wines with bentonite, 102–4
  KMBS addition, 120
  lysozyme and use in winemaking, 124–5
  molecular sulphur dioxide and its relation to free and total SO_2, 117–18
  prevention of malolactic bacteria or Brett growth, 121
  protection from spoilage during transport, 119–20
  role of vineyard nitrogen management in juice processing and wine instabilities, 98–9
  free SO_2 required at wine pH, 122
  malolactic fermentation, 77–97
  pH change effect on protein stability, 106
  sodium or calcium bentonite, 105–7
   bentonite slurry additions for bench trials, 106

effect of blending two protein-stable wines, 106
influencing factors on protein stability, 106–7
proper mixing, 106–7
rehydration of bentonite, 105–6
settling of bentonite lees, 107
use of sorbate, 111–12
dose, 114
geranium tone, 111
potassium sorbate, 113
sorbic acid, 113
yeast fermentation, 52–76
see also specific wine
wine bottling
control of oxygen uptake, 172–5
occasional empty bottles, 233
oxygen in bottles at filling, 176–7
oxygen pick up, 170–1
post-bottling storage conditions on package performance, 189–90
underfilled bottles, 232–3
without filtration, 149
see also sterile bottling
wine clarification, 98–125
wine colour, 356–8
wine diamonds, 100
wine expansion, 189, 191
wine filtration, 127–49
causes of difficult filtration
colloids and polysaccharides, 141
heavy solids load, 141
pectins, 141
cross- or tangential-flow filtration, 132–4
effect on wine quality, 143
integrity testing, 146
membrane flushing schemes, 147–8
back-flushing, 147
chemical regeneration, 147
membrane storage, 148
sanitisation, 148
water regeneration, 147
membrane pore size, 138
minimising filtration, 140
monitoring operational parameters, 145
options, 130–4
osmotic distillation, 135
perpendicular flow filtration, 130–4
depth filtration, 130–1
membrane filtration, 131–2
screen filtration, 130
size interpretation, 127–9
sterile fertile pads, 139
types of filters, 136–7
cartridge, 136
centrifuge, 137
cross-flow, 137
diatomaceous earth, 136
lees press, 137
plate and frame, 136
vacuum drum, 137
why filtration is necessary, 144
wine bottling without filtration, 149

wine particulates on filtration rate and volume, 142
wine oxidation, 361–2
wine packaging and storage, 150–98
bag in box packaging, 184–5
bottling line disinfection, 154–5
bottling line sterilisation, 150–2
importance of water quality, 152
quality control monitoring, 152
sources and causes of reinfection during bottling, 151
steam and hot water benefits, 150
sterilisation protocol and identifying sources of infection, 150–2
changes in internal bottle pressure, 187
chemical additives for refermentation or microbiological instability prevention, 164–9
dimethyldicarbonate, 167–8
residual sugars by Clinitest, 169
sorbic acid, 166–9
sulphur dioxide, 164–6
control of oxygen uptake at bottling, 172–5
benefits of limited oxygen, 172
elimination at bottling, 173–5
elimination prior to bottling, 172–3
inline oxygen sparger, 173
cork sampling, 178–9
national and global shipments temperature and storage, 195–8
bottle position, 198
recommended global wine shipping and storage temperature, 198
optimum environmental parameters and their effect on wine quality, 191–4
humidity, 193–4
light, 193
relative rates of selected reactions in wines, 192
temperature, 191–3
temperature fluctuation, 193
white wine sensory evaluations, 192
oxygen in bottles at filling, 176–7
oxygen pick up during bottling, 170–1
post-bottling storage conditions on package performance, 189–90
screw caps, 180–1
steaming the bottling line, 156–9
procedure, 156–8
purpose, 156
sterile bottling, 160–3
bottling room and environmental factors, 162–3
depth filters, 160
important factors, 162
integrity test, 161–2
membrane filter integrity testing, 161
membrane filters, 160–1
sterile bottling system, 162
synthetic closures, 182–3
wine bottle storage, 186–8
wine preservation, 98–125
wine press, 15–16, 255
wine quality, 345–86

astringency, 373–5
Burgundian bâtonnage, 385–6
effect of filtration, 143
factors of ten for blend trials, 349–52
   calculation, 351
   helpful tips, 352
   location, 349
   process, 350–1
   reason, 349
   results, 351–2
   time, 349
   tools needed, 349–50
   trials by ten, 349
film formation, 380
formation of thin, dry, white, filamentous film, 381–2
high volatile acidity from lactic acid bacteria, 377
improvement of colour, 356–8
old fish smell, 371–2
oxidative vs non-oxidative browning, 363
pH and titratable acidity, 354–5
presence of disulphides, 367–8
residual sugar causing yeast growth and/or carbonation, 379
sampling, 345–8
   applications, 348
   barrel sampling, 347
   blending sessions, 347–8
   representative sampling, 345–6
   tank sampling, 346
silver alloys to bind hydrogen sulphide and mercaptans, 369–70
sulphur dioxide resistant yeast, 383–4
sulphur off-odours
   identification, 364
   prevention, 366
   treatment, 364
types of hazes, 378
wine oxidation, 361–2
wine stabilisation, 98–125
wine tanks cleaning, 278–82
   chemistry, 278
   needed materials, 278–9
   preparation, 279
   procedure, 279–81
winemaking
   grape analysis, 1–14
      berry ripeness measurement, 9–11
      grape sampling, 4–5
      importance of effective grape sampling and analysis, 1–2
      sensory ripeness assessment, 12–14
      storage, transport, and processing, 6–8
   juice and must preparation, 15–51
      achieving desired sugar and alcohol concentrations, 47–8
      avoiding oxygen exposure with white must, 35
      botrytis or rot management, 17–18
      calculation and addition of grape concentrate, 24–5
      cold settling of white juices, 31–2
      enological tannins, 45–6
      facilitating fermentation of red musts, 37–8
      ideal temperature to press ice wine grapes, 50–1
      must clarification, 29–30
      pectinase use in white must preparation, 28
      pH adjustments, 19–20
      pH and titratable acidity adjustment before fermentation, 21
      saignée and effect on red wine, 39–40
      skin contact on white wine style, 26–7
      sugar content nomenclature, 22–3
      thermovinification, 41–3
      treating must containing laccase, 33
      wine press size, 15–16
   lysozyme and use in winemaking, 124–5
   Victoria grape PPO thermal stability, 42
winemaking equipment
   bubble point membrane filter integrity testing, 218–22
      advantages, 222
      definition and principle, 218–19
      in-process bubble point test illustration, 221
      procedure, 220–2
   maintenance and trouble shooting, 199–256
      barrel, 223–31
      corrosion, 237–40
      electric motor, 202–5
      lubrication, 253–6
      passivated materials corrosion, 246–8
      pumps, 206–16
      underfilled bottles, 232–3
   preventative maintenance, 234–6
      considerations, 234–6
      plan making, 234
   stainless steel, 241–5
      protection, 249–52
   stemming/crushing operation, 199–201
      centrifugal stemmer/crusher, 199–200
      common problems, 200–1
      roller and wiper type stemmer/crusher, 200
winery microbiology, 257–89
winery preventative maintenance, 234–6
   considerations, 234–6
   lubrication, 235–6
   special considerations, 236
   tank gaskets, 235
   tank temperature controllers, 235
   winery hoses and pumps, 235
   winery refrigeration system, 234–5
   plan making, 234
winery sanitation, 257–89
   alternatives for chlorine bleach, 274–5
   citric acid solutions, 275
   hot water, 274
   iodine, 274
   mechanical cleaning, 274
   ozonated water, 275
   per(oxy)acetic acid, 274
   per(oxy)carbonate-based products, 274
   potassium metabisulphite solutions, 274

quaternary ammonium compounds, 275
trisodium phosphate, 275
associated safety issues, 289
biofilms, 262
bleach or chlorinated cleaners, 272
chlorine bleach, 273
cross-contamination, 263–7
   airlocks and bungs, 265–6
   blending, 264
   chemicals and dry goods, 265
   filter pads and diatomaceous earth, 266
   hands and clothing, 265
   insects and creatures, 265
   pomace, 266
   sampling, 263–4
   tanks, 266
   topping, 264
   transfers, 264
   winery cellar, 263
effective sanitation program, 288
environmental TCA, 276
operational sanitation program essential elements, 257–61
   acidulated sulphur dioxide, 260
   built detergents, 259
   finishing operations, 261
   hot water and steam, 260
   iodine-based formulations, 259
   monitoring, 261
   ozone, 260
   peroxides, 259–60
   quaternary ammonium compounds, 259
   ultraviolet light, 261
ozone, 277
use of microscope, 270–1
viable but non-culturable organisms, 268–9
wine tanks cleaning, 278–82
   chemistry, 278
   needed materials, 278–9
   preparation, 279
   procedure, 279–81
winery transfer hoses cleaning, 283–7
   chemistry, 283
   needed materials, 283–4
   preparation, 284
   procedure, 284–6
wrought iron, 241

yeast assimilable nitrogen, 40, 62–4
   adjustment, 67–70
      2008 limit for DAP additions to must, 67
      adding DAP before fermentation mid point, 68
      case studies and applied response in nitrogen management, 68–9
      red must, 69–70
   Brix adjusted concentrations, 63
   data interpretation, 66
   deficient YAN, 63–4
   excessive YAN, 64
   optimal levels, 63
   required levels, 62
   variability in must and juices, 62
yeast available nitrogen, 330, 385
yeast fermentation, 52–76
   active dry yeast rehydration, 71
   carbonic maceration and whole berry fermentation, 76
   effect of fermentors on fermentation and storage, 55–6
   fermentor in red and white wines, 52–4
      material of construction, 52
      shape and size, 52–3
   fermentor types and use in red and white wines
      other factors to consider, 53–4
      type of wine to be made, 53
   grape inoculation with yeast before cold soaking, 75
   influence of oxygen addition on yeast, 74
   materials used in constructing fermentors, 55–6
   native flora fermentation, 60
   native flora vs yeast inoculum, 61
   yeast assimilable nitrogen, 62–4
      adjustment, 67–70
      Brix adjusted concentrations, 63
      data interpretation, 66
      deficient YAN, 63–4
      excessive YAN, 64
      measurement, 65
      optimal levels, 63
      required levels, 62
      variability in must and juices, 62
   yeast population homogeneity, 73
   yeast strains, 57
      influence on fermentation kinetics, wine aromas and flavours, 58
yeast strain, 88
yeasts, 57
YM agar, 323
YM Green, 323

Z-Brett, 320
ZEP-I-Dine, 274
zerk fitting, 214
Zeta potential, 130
zinc, 236
zinc plating, 240
Zinfandel, 356
Zygosaccharomyces, 166